THE CHANGING SHAPE OF GEOMETRY

Celebrating a century of geometry and geometry teaching, this book will give the reader an enjoyable insight into all things geometrical. There is a wealth of popular articles including sections on Pythagoras, the golden ratio and recreational geometry. Historical items, drawn principally from the *Mathematical Gazzette*, are authored by mathematicians such as G. H. Hardy, Rouse Ball, Thomas Heath and Bertrand Russell as well as some more recent expositors.

Thirty 'Desert Island Theorems' from distinguished mathematicians and educationalists give light to some surprising and beautiful results. Contributors include H. S. M. Coxeter, Michael Atiyah, Tom Apostol, Solomon Golomb, Keith Devlin, Nobel Laureate Leon Lederman, Carlo Séquin, Simon Singh, Christopher Zeeman and Pulitzer Prizewinner Douglas Hofstadter. The book also features the wonderful Eyeball Theorems of Peruvian geometer and web designer, Antonio Gutierrez.

For anyone with an interest in mathematics and mathematics education this book will be an enjoyable and rewarding read.

The Spectrum Series of the Mathematical Association of America was so named to reflect its purpose: to publish a broad range of books including biographies, accessible expositions of old or new mathematical ideas, reprints and revisions of excellent out-of-print books, popular works, and other monographs of high interest that will appeal to a broad range of readers, including students and teachers of mathematics, mathematical amateurs, and researchers.

THE CHANGING SHAPE OF GEOMETRY

Celebrating a Century of Geometry and Geometry Teaching

Edited on behalf of The Mathematical Association by

CHRIS PRITCHARD

MATHEMATICAL ASSOCIATION

supporting mathematics in education

THE MATHEMATICAL ASSOCIATION OF AMERICA

CAMBRIDGE
UNIVERSITY PRESS

PUBLISHED BY THE PRESS SYNDICATE OF THE UNIVERSITY OF CAMBRIDGE
The Pitt Building, Trumpington Street, Cambridge, United Kingdom
MATHEMATICAL ASSOCIATION OF AMERICA
1529 Eighteenth Street, NW Washington DC 20036

CAMBRIDGE UNIVERSITY PRESS
The Edinburgh Building, Cambridge CB2 2RU, UK
40 West 20th Street, New York, NY 10011-4211, USA
477 Williamstown Road, Port Melbourne, VIC 3207, Australia
Ruiz de Alarcón 13, 28014 Madrid, Spain
Dock House, The Waterfront, Cape Town 8001, South Africa

http://www.cambridge.org

First published 2003

Printed in the United Kingdom at the University Press, Cambridge

Typeface Times 11/14 pt   *System* LATEX 2_ε [TB]

A catalogue record for this book is available from the British Library

Library of Congress Cataloguing in Publication data

ISBN 0 521 82451 6 hardback
ISBN 0 521 53162 4 paperback

Contents

Foreword

My foreword to this inspiring book unabashedly tells the very personal tale of one individual's various encounters with geometry and geometries over a good number of decades, but the hope behind this act of self-indulgence is that my narration's idiosyncratic aspect is counterbalanced by something more universal in the powerful emotions and the significant experiences recounted.

I grew up deeply in love with mathematics. The lure of graceful patterns and the fascination with the mysterious were simply part of my basic makeup. I first heard about the irrational square root of 2 when I was perhaps eight years old, about imaginary numbers when I was perhaps nine, about matrix multiplication when I was perhaps ten, about Euler's constant e and its connection with i, π, and -1 when I was perhaps eleven, and so forth. All of these things excited me enormously, although, to be sure, I had but a childish grasp of them.

Somehow, though, I did not get much exposure to geometry. I recall that when I was in fourth grade or so, a slightly older friend showed me how to construct a regular pentagon inscribed in a circle with ruler and compass, and for weeks I went around constructing perfect pentagons on every blank piece of paper that came my way, but today, sad to say, I have no more than a blurry memory of how that beloved construction went. As another friend of mine wryly observed a few years ago, "Pentagons are a young man's game."

I also entirely missed out on geometry in school. When I was 13 – the year when I normally would have had a geometry course – my family spent the year in Switzerland, and for various reasons I was exempted from taking mathematics altogether. I recall seeing my classmates work on geometry problems, and being turned off by the way they were forced to acquire the dubious skill of finding absurdly complex 'proofs' of trivially- as well as visually-obvious facts (the two tangents to a circle from an exterior point have the same length, things like that). Oddly enough, I had never encountered the notion of 'proof' in mathematics before, and because I only saw uninspiring examples of the phenomenon, geometry struck me as a set of pedantic exercises, much like counting the number of angles that can dance on the head of a pin (the answer, of course, depends on whether they are acute or obtuse).

After our Genevan year, we returned to America and I started a lifelong habit of avidly browsing the mathematics sections of bookstores. Once in a while I would encounter books with titles like *Advanced Euclidean Geometry* and would flip through them. I vividly recall how I perceived them: as being filled with hugely dense and complicated figures crammed with tightly overlapping circles and lines and far too many letters of the alphabet for my taste. In my young mind, the word 'geometry' took on a stale attic-exuding odor, and the practitioners of the art seemed to be detail-obsessed old fogeys who loved only the most obscure and unimportant nooks and crannies of mathematics.

Something changed a bit when I went to Stanford University. I read about point-line duality and projective geometry in Courant and Robbins' famous tome *What is Mathematics?* as well as in W. W. Sawyer's little gem of a book *Prelude to Mathematics*, and somewhere or other I also ran into Morley's abracadabric manu-facture of equilaterality out of nowhere (which I struggled to prove but could not). I encountered the inside-out universes engendered by inversion in a circle when I studied complex variables under Prof. Gordon Latta, and I adored the fact that circles mapped to circles. Perhaps I was even on the verge of falling in love with that musty old game played by musty old fogeys, but then something unfortunate happened.

In those days – the mid-1960s – virtually no courses on geometry were offered at Stanford; however, there was one professor, Harold Bacon, who was known to love geometry, and so I approached him with the idea of doing a reading course with him on projective geometry. I expected to encounter all sorts of magical diagrams like the ones I had seen of Pascal's and Brianchon's theorems, but to my surprise, the book he selected contained very few if any diagrams and seemed to be almost exclusively about linear algebra, with the points of the plane represented by homogeneous coordinates. Although I could appreciate the elegance and symmetry of the matrix equations, I was so disappointed by the abstractness and apictoriality of the course that for many years thereafter I never gave a further thought to geometry.

Convinced nonetheless that I was destined to be a professional mathematician, I started graduate school in mathematics in 1966, but in the courses that I was required to take, I once again encountered that same terrible barrier of enormous abstraction coupled with an utter lack of pictures, and as a result, I became so disillusioned with mathematics that I dropped out, and eventually I wound up doing a doctorate in theoretical physics instead, where I was at least able to keep in touch with certain less forbidding, more concrete areas of mathematics.

One morning in 1992, long after my doctorate and nearly 30 years after my brief brush with geometry at Stanford, I woke up with an image of circles inverting into circles in my mind's eye. How or why this ancient echo of Prof. Latta's amazingly precise and elegant blackboard diagrams flashed into my sleepy head that day, I have no idea, but I remember how much it charmed me and intrigued me, and I decided to prove it for myself that very morning. I tackled the challenge the only way I knew – purely algebraically – and after some work, I managed to get the equations to jump through the hoops I had set up, and lo and behold, the equation of one circle metamorphosed, *mirabile dictu*, into the equation of another circle. Eureka!

I felt so pleased with myself that I asked some mathematically-inclined friends if they, too, could find the proof of this result, and most of them attacked it, as had I, with an algebraic toolkit; one of them, however, showed me a purely synthetic, geometric proof that dazzled me, a proof whose clarity knocked me off my seat. Whereas none of the algebraic proofs offered any insight into WHY one circle's image is another circle, the geometric proof was so direct and so much more to the point that one had the sense of genuinely GRASPING this heretofore elusive fact's *raison d'être*.

This was a beautiful sensation and thus brought me joy, but at the same time it brought me pain, for I felt a sense of shame that I had never experienced this pure brand of clarity before. I realized that for many years, I had confused the reading of any proof whatsoever in a book (or one's own generation of a proof, no matter how jumbly and awkward) with coming to understand WHY, but now I saw that there could be a huge gulf between the two types of experience. Inspired by this new vision, I set out to find or invent sharp, clean, purely geometrical, WHY-revealing proofs for various other geometrical facts that I knew, and within short order, a deep love for geometry – for pure geometry without any algebra – was born in me.

As my initial spark grew into a bonfire, I wanted to share this new passion with others, so I decided to teach a personal seminar on geometry at Indiana University. One day during the first of these 'CaT:DoG' courses ('Circles and Triangles: Diamonds of Geometry'), it occurred to me that I could probably write a computer program to let me construct interrelated circles and triangles and so forth in such a way that I could drag points or lines around on the screen while watching the deep relationships remain invariant. Years earlier, I had seen a computer program like this, in which the images of physical objects could be connected together in intricate ways and then one could move one virtual object on the screen and watch the others react according to the laws of mechanics, so I had a pretty clear image in my head of how my hypothetical geometry program would look. When I mentioned this hope to a very visually-minded friend, he caught me off guard, saying, "I just saw a program exactly like what you've described – I even know the people who designed it, and I'm sure they'd be happy to meet you!" It was in this way that I met the designers of *Geometer's Sketchpad* (who kindly gave me a free copy of their product!), and I was thereby spared the effort of writing such a program myself (or getting a student to do it for me).

Over the next few years, I delved into many areas of geometry, making quite a few discoveries of my own and encouraging my students to do the same. Crystal-clarity of proofs became a religion for me, and I was ever in search of simpler and simpler ways of understanding facts for which I already had a decent, perhaps simple, proof. Moreover, the fascinating way in which the use of analogy and the extrapolation of patterns allow mathematicians to generate fresh new concepts and to reperceive and reconceive familiar ideas and patterns in radically new ways became central to my personal thinking and to my CaT:DoG courses. (In this volume, there are many stunning examples of this deep phenomenon of flexible reperception and analogy-based creativity, but I personally was struck by two lovely variations on the theme of Pythagoras – one in the article by Hazel Perfect, the other in an article by Larry Hoehn.)

As my experiences with geometry grew deeper and wider, I started exploring not just geometry but geometries, and my goal of understanding, in as clear a manner as possible, the interconnections between different sorts of geometries became another passion. For instance, I knew very well that deleting one line from the projective plane leads to Euclidean geometry, and so one day, when in a playfully dualistic mood I asked myself what would happen if instead I were to delete a point from the projective plane, I was rapidly and inevitably led to a very disorienting new geometry that I dubbed 'Euclidual' (meaning 'dual to Euclidean' and rhyming with 'residual'), in which lines are closed and have finite length, and points, too, have bizarre new counterintuitive properties, including the existence of 'parallel points'. I investigated the shape of 'circuals' in this new geometry, I found that the sum of the three 'internal slides' of a 'trislide' is π, and so forth.

I soon discovered that deleting not just one but a whole family of lines from the projective plane leads to the non-Euclidean geometry discovered by Janos Bolyai and Nikolai Lobachevsky, and that the dual act of deleting a family of points instead leads to what I naturally dubbed 'non-Euclidual geometry'. And then, mirroring the marvelous self-duality of projective geometry, there were even new self-dual geometries obtained by simultaneously deleting a line plus a point, or else a family of lines plus a family of points. Altogether, then, I wound up discovering an elegant family of exactly nine different two-dimensional geometries obtained by various combinations of simple deletions from the projective plane. I then had the chutzpah to write to one of my heroes, Donald Coxeter, and ask him if any of this was new.

Coxeter's reply to me, extremely cordial and gracious, led me to the realization, after some bibliographical research, that my personal 'new' geometries had in fact been found by Arthur Cayley over one hundred years earlier – a fact that was rather a crushing blow. A year or two later, on a visit to Toronto where I met Coxeter in person, I also made the acquaintance of Abe Shenitzer, who gave me a wonderful book on non-Euclidean geometry by the Russian mathematician I. M. Yaglom (Shenitzer had translated it into English), in which I found 'my own' 3×3 family of plane geometries explored in an extraordinarily systematic and incisive fashion – another blow to my ego. My only solace lay in the thought that I had come across these geometries via a different and somewhat more down-to-earth route.

Since then, despite occasional setbacks in which I have realized that some exquisite recent 'discovery' that I think of as mine alone is in fact someone else's rather old hat, I have continued my avid explorations of Euclidean geometry with the aid of *Geometer's Sketchpad*, not to mention the aid of many students and friends. I can safely say that I have developed as profound a love for geometry as I have ever had for any branch of mathematics. But how did my perspective on it change so radically? After all, those old books on 'advanced Euclidean geometry' that once struck me as infinitely dull and musty trunks in the attic now strike me as treasure boxes filled with deeply intoxicating magic.

All I can say is that geometry, like fine wine, is an acquired taste. Every theorem and every idea has its own unique bouquet. Some appeal to certain individuals, others to other individuals. There are fine French wines, fine Italian wines, fine German wines, and so forth. Poncelet, Gergonne, Brianchon, Beltrami, Klein, Plücker, von Staudt, Feuerbach ... The list goes on virtually forever, and I must admit that I occasionally fantasize about how such old-time geometers would be thrilled if they could only watch over my shoulders as I drag a point about on my laptop's screen and reveal such magical phenomena as the fourfold kiss of Feuerbach's theorem or the uncanny precision of Poncelet's porism. But such things are not to be. The dead are dead, and they will never know that particular joy, though I like to think that in some sense they knew these phenomena so intimately inside their heads that they might not even have been surprised to witness them taking place in dynamic color before their eyes.

Each generation is a product of its age and its culture, and ours is of course a computer culture. We few who love geometry are privileged to be able to witness great depth and visual magic on our screens – and yet, how are computers used by most school-aged children? Mostly for playing video games and watching movies – only very rarely for learning mathematics, let alone exploring it on their own. It's the same story as for television, just retold an octave higher: a marvelous technology is invaded and possessed by the money-driven entertainment industry, and what once was the most promising of educational tools becomes instead a bottomless pit of banality and triviality.

And yet there is still hope. With Euclidean and even non-Euclidean geometry now directly visualizable on screens, with visual treatments of four-dimensional special relativity and even of Riemannian manifolds now accessible with a few flicks of a cursor, high school and university students are going to be able to penetrate rapidly into conceptual universes that previous generations could barely imagine. In my opinion, thanks to these new sorts of visualization tools, geometry is poised on the threshold of a new golden era.

But what is this thing called 'geometry'? As Michael Atiyah suggests in this volume, it may be that geometry is less a specific branch of mathematics than a way of looking at any or all branches of mathematics – it may be that geometry is simply that which one does when one visualizes, as opposed to what one does when one calculates or cogitates. And it is certainly undeniable that there has been a great turnaround since the incredibly depressing picture-barren days when I was an undergraduate mathematics student. In those days, whole textbooks on such undeniably geometrical topics as topology were standardly written with nary a figure anywhere. Today, austerity that grim would be unthinkable. Indeed, just a few years ago, a book with the title *Visual Complex Analysis* (by Tristan Needham) garnered high prizes for its intuitive manner of presenting ideas that formerly were presented in texts that scorned intuition, that revered rigor and arcane symbolism, and that were totally or almost totally devoid of visual explanations. By contrast, Needham's work of art, with its hundreds and hundreds of beautiful figures *à la* Latta, brings complex analysis alive in an unprecedented manner.

Along these same lines, a horribly opaque, utterly algebraic, mind-numbingly subscript-studded proof of Van der Waerden's exquisite theorem ('If you color every positive integer either red or green, then no matter how you do it, one or the other color will contain arithmetic progressions of arbitrary length'), such as is given by Alexander Khinchin in his poetically titled but antipoetically written little book *Three Pearls of Number Theory*, is now, thank God, extremely outmoded, having yielded its place to a highly diagrammatic and transparently intuitive proof such as can be found in *Ramsey Theory* by Ronald Graham *et al*. Though Van der Waerden's theorem is not usually thought of as geometry, such a highly visual proof would bring it under that rubric, at least according to Atiyah's characterization.

There is an ironic aspect of Atiyah's characterization, though, which is the fact that much of what has traditionally been labeled 'geometry' would not count as geometry, as it is usually presented in an unapologetically algebraic and strictly apictorial manner; conversely, areas of mathematics that traditionally have never been considered 'geometry' could acquire that label as they are rendered more and more pictorial and accessible to the mind's eye.

This recent movement toward the invitingly visual and away from the forbiddingly formal is a marvelous trend, and I hope that it continues to flourish. I must nonetheless admit that when I flip through contemporary schoolbooks that purport to teach geometry, I find an appalling lack of rigor and an appalling lack of visual beauty. To be sure, I find plenty of bright colors on every page, plenty of different typefaces and distracting typographical virtuosity, and plenty of allusions to the worlds of advertising and clothes and engineering and space travel and sports, but almost no links whatsoever to the world of – well, the world of mathematics.

It would be in vain that one would search today's high school geometry texts for such beautiful concepts as the Euler line, the nine-point circle, Feuerbach's theorem, Desargues' theorem, Pappus' theorem, Pascal's theorem, Ceva's theorem, Menelaus' theorem, inversion in a circle, and on and on and on. Where did it all go? Why were such things commonplace in texts 80 years or so ago, whereas now they are totally unheard of? Why was the initially austere and perhaps pointless-seeming but subsequently deeply rewarding activity of erecting an intellectual edifice by means of logical deduction taken for granted as part of school curricula back then? Why does mathematics today have to be 'relevant' and 'fun', have to prove its worth by chewing gum like a sports star, acting sexy like a movie star, spouting cutesy sound-bites like a with-it journalist, displaying itself as eye candy like a top model – but, heaven forbid, not by exploring unsuspected symmetries and subtle patterns purely for their own sake, like a scientist (let alone a mathematician!)?

I have no final answers to offer, ultimately, to these conundra. I am baffled and troubled by the widespread societal loss of interest in and respect for the beauty of pure mathematics. All I can say is that each age has its pluses and its minuses, its own unique points of view and its own unique blind spots. Today we find ourselves somewhere along a vast swing of a vast pendulum, and its arc may not be a periodic one. Who knows where the swing of mathematics education will end up? All we can do is try our best to influence it in the manner we feel is most useful and most beautiful – and it is indubitably in the spirit of high reverence for hidden beauty that Chris Pritchard has fashioned this book, his contribution to the ongoing debate.

May this book have a profound effect on the trajectory of the swinging pendulum! May geometry soon enjoy a grand reflowering! May our children come to appreciate the subtle and sublime harmonies that link triangles and quadrangles, conic sections and cross ratios, polygons and polyhedra, tessellations and tori, curvatures and conjugacies, poles, polars, and polarities! Long live the ancient art and the timeless poetry of geometry!

Douglas Hofstadter
Center for Research on Concepts and Cognition
Indiana University

Acknowledgements

This book is the result of the ideas and the practical support of numerous people. Late in 1999, at the instigation of its secretary, Peter Thomas, The Mathematical Association Teaching Committee decided that it would be fitting to mark its centenary, then three years away. The committee's chair, Doug French, and former chair, Janet Jagger, promoted the idea of putting together a book on geometry and its teaching, taking articles previously published in the association's journals, and the case for seeking partnership with Cambridge University Press was strongly put by Charlie Stripp. Much of the material was reformatted by Bill Richardson, Chair of The Mathematical Association's Council and production editor of the *Mathematical Gazette*, and this enabled the project to proceed at crucial stages. The support of these and other members of Teaching Committee proved invaluable.

The editing process has been eased by the assistance of a number of specialists in their fields – Doug French, Jeremy Gray, Janet Jagger, Ruth Lawrence, Adam McBride, Michael Price and Robin Wilson. The points they raised or addressed were supplemented by the valuable comments of the mathematicians who previewed some of the 'Desert Island Theorems' on behalf of Cambridge University Press and the Mathematical Association of America. Contributions have come in from the four corners of the world but not without the occasional glitch and the project would have foundered without the expertise of Lewis Paterson and Kellie Gutman. They provided vital assistance in overcoming hardware and software problems. Barry Lewis, President of The Mathematical Association, did much liaison work behind the scenes. We are grateful for all their help, so willingly given.

The final phase from manuscript to book has been undertaken with enormous skill and professionalism by the staff of Cambridge University Press. Our thanks go in particular to the commissioning editor, Jonathan Walthoe.

Finally, the editor would wish to offer thanks to his colleagues in the Mathematics Department of McLaren High School for their indulgence and unfussy help at a time when this project has been the main focus of his attention, and particular to his wife, Audrey, for the constancy of her support and encouragement.

General Introduction: Simplicity, Economy, Elegance

CHRIS PRITCHARD

As children we build sandcastles and snowmen, construct buildings with LEGO and play computer games that create the impression of rapid movement through three-dimensional space. In later life, we hang wallpaper, negotiate narrow spaces in our cars and tease furniture through doorways. A surprising number of us use specially-honed spatial skills to earn a living, such as barbers and sculptors, interior designers and footballers, bus-drivers and architects, supermarket shelf-stackers, couturiers and civil engineers. Occasionally a Pelé, an Yves Saint Laurent or a Barbara Hepworth comes along. But regardless of the extent to which our spatial talents are developed, from the cradle to the grave, we are all geometers.

This book celebrates the best of geometry in all its simplicity, economy and elegance. Such a wonderful tray of attributes might have formed a 'splendid title' for the book as a whole, in the view of H. S. M. Coxeter, but instead it serves as a theme for the Desert Island Theorems and as the title of this general introduction. The source is a lovely anecdote about Peter Frazer, related by H. E. Huntley in his book *The Divine Proportion* [1, p. 5]:

Peter Frazer . . . a lovable man and a brilliant teacher, was discussing cross ratios with a mathematics set. Swiftly, he chalked on the blackboard a fan of four straight lines, crossed them with a transversal and wrote a short equation; he stepped down from the dais and surveyed the figure. . . . Striding rapidly up and down between the class and the blackboard, waving his arms about excitedly, with his tattered gown, green with age, billowing out behind him, he spoke in staccato phrases: 'Och, a truly beautiful theorem! Beautiful! . . . Beautiful! . . . Look at it! *Look at it!* What simplicity! What economy! Just four lines and one transversal . . . What elegance!'

It appears that Frazer saw simplicity, economy and elegance as complementary essences of beauty in geometry. If I were to add a fourth attribute, it would be 'surprise'. To elaborate on this theme consider two results from elementary Euclidean geometry. Firstly, in the figure, $ABCD$ is a square and E, F, G, H are the midpoints of AB, BC, CD, DA respectively. Then:

$$\text{Area } IJKL = \frac{1}{5} \text{ Area } ABCD.$$

A simple dissection proof consists of rearranging the figure into a pentomino in the form of a cross by rotating $\triangle EJB$ through 180 degrees about E, and rotating the other three small triangles in like manner.

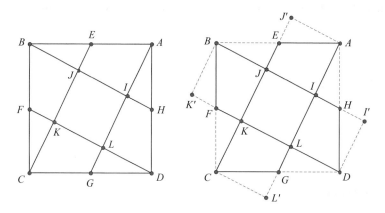

Secondly, take two maps of the same area, one a pocket map and the other a comprehensive map on a larger scale. (Alternatively, take two differently-sized prints of the same photograph.) Place the pocket map on the detailed map at any angle. Then there is one and only one point on the pocket map lying immediately on top of the point corresponding to it on the map beneath. This can be demonstrated using simple transformation geometry. First rotate the pocket map $ABCD$ about the common point, E, to give $MNPQ$ with the same orientation as the larger map

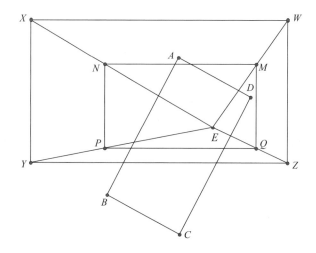

W X Y Z. The common point then acts as the centre of a dilatation (or enlargement) of the pocket map. Now let us reverse the argument. Given the two maps with the smaller cast carelessly onto the larger, how would we go about finding the common point? (A solution using cyclic quadrilaterals is shown at the end of the introduction.)

The Mathematical Association came into existence in 1871 at a time when geometry teaching was in something of a tumult. The school and university geometry curricula and Euclid's *Elements* were as two peas in a pod. So completely was the curriculum determined by the standard text that Sylvester sarcastically referred to the *Elements* as 'one of the advanced outposts of the British Constitution'. Calls for reform in the universities met with the antipathy of De Morgan, Cayley and Kelland and with positions entrenched, it fell to schoolteachers to advance the case for 'loosening the shackles' of Euclid [2, ch. 2].

The sort of issues that challenged teachers at the time, concerned the choice and order of the Euclidean propositions to be taught, to whom and at what age? The role of riders in geometry education was also considered. Geometry continued to be viewed as the ideal vehicle for developing an understanding of formal proof. It is only in recent decades, with the advent of modern mathematics and its emphasis on transformation geometry, that the status of formal proof has fallen away and its 'natural habitat' relocated to algebra. We stand on the threshold of a possible reversal of this trend, though as ever it will be difficult to reach a consensus.

In its first thirty years, the movement to improve geometry teaching had only limited success. But by the turn of the century there was a more general acknowledgement of the problems by both educationalists and administrators and the time was ripe for significant change. A growing understanding of the need to broaden the scope and methods of geometry in schools saw the first use of measuring instruments such as protractors and the adoption of four-figure tables.

To digress a little, at high school in the late 1960s, my classmates and I made use of such four-figure tables, compiled according to the frontispiece by 'the late C. Godfrey and the late A. W. Siddons'. We had the great good fortune to be taught geometry by Wil Williams, a teacher of unusual clarity and not a little humour – "they died calculating the entries", was his quip. At the time, little did he or we know of the enormous roles Godfrey and Siddons played in The Mathematical Association. Twenty-three of its early members formed the first Teaching Committee in 1902 and Godfrey and Siddons were prominent among them. This book celebrates the centenary of Teaching Committee, a body which in its early years helped to subjugate the geometry text to the requirements of the designed curriculum and which has continued ever since to provide advice and resources for mathematics teachers and to seek to influence national policies on mathematics education.

At the Annual Meeting of the National Council of Teachers of Mathematics in Florida in April 2001, I sought the views of one of its senior figures on The Mathematical Association. It was no great surprise that the first thing he said in reply was that the association has very good journals. And indeed, the journals attract contributions from all parts of the world. Since it is from two of these journals, the *Mathematical Gazette* and *Mathematics in School*, that a goodly proportion of the material for this book is taken, a little more information about them is in order.

The *Mathematical Gazette*, the premier journal of The Mathematical Association, was established in April 1894 under the editorship of Edward Mann Langley. The opening editorial made it plain that it would enable mathematics teachers to share successful approaches to their art, other than those to be found in the texts of the day [2, p. 40]. In this role it has met with fluctuating success during its 108-year history, whilst maintaining greater consistency as a minor mathematical serial. It was Langley, incidentally, who as a teacher at Bedford Modern School converted E. T. Bell to mathematics and quite likely helped shape that accomplished number theorist and somewhat unreliable mathematical biographer [3, ch. 2]. Bell duly contributed an article to the *Mathematical Gazette*'s 250th number in July 1938 and, with its reputation soaring both in Britain and internationally, the 500th number was published in July 2000. The launch of *Mathematics in School* in November 1971 provided members with a journal designed to support the teaching of mathematics to younger children. Its style remains to this day rather more informal than that of the *Mathematical Gazette*, with a larger proportion of articles focussing on classroom practice, often with a hands-on flavour. Elementary geometry and its teaching have formed the subject of several thousand articles which have appeared in these two journals over the last century; many of the best get a second airing in this volume.

Following an initial trawl some twenty geometrical themes were identified for possible inclusion in this book. Each perhaps merited a separate part but six were finally selected for inclusion. The first two parts, on the nature and history of geometry, and the last part, on its teaching, are somewhat weightier than the other three on Pythagoras' Theorem, the golden section and recreational geometry. They tend to set the scene, to help the reader to understand the contexts within which geometry and its teaching developed. A number of articles in these sections refer to Euclidean definitions and theorems simply by taking the book number and the proposition number from the *Elements*. Thus, Pythagoras' Theorem is Euclid I, 47 because it is the forty-seventh proposition of the first book of the *Elements*. Note that Euclid and his book are synonymous. Since a large proportion of a modern audience will be unfamiliar with these tags, each of the propositions referred to in this way has been stated in full in the second appendix.

Two of the most illustrious mathematicians of this century, G. H. Hardy and Michael Atiyah offer views on what constitutes geometry. Hardy confesses from the outset that 'I do not claim to know any geometry, but I do claim to understand quite clearly what geometry is' and comes to a pessimistic conclusion about ever reaching a consensus on the geometry curriculum. Atiyah takes an historical approach to tease out the nature of the subject.

Following a general overview of the history of geometry, the history articles open with an essay on Greek mathematics by Thomas Heath. A notable feature of this section is a run of items highlighting non-European contributions to the development of geometry, especially one by the author of *The Crest of the Peacock*, George Gheverghese Joseph. It is rounded off with a detailed review of nineteenth century geometry by Gaston Darboux, penned at the turn of the century without the benefit that a time lapse often affords. The reader may judge whether Hardy is harsh in suggesting in his article that Darboux was a great geometer with little feel for the nature of his discipline. The section on the teaching of geometry is introduced by a new and comprehensive essay from The Mathematical Association's historian, Michael Price. It is followed by what Price describes as a 'concise and stylish attack on Euclid from an advanced pure mathematical standpoint' [2, p. 55]. Its author is Bertrand Russell, the year 1902, the very year that Teaching Committee came into existence.

There is a greater focus on elucidating elementary geometry in the other three parts of the book. Articles on Pythagoras' Theorem are introduced by Janet Jagger, an educationalist and former Chair of Teaching Committee. Ron Knott casts his expert eye over the articles on the golden section. Incidentally, his peerless Fibonacci website is a treasury of information on the sequence and the associated ratio. Articles on recreational geometry are prefaced by the thoughts of Brian Bolt, whose numerous books on recreational mathematics are much in the vein of Martin Gardner. It is a real pleasure to see reproduced here an extract of an early *Mathematical Gazette* article on 'rep-tiles' (replicating figures in the plane) by another hugely influential recreational mathematician, Solomon Golomb. His enthusiastic followers will also be pleased to find among the Desert Island Theorems previously unpublished proofs of a tromino theorem.

Since 1942, the BBC has broadcast a popular weekly radio programme called *Desert Island Discs*. These days its guests are invited to select eight pieces of music that they would take with them to a desert island. They also nominate a favourite book though, presumably to guarantee variety week on week, the key works of the great religions and the works of Shakespeare are excluded. Thirdly, a luxury inanimate object of no practical use is selected. The choices made by guests are little more than devices to prompt them to discuss their lives.

For this book, the 'desert island theme' has been developed in a different direction and for a different reason. Imagine you are a mathematician, cast away on a desert island. You have no access to the modern world and little prospect of being reunited with the rest of humanity. Which geometrical theorem would you least wish to be without, in some sense; which perhaps are you truly glad was discovered. The focus is more on the theorem but might well have something of the personal too. No claim is made to novelty in this shift from desert island music to desert island mathematics. Those who have previously written or spoken on desert island mathematics include David Burghes, Tony Crilly and Colin Fletcher and it is appropriate that their theorems should be included in this book.

Prompted by an early success in persuading Coxeter to offer his Desert Island Theorem – he actually took no persuading at all, such is his love of geometry – a number of eminent mathematicians and physicists, teachers and educationalists were invited to follow suit. They were asked to nominate an elementary, geometrical theorem, to try to keep to a maximum of about 500 words plus a diagram and to adopt where possible what might be termed a 'popular science' style. As you will see, some found the brief restrictive but the brief was always intended to guide rather than stifle. Only in a very small number of cases was the same theorem chosen by two contributors.

The response from those approached was overwhelmingly positive and enthusiastic. The Mathematical Association notes with gratitude the affection shown towards it by the community of mathematicians, especially in Britain and North America during the preparation of this book. Among those in higher education and research there would appear to be fulsome recognition of the valuable support and advice given to mathematics teachers by The Mathematical Association and similar organizations worldwide.

The informality and brevity of the contributor's designations has raised the eyebrows of some who have previewed material in this book. Leaders in their fields usually need little introduction and certainly no puffing up. Indeed, what needs to be added to an author's name tends to vary inversely with eminence. In the minds of thousands of students or former students, Tom Apostol is the author of the two best calculus texts of all time. It is fitting therefore to refer to him as a teacher of the calculus, especially in a book put together to celebrate a centenary of relevance to mathematics teachers. Further, in most cases, academic positions (though not institutions) have been omitted as have titles and other honours. This is an attempt – possibly a vain attempt, human nature being what it is – to focus on the Desert Island Theorems themselves rather than the eminence of the writer. Several of them are beautifully crafted. Where a contributor is or has been a servant of The Mathematical Association, this is noted especially for the benefit of members. Each castaway has either suggested the designation provided or else expressed contentment with it.

There are thirty Desert Island Theorems in total, arranged very loosely into five groups:

A. ancient Greek geometry
B. elementary Euclidean geometry (of the last four centuries)
C. advanced Euclidean
D. spherical geometry and topology
E. geometrical physics.

As a rule of thumb, the difficulty level of the theorems tends to increase as the book proceeds. Arguably, there is a move further and further from pure geometry at the same time. The vast majority of the theorems should be accessible to those who have studied mathematics to the age of eighteen and possess a willingness to wrestle with some of the geometrical arguments. Readers with an interest in the development of geometry will find much to savour.

If it can ever be said that a single person has pushed back the boundaries of a branch of mathematics, particularly in recent times, then it might justifiably be said of Coxeter in relation to geometry. Coxeter's sixty-six year association with the University of Toronto may have obfuscated the fact that he was born and raised in England and received his university education at Cambridge. The first recorded evidence of his prodigious grasp of geometry is to be found in the *Mathematical Gazette*, where in the issue of October 1926, the young 'Donald' Coxeter asked the readership via Alan Robson, his teacher at Marlborough School, if it knew of an 'elementary verification' of $\int_0^{\frac{\pi}{2}} \sec^{-1}(\sec x + 2)\,dx = \frac{5\pi^2}{24}$ and two similar results 'suggested by a geometrical consideration and verified graphically' [4]. At the time, Robson was already prominent in The Mathematical Association. He would go on to hold high office as Chair of Teaching Committee and as President, either side of the Second World War [2, pp. 151, 161]. The direct influence of Robson on Coxeter is undoubted, the indirect influence of the Mathematical Association and its Teaching Association rather more speculative.

Finally, let me return to my previous theme to consider the simple, economical, elegant and surprising geometry of the cauliflower. Almost any elementary treatment of the Fibonacci sequence and the golden section draws attention to the structure of the spiralling segments of the pineapple. Yet the cauliflower appears to have exactly the same structure and it seems a shame that this humble vegetable should have been so thoroughly eclipsed by an exotic fruit. If you look carefully enough the geometrical structure is certainly evident in the common white variety, though it is striking in the romanesco variety of Fibonacci's homeland of northern Italy. A few years ago I was delighted to find that the cauliflower was not without its champion. Touring the Cité des Sciences et de l'Industrie at La Villette in Paris, I came across an exhibit which highlighted both the cauliflower's spirals and

The spirals and fractals of the romanesco cauliflower

its self-replicating fractal form. But whether we invoke the pineapple, the nautilus shell, or Fibonacci's romanesco cauliflower one thing seems clear – the attributes of the most attractive and pleasing geometry remain the same irrespective of whether the geometry is of our invention or of our discovery, crafted by man or designed by nature. In the words of John Keats: *A thing of beauty is a joy for ever*.

Solution to the Maps Question (by Doug French)

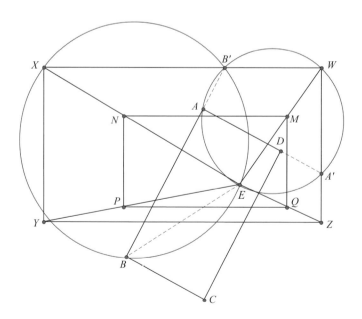

First, let BA produced meet XW at B'.

$$\angle BEX = \angle BB'X = \text{ angle of rotation}$$

These angles are subtended by the same chord, XB. So $XBEB'$ is a cyclic quadrilateral.

Similarly, let AD produced meet WZ at A'.

$$\angle AEW = \angle AA'W.$$

These angles are subtended by the same chord, AW. So $WAEA'$ is a cyclic quadrilateral.

Since E lies on the circumference of both circles, it is located at one of the points of intersection.

References

1. H. E. Huntley, *The Divine Proportion: A Study in Mathematical Beauty* (Dover Publications, 1970).
2. Michael H. Price, *Mathematics for the Multitude: A History of the Mathematical Association* (The Mathematical Association, 1994).
3. Constance Reid, *The Search for E. T. Bell, Also Known as John Taine* (Washington DC: Mathematical Association of America, 1993).
4. D. Coxeter (per A. R.), Note 853, *Mathematical Gazette* **13** (October 1926), 205. This was Coxeter's first publication. [See Coxeter, H. S. M. *Twelve Geometry Essays* (Southern Illinois University Press, 1968) (Preface, p. vii).]

I

The Nature of Geometry

1.1

What is Geometry?

G. H. HARDY

I have put the title of my address in the form of a definite question, to which I propose to return an equally definite answer. I wish to make it quite plain from the beginning that there will be nothing in the least degree original, still less anything paradoxical or sensational, in my answer, which will be the orthodox answer of the professional mathematician.

I expect that you, as members of an association which stands half-way between the ordinary mathematical teacher and the professional mathematician in the narrower sense, will probably agree with me that I am wiser to avoid topics of what is usually called a 'pedagogical' character. I am sorry to be compelled to use the unpleasant word 'pedagogical' and I am sure that you will believe me when I say that I do not use it in any contemptuous sense, and that I am enough of a pedagogue myself to realise the very genuine interest of many 'pedagogical' questions. But I do not regard it as the business of a professional mathematician to concern himself primarily with such questions, and, even if I did, I should have very little to say about them. It has always seemed to me that in all subjects, and most of all in mathematics, questions concerning methods of teaching, whether this should come before that, and how the details of a particular chapter are best presented, however interesting they may be, are of secondary importance; and that in mathematics at all events there is one thing only of primary importance, that a teacher should make an honest attempt to understand the subject he teaches as well as he can, and should expound the truth to his pupils to the limits of their patience and capacity. In a word, I do not think it matters greatly what you teach, so long as you are really certain what it is; and I feel that you might reasonably be impatient with me, whether you agreed with me or not, if I occupied your attention for an hour and had nothing more to say to you than that. It is obviously better that I should take some definite chapter of mathematical doctrine, a chapter which is at any rate of

The 1925 Presidential address to The Mathematical Association.
First published in *Mathematical Gazette* **12** (March 1925), pp. 309–316.

the most obvious and direct educational interest, and expound it to you as clearly as I can.

It is, however, quite likely that some of you, and particularly any genuine geometer who may be present, will criticise my choice of a subject in a manner which I might find a good deal more difficult to meet. You might object that it would be reasonable enough for me to try to expound the differential calculus, or the theory of numbers, to you, because the view that I might find something of interest, to say to you about such subjects is not *prima facie* absurd; but that geometry is, after all, the business of geometers, and that I know, and you know, and I know that you know, that I am not one; and that it is useless for me to try to tell you what geometry is, because I simply do not know. And here I am afraid that we are confronted with a regrettable but quite definite cleavage of opinion. I do not claim to know any geometry, but I do claim to understand quite clearly what geometry is.

I think that this claim is in reality not quite so impertinent as it may seem. The question 'What is geometry?' is not, in the ordinary sense of the phrase, a geometrical question, and I certainly do not think it absurd to suppose that a logician, or even an analyst, may be better qualified to answer it than a geometer. There have been very bad geometers who could have answered it quite well, and very great geometers, such as Apollonius, Poncelet, Darboux, who would probably have answered it extremely badly. It is a comfort, at any rate, to reflect that my answer can hardly be worse than theirs would in all probability have been.

I propose, then, to cast doubts of this sort aside, and to proceed to answer my question to the best of my ability. There are two things, I think, which become quite clear the moment we reflect about the question seriously. In the first place, there is not one geometry, but an infinite number of geometries and the answer must to some extent be different for each of them. In the second place, the elementary geometry of schools and universities is not this or that geometry, but a most disorderly and heterogeneous collection of fragments from a dozen geometries or more. These are, or should be, platitudes, and I have no doubt that they are to some extent familiar to all of you; but it is a small minority of teachers of geometry that has envisaged such platitudes clearly and sharply, and it is probably desirable that I should expand them a little.

I begin with the second. It is obvious, first, that a great part of what is taught in schools and universities under the title of geometry is not geometry, or at any rate mathematical geometry, at all, but physics or perhaps philosophy. It is an attempt to set up some kind of ordered explanation of what has been humorously called the real world, the world of physics and sensation, of sight and hearing, heat and cold, earthquakes and eclipses; and earthquakes and eclipses are plainly not constituents of the world of mathematics.

It is dangerous to repeat truisms in public, and the particular truism which I have just stated to you is one which I have often expressed before, and which has sometimes been received in a manner very different from that which I had anticipated. But I am not speaking now to an audience of rude and simple physicists, or of philosophers dazed by centuries of Aristotelian tradition, but to one of mathematicians familiar with common mathematical ideas. I find it difficult to believe that any mathematician of the twentieth century is quite so unsophisticated as to suppose that geometry is primarily concerned with the phenomena of spatial perception, or the physical facts of the world of common sense. It is, however, perhaps unwise to take too much for granted, and I will therefore try to drive home my point by a simple illustration.

Imagine that I am giving an ordinary mathematical lecture at Oxford, let us suppose on elementary differential geometry, and that I write out the proof of a theorem on the blackboard. John Stuart Mill would have maintained that the theorem was at the best approximately true, and that the closeness of the approximation depended on the quality of the chalk; and, though Mill was a man for whom I feel in many ways a very genuine admiration, I can hardly believe that there is anybody quite so innocent as that today. I want, however, to push my illustration a stage further. Let us imagine now that a very violent dynamo, or an extremely heavy gravitating body, is suddenly introduced into the room. Einstein and Eddington tell us, and I have no doubt that they are right, that the whole geometrical fabric of the room is changed, and every detail of the pattern to which it conforms is distorted. Does common sense really tell us that my theorem is no longer true, or that the strength or weakness of the arguments by which I have established it has been in the very slightest degree affected? Yet that is the glaring and intolerable paradox to which anyone is committed who supports the old-fashioned view that geometry is 'the science of space'.

The simple view, then – the view which I will call for shortness the view of common sense, though there is uncommonly little common sense about it – the view that geometry is the science which tells us the facts about the space of physics and sensation, is one which will not stand a moment's critical examination; and this, of course, was plain enough before Einstein, though it is Einstein who, by enabling us to exhibit its paradoxes in so crude a form, has finally completed the demonstration. The philosophers, of course, have tried to restate the view of common sense in a more sophisticated form. Geometry, they have explained to us, tells us, not exactly the facts of physical or perceptual space, but certain general laws to which all spatial perception must conform. Philosophers have been singularly unhappy in their excursions into mathematics, and this is no exception. It is, as usual, an attempt to restrict the liberty of mathematicians, by proving that it is impossible for them to think except in some particular way; and the history of mathematics shows conclusively that mathematicians will never accept the tyranny of any philosopher. The moment a philosopher has demonstrated the impossibility of any mode of thought, some rebellious mathematician will employ it with unconquerable energy and conspicuous success. No sooner was the apodeictic certainty of Euclid firmly established, than the non-Euclidean geometries were constructed; no sooner were the inherent contradictions of the infinite finally exposed, than Cantor erected a coherent theory. I do not think, then, that we need trouble ourselves with the views of the philosophers concerning geometry. They are, indeed, of much less interest than those of the man in the street, which do possess some interest, since there are valid reasons for supposing that others may share them.

It will be more profitable to leave the philosophers alone, and to consider what the mathematicians themselves have to say. We shall then have reasonable hope of making some substantial progress, since mathematicians, or those of them who are at all interested in the logic of mathematics, hold fairly definite views, and views which are in tolerable agreement, concerning this question of the relation of geometry to the external world. The views of the mathematicians are also much more modest than those which the philosophers have tried to impose upon them.

A geometry like any other mathematical theory, is essentially a map or scheme. It is a picture, and a picture, naturally, of *something*; and as to what that something is opinions do and well may differ widely. Some will say that, it is a picture of something in our minds, or evolved from them or constructed by them, while others, like myself, will be more disposed to say that it is a picture of some independent reality outside them; and personally I do not think it matters very much which type of view you may prefer to adopt. What is much more important and much clearer is this, that there is one thing at any rate of which a geometry is not a picture, and that that is the so-called real world. About this, I think that almost all modern mathematicians would agree.

This is only common mathematical orthodoxy, but it is an orthodoxy which outsiders very frequently misunderstand or misrepresent. I need hardly say that it does not mean that mathematicians regard the world of physical reality as uninteresting or unimportant. That would be on a par with the view that mathematicians are peculiarly absent-minded, always lose at bridge, and are habitually unfortunate in their investments. Still less does it mean that they regard as uninteresting or unimportant the contribution which mathematics can make to the study of the real world. The Ordnance Survey suggests to me that Waterloo Station, and Piccadilly Circus, and Hyde Park Corner lie roughly in a straight line. That is a geometrical statement about reality, and it enables me to catch my train at Paddington. Einstein is more daring, and issues his orders to the stars, and the stars halt in their courses to obey him. Einstein, and the Ordnance Survey, and even I, can all of us, armed with our mathematics, put forward suggestions concerning the structure of physical reality, and our suggestions will continually prove to be not merely interesting, but of the most direct and practical importance. We can point to this or that mathematical model, Euclidean or Lobatschewskian or Einsteinian geometry, and suggest that perhaps the structure of the universe resembles it, or can be correlated with it in one way or another; that that is a possibility at any rate which the physicists may find it worth their while to consider. We can offer these suggestions, but, when we have offered them, our function as mathematicians is discharged. We cannot, do not profess to, and do not wish to *prove* anything whatsoever. There is not, and cannot be, any question of a mathematician proving any thing about the physical world; there is one way only in which we possibly discern its structure, that is to say the laboratory method, the method of direct observation of the facts.

I will venture here on an illustration which I have used before. If one of you were to tell me that there are three dimensions in this room, but five for Southampton Row, I should not believe him. I would not even suggest that we should adjourn our discussion and go outside to see. The assertion would of course, be one of an exceedingly complicated character, and a very painstaking analysis might prove necessary before we were quite certain what it meant. However, I could attach a definite meaning to it. I should understand it to imply that, owing to particularities in the geography of London which had up to the present escaped my attention, the common three-dimensional model, sufficient for our purposes in here, becomes inadequate when we pass out into the street. And, however sceptical I might feel about such a theory, I should certainly not be so foolish as to advance mathematical arguments against it, for the all-sufficient reason that I am quite certain that there are none. I should be sceptical, not as a geometer but as a citizen of London, not because I am a mathematician, but in spite of it; and, indeed, I am sure that, if you appealed from me to the nearest policeman, you would find him not less but far more obstinately sceptical than me.

I must pass on, however, to what is really the proper subject matter of my address. Geometries, I will ask you to agree provisionally, are *models*, and models of something which, whatever it may be in the last analysis, we may allow for our present purposes to be described as mathematical reality. The question which we have now to consider is that of the nature of these models; and the characteristics which distinguish one from another; and there is one great class of geometries for which the answer is immediate and easy, namely, that of the *analytical* geometries.

An analytical geometry, whether of one, two, three, four, or n dimensions, whether real or complex, projective or metrical, Euclidean or non-Euclidean, and it may, of course, be any of these, is a branch of analysis concerned with the properties of certain sets or classes of sets of numbers. I will take the simplest example, the two-dimensional Cartesian geometry which resembles very closely, though it is by no means the same as, the elementary 'analytical geometry' taught in schools. I will call it, as I usually call it in lectures, *Common Cartesian Geometry*.

In Common Cartesian Geometry, a *point* is, by definition, a pair of real numbers (x, y), which we call its *coordinates*. A line is, again by definition, a certain class of points, viz. those which satisfy a linear relation $ax + by + c = 0$, where a, b, c are real numbers and a and b are not both zero. The relation itself is called the *equation* of the line. If the coordinates of a point satisfy the equation of a line, the line is said to *pass through* the point, and the point to *lie on* the line. And that is the end of Common Cartesian Geometry, in so far as it is projective, that is to say in so far as it does not use the so-called metrical notions of distance and angle, and in so far as it is concerned only with equations of the first degree. What remains is just algebraical deduction from the definitions.

Common Cartesian Geometry, as I have defined it, is a very simple and not a very interesting subject. It gains a great deal in interest, as you will readily imagine, when 'metrical' concepts are introduced. We define the *distance* of two points (x_1, y_1) and (x_2, y_2) by the usual formula

$$d = \sqrt{\{(x_1 - x_2)^2 + (y_1 - y_2)^2\}},$$

and the *angle* between two lines by another common formula, which I need not repeat. We have still, however, only to explore the algebraical consequences of our definitions, and no new point of principle arises, so that I can illustrate what I want to say quite adequately from the projective and linear system. This system, trivial as it is, has certain features to which I wish to call your attention as characteristic of analytical geometries in general.

The first feature is this, that a point in Common Cartesian Geometry is *a definite thing*. This is so in all analytical geometries. Thus in any system of two-dimensional and homogeneous analytical geometry a point is a class of triads (x, y, z), those triads being classified together whose coordinates are proportional, and in the geometry of Einstein a point is a set of four numbers (x, y, z, t). This is a very obvious observation, but it is of fundamental importance, since it marks the most essential difference between analytical geometries and 'pure' geometries, in which, as we shall see, a point is not a definite entity at all.

The next point which I ask you to observe is the absence of *axioms*. There are no axioms in any analytical geometry. An analytical geometry consists entirely of *definitions* and *theorems*; and this is only natural, since the object of axioms is, as we shall see, merely to limit our subject matter, and in an analytical geometry our subject matter is known.

It is most important to realise clearly that, in different geometrical systems, propositions verbally identical may occupy entirely different positions. What is an axiom in one system may be a definition in another, a true theorem in a third, and a false theorem in a fourth. You are accustomed, for example, to *proving* that the equation of a straight line is of the first degree, and I am not suggesting that the 'proof' to which you are accustomed is meaningless, trivial, or false. You profess to be proving a theorem, and you are, in fact, genuinely proving something, though it might take us some time to ascertain exactly what it is. There is one thing, however, that is quite plain, and that is that the something which you are proving is not a theorem of analytical geometry, for your supposed theorem is, as a proposition of analytical geometry, not a theorem at all but the definition of a straight line.

Let us take another simple illustration, the 'parallel postulate' of Euclid. *If L is a line, and P is point which does not lie on L, then there is one and only one line through P which has no point in common with L.* This, in school geometry, is sometimes called an 'axiom' and sometimes, I suppose, an 'experimental fact'. It cannot be either of these in analytical geometry, where there are neither axioms nor experimental facts, and it is obviously not a definition. It is, in fact, a theorem, which in Common Cartesian Geometry is true, though in other systems it may be false; and it is a theorem which any schoolboy can prove. It is the algebraical theorem that, given an equation $ax + by + c = 0$, and a pair of numbers, x_0, y_0, which do not satisfy this equation, then it is possible to find numbers A, B, C, such that

$$Ax_0 + By_0 + C = 0 \tag{1}$$

and the equations

$$ax + by + c = 0, \quad Ax + By + C = 0 \tag{2}$$

are inconsistent with one another; and that the ratios $A : B : C$ are determined uniquely by these conditions.

These are the characteristics of Common Cartesian Geometry which it is most essential for us to observe at the moment. There are others which I should like to say something about if I had time. There is no infinite and no imaginary in this geometry; there are imaginaries, naturally, only in complex systems, and infinites in homogeneous systems. Further, the principle of duality is untrue. All these topics call for comment; and I should have liked particularly to say something on the subject of the geometrical infinite, since the tragical misunderstandings which have beset many writers of text-books of analytical geometry, and which have generated such appalling confusion in the minds of university students, are misunderstandings for which writers like myself of text-books on analysis have been largely though innocently responsible. The geometrical infinite, however, is a subject which would demand at least a lecture to itself. Apart from this, there is nothing in analytical geometry which presents any logical difficulty whatever, and I may pass to the slightly more delicate topic of pure geometry.

The nature of a system of pure geometry, such as the ordinary projective system, is most easily elucidated, I think, by contrast with analytical systems. The contrasts, which I have made by implication already, are sharp and striking, and when once they have been clearly observed the road to the understanding of the subject is open. I observed, first, that the points and lines of analytical geometry were *definite objects*, such as the pair of numbers (2, 3). Secondly, I observed that there were no *axioms* in an analytical geometry, which consists of definitions and theorems only; and that it is the definitions which differentiate one system of analytical geometry from another. The business of an analytical geometer is, in short, to investigate the properties of *particular systems of things*. The standpoint of a pure geometer is entirely different. He is not, except for incidental and subsidiary purposes, concerned with particular things at all. His function is always to consider *all things which possess certain properties*, and otherwise to be strictly indifferent to what they are. His 'points' and 'lines' are neither spatial objects, nor sets of numbers, nor this nor that system of entities, but *any* system of entities which are subject to a certain set of logical relations. The particular system of relations which he studies is that which is expressed by the *axioms* of his geometry. It is the axioms only which really matter; it is they which discriminate systems, and the definitions play an altogether subsidiary part.

Suppose, for example, if I may take a frivolous illustration, that a pure geometer and an analytical geometer were to go together to the zoo. The analytical geometer might be interested in tigers, in their colour, their stripes, and in the fact that they eat meat. A point, he would say, is by definition a tiger, and the central theorems of my geometry are that 'points are yellow,' that 'points are striped,' and above all that 'points eat meat.' The pure geometer would reply that he was quite indifferent to tigers, except in so far as they possessed the properties of being yellow and striped; that *anything* yellow and striped was a point to him; that 'points are yellow' and 'points are striped' were the *axioms* of his geometry, and that all he wanted to know was whether 'points eat meat' is a logical deduction from them.

You will, in fact, find, if you consult any standard work on pure geometry, such as Hilbert's *Grundlagen* or Veblen and Young's *Projective Geometry*, that a pure geometer begins somewhat as follows. We consider a system S of objects $A, B, C, \ldots$. We call these objects *points*, and their aggregate *space*; the *plane*, I may say, if I confine myself for simplicity to geometries of two dimensions. From the complete system S which constitutes space we pick out certain partial aggregates $L, M, N, \ldots$, which we call *lines*. If a point A belongs to the particular partial aggregate L, we say that A *lies on* L and that L *passes through* A. These are the *definitions*, and you will observe the quite subsidiary part they play. They are, in fact, purely verbal, and common to all systems; and they do not indicate or imply any special property whatever of the objects which they are said to define, which are indeed often called the *indefinables* of the geometry. The function of the definitions, in fact, is merely to point to the indefinables.

The serious business of the geometry begins when the axioms are introduced. We suppose next that our points and lines are subject to certain logical relations. These suppositions are assumptions, and we call them axioms. To construct a geometry is to state a system of axioms and to deduce all possible consequences from them.

Let us take an actual example. I select the following system of axioms:

Axiom 1. *There are just three different points.*
Axiom 2. *No line contains more than two points.*
Axiom 3. *There is a line through any two points.*

These axioms are consistent with one another, for it is easy to construct a system of objects which satisfy them. We might, for example, take the numbers 1, 2, 3 as our points and the pairs of numbers 2 3, 3 1, 1 2 as our lines, in which case all our axioms are obviously satisfied. Further, the axioms are independent of one another. If the numbers 1, 2, 3 were still our points, but the pairs 2 3, 3 1 alone, and not the pair 1 2, were taken as lines, then the first and second axioms would be satisfied but not the third, and it naturally follows that Axiom 3 is incapable of deductions from the other two. You will have no difficulty in proving in a similar manner, if you care to do so, that each of the three axioms is logically independent of the others. I do not profess to have stated the axioms in the best form possible, but at any rate they are consistent and independent.

It is easy to deduce from our axioms:

Theorem 1. *There are just three lines.*
Theorem 2. *There are just two lines through any point.*

The state of affairs in this geometry is, in short, that suggested by a figure consisting of three points on a blackboard and three lines joining them in pairs. With this, our geometry appears to be exhausted.

The geometry which I have constructed is not an interesting system, since it has no particular application and virtually no content. For our present purpose, however, that is an advantage, as it makes it possible for me to exhibit the system to you in its entirety. However little interest, it may possess, it is a perfectly fair specimen of a pure geometry. All systems of pure geometry, projective geometry, metrical geometry Euclidean or non-Euclidean, are constructed in just this way. They are usually very much more complicated, for you must naturally be prepared to sacrifice simplicity to some extent if you wish to be interesting; but their differences from my trivial geometry are differences not at all of principle or of method, but merely of richness of content and variety of application.

I have now given to you the substance of the orthodox answer to the question which I started by asking. I might expand it indefinitely in detail, but I should add nothing essentially new. Geometry is a collection of logical systems. The number of systems is infinite, and any of you can invent as many new systems as you please; I have myself, with the aid of a few pupils, constructed seven or eight in the course of an hour. There are two kinds of systems, analytical geometries and pure geometries. An analytical geometry attaches the usual geometrical vocabulary to more or less complicated systems of numbers, and investigates their properties by means of the ordinary machinery of algebra and analysis. A pure geometry, on the other hand, considers all possible fields of certain logical relations, and explores their connections without reference to the nature of the objects among which they hold.

I said when I started that I did not propose to offer any very definite suggestions about the teaching of mathematics; but I should like to conclude with a few words about some of the practical problems with which members of this association are primarily concerned. It should be obvious to you by now, I think, that school geometry is, as I stated early in my address, not a well-defined subject, a rational exposition of a particular geometrical system, but a collection of miscellaneous scraps, a selection of airs from different pieces, strung together in the manner which experience shows to be the most enlivening. It would be very easy for me to illustrate my thesis by examining a few passages from current text-books of geometry. What is taught as projective geometry, for example, is not projective geometry, and makes very little pretence of being so; since it is based quite frankly on ratios of lengths and other obviously metrical concepts. Indeed, so far as I know, no English book on projective geometry proper exists, except Mathews' *Projective Geometry*, Dr. Whitehead's tract, and parts of Prof. Baker's treatise. On the other hand, a great deal of what is taught as analytical geometry is not analytical geometry, but an attempt to apply the methods of analytical geometry in other fields, partly to some rough kind of physical geometry supposed to be given intuitively, partly to some system of hybrid pure geometry of which some previous knowledge is assumed. But I must not enter into detail, since detail would mean criticism, and criticism of particular books and particular passages, which I have no time for, and am in any case anxious to avoid.

It is not my object now to offer criticisms of the present methods of geometrical teaching. There are a good many very obvious criticisms suggested by the doctrines which I have tried to explain to you, but I recognise that most of these criticisms would be to a very great extent unfair. It is obvious that the teaching of geometry must be based on what is at best a very illogical compromise, and I am prepared to believe that the compromise evolved by experience, and applied by people who know a good deal more about the practical necessity of compromise than I do, is in substance as reasonable a compromise as the difficulties of the problem permit. My object, so far, has been one not of criticism but of explanation.

I do propose, however, to conclude with one word of criticism, directed only to those of you whose pupils are comparatively able and comparatively mature. There is no doubt that the standard of teaching of analysis has improved out of all knowledge during the last twenty years. The elements of the calculus, even the elements of what foreign mathematicians call algebraical analysis, are taught in a manner with which I personally have comparatively little fault to find. The stupid old superstition that falsehood is always easy and attractive, the truth inevitably repulsive and dull, is almost dead, and it is no longer supposed that ignorance of analysis is in itself a proof either of superior intelligence, or high moral character, or profound geometrical or physical intuition. The teaching of higher geometry does not seem to me to have advanced in the same degree.

I think that it is time that teachers of geometry became a little more ambitious. Geometry in its highest developments may be, for all I know, a more difficult subject than analysis; it is not for me as an analyst to deny it. But what may be true enough of the theory of deformation of surfaces, or of algebraical curves in space, is not even plausible of the elements of higher geometry. Those stages of the subject are surely very much easier than the corresponding stages of analysis. There is something hard and prickly about the basic difficulties of analysis, definite stages on the road where definite types of mind seem to come to an inevitable halt. The difficulties of geometry seem to me a little softer and vaguer; knowledge and general intelligence will carry a student appreciably further on the way. And, if this is so, it seems to me regrettable that students are not given the opportunity, while still at school, of learning a good deal more about the real subject matter out of which modern geometrical systems are built. It is probably easier, and certainly vastly more instructive than a, great deal of what they are actually taught. Anyone who can investigate properties of six or eight points on a conic is capable of understanding what projective geometry is. Anyone who has the faintest hope of a scholarship at Oxford or Cambridge could learn the nature of an axiom, and how a system of axioms may be shown to be consistent with, or independent of, one another. And anyone who can be taught to project two arbitrary points into the circular points at infinity could learn, what he certainly does not learn at present, to attach some sort of definite meaning to the process he performs. Small as my own knowledge of geometry is, and slight as are my qualifications for teaching it to anybody, I have not yet encountered the student who finds difficulty with such ideas when once they are put before him clearly. I am well aware of the very great services which the Association has rendered in the improvement of geometrical teaching. I think that it might well now concentrate its efforts on a general endeavour to widen the horizon of knowledge, recognising, as regards niceties of logic, sequence, and exposition, that the elementary geometry of schools is a fundamentally and inevitably illogical subject, about whose details agreement can never be reached.

1.2

What is Geometry?

MICHAEL ATIYAH

1. History

Of all the changes that have taken place in the mathematical curriculum, both in schools and universities, nothing is more striking than the decline in the central role of geometry. Euclidean geometry, together with the allied subject of projective geometry, has been dethroned and in some places almost banished from the scene. While educational reform was certainly needed there is always the danger that the pendulum may swing too far the other way and that insufficient attention may be paid to geometry in its various forms. Much of the difficulty here centres round the elusive nature of the subject: What is geometry? I would like to examine this question in a very general way in the hope that this may clarify the educational reasons for teaching geometry, and for deciding what is appropriate material at different levels.

Let me begin by taking a historical look at the development of mathematics. It is I think no accident that geometry, in the hands of the Greeks, was the first branch of mathematics to reach maturity. The fundamental reason is that geometry is the least abstract form of mathematics: this means that it has direct applicability to everyday life and also that it can be understood with less intellectual effort. By contrast algebra is the essence of abstraction, involving a dictionary of symbolism which has to be mastered by great effort. Even arithmetic, based as it is on the process of counting, depends on its own dictionary such as the decimal system, and took longer to evolve.

Of course, at a sophisticated level, geometry does involve abstraction. As the Greeks recognised, the points and lines which we meet in the real world are only approximations to some 'ideal' objects, in an 'ideal' world where points have no magnitude and lines are perfectly straight. These philosophical reflections do not

The 1982 Presidential Address.
First published in *Mathematical Gazette* **66** (October 1982), pp. 179–184.

however worry the practitioner of geometry, be he a school-child or a civil engineer and geometry at this level remains the practical study of physical shapes.

For many centuries Euclidean geometry dominated the mathematical scene, but the emergence of algebra, its application by Descartes to geometry and the subsequent development of calculus altered the whole character of mathematics. It became much more symbolic and abstract. Inevitably geometry came to be regarded as primitive and old-fashioned.

While these rival branches of mathematics were being developed the foundations of geometry and its relation to the physical world were being re-examined. In the nineteenth century the famous 'parallel postulate' of Euclid, asserting the existence of a unique line parallel to a given line and passing through a given point, was shown to be independent of the other axioms. Non-Eudidean geometries, in which this postulate does not hold, were discovered. This had the profound if disturbing effect of liberating geometry from physics. While there is only one physical universe there are many different geometries and it is not clear which one is most relevant to our universe. For a while algebra attempted to take advantage of this division of ranks in the geometrical field. Felix Klein in his famous *Erlanger Programm* attempted to define geometry as the study of those properties which are invariant under a given group of symmetries: different geometries corresponding in this way to different symmetry groups. While this has been a very fruitful point of view in connection with non-Euclidean geometries its scope had already been undermined by the earlier far-reaching ideas of Bernhard Riemann. For Riemann space did not have to be homogeneous, its curvature could vary from point to point and there might be no symmetries at all. Instead of group theory Riemann based his geometry on the differential calculus and, as we know, his view-point was ultimately vindicated by Einstein's general theory of relativity.

The outcome of all this introspection by geometers showed that geometry is not just the study of physical space. In particular it is not restricted to 3 (or 4) dimensions. But in that case what is the use of all these other geometries? Are they simply abstract games played by mathematicians? I shall attempt to answer these questions by showing how abstract 'spaces' and geometries occur naturally in a wide variety of ways. Some of my examples will be blindingly familiar but others may be more novel.

2. Examples of Non-Physical Spaces

2.1. Graphs

At all levels in mathematics graphs are familiar and widely used. The simplest is perhaps the distance/time graph for a moving object. Of course the (x, t) plane is part of Einstein's 4-dimensional space-time, but in our graph the time variable t is replaced by a second space variable y. For other examples, however, such as those used in economics, the variables may have no relation to space-time at all. The plane in which we draw our graph is an abstract plane, but the practical advantage of such pictorial representations is clearly enormous. The advantage rests on the capacity of the brain to see two-dimensional patterns literally at a glance.

2.2. The Complex Plane

The representation of complex numbers $x + iy$ by points of a plane is again an almost too familiar example of an 'abstract' plane. If for instance x represents distance along a fixed line, and if an algebraic problem involving this distance leads to a complex solution $x + iy$, then y does not correspond to any real direction. Familiarity breeds contempt and long exposure to complex numbers makes the complex plane almost tangible. However, as the great Gauss said 'the true metaphysics of $\sqrt{-1}$ is elusive'. Those who have to introduce students to complex numbers for the first time may well agree.

2.3. Riemann Surfaces

While the two previous examples are familiar, when we combine them the geometry becomes more serious. Consider for example the graph of the (two-valued) function $y^2 = f(x)$ where f is some polynomial. When x, y are real we can draw this in an ordinary real plane, but if we take x and y both to be complex the graph becomes a real surface in a four-dimensional real space. This is the Riemann surface of the function and its geometric (or topological) properties are of fundamental importance in the analytic study of the function. This illustrates the importance of 'abstract' geometrical ideas in the study of polynomials or analytic functions of several variables. In fact complex algebraic (and analytic) geometry are now flourishing branches of mathematics.

2.4. Dynamics

In Newtonian mechanics the motion of a particle in a given field of force is determined if we know its position and velocity at one instant. To describe its subsequent motion it is then convenient to introduce the 'phase-space' of pairs (x, v), where both components are 3-vectors and represent position and velocity respectively. The motion will then be represented by a curve $((x(t), v(t))$ in this 6-dimensional space. For instance if motion is on a line rather than in 3-space, the phase space is then 2-dimensional and simple harmonic motion corresponds to circles in this phase-plane. These phase pictures are extremely useful in general dynamical problems.

2.5. *Rigid Bodies*

Suppose, instead of a particle, we have a rigid body. Before proceeding to look at the motion of such a body consider first the static problem of just describing its position. Assuming its centre of mass fixed at the origin we are left with the rotations about the origin. Such rotations have three degrees of freedom but these do not correspond to specifying three Cartesian coordinates (x, y, z). The 'space' of rotations is actually the non-Euclidean (elliptic) 3-space, the quotient of the 3-sphere by the anti-podal map. (The most elegant way of seeing this is to use quaternions. The quaternions q of unit norm form the 3-sphere and act on the 3-space of imaginary quaternions x by $q \times q^{-1}$. This gives all rotations of 3-space but $\pm q$ give the same rotation and are anti-podal pairs on the 3-sphere.) Thus non-Euclidean geometry appears here out of Euclidean geometry.

2.6. *Line Geometry*

If our rigid body is replaced by a long thin rod idealised to a line of infinite length then we cannot describe its position in quite the same way because its centre of mass is not now defined. Instead we can pick two points $X = (X_1, X_2, X_3)$ and $Y = (Y_1, Y_2, Y_3)$ on the line and consider the 6-vector $(X - Y, X \wedge Y)$. If we pick any other pair on the line this 6-vector gets multiplied by a scalar. Moreover its components satisfy the quadratic relation

$$(X - Y) \cdot (X \wedge Y) = 0.$$

This means that lines in 3-space can be parametrized by points of a quadric in projective 5-space. This is the famous Klein representation. When I first encountered it as a young student I thought it one of the most beautiful ideas in mathematics. To illustrate its properties let me recall that in 3-space a hyperboloid of one sheet has two systems of generating lines. Similarly in 5-space the Klein quadric has two systems of generating planes. A plane of one system parametrizes all lines in 3-space through a fixed point, while a plane of the other system parametrizes all lines in 3-space lying in a fixed plane. From this we can immediately deduce the incidence properties of the generating planes. Namely two planes of the same system always meet in a point (e.g. because two points in 3-space lie on a unique line), while planes of opposite systems usually do not meet (because there is in general no line in 3-space which goes through a given point p and lies in a given plane π. However exceptionally planes of opposite systems can meet and then they meet in a line (if P lies in π there is a whole pencil of lines through P lying in π).

It is interesting that this Klein representation has in recent years played a fundamental role in the work of Roger Penrose in theoretical physics. Roughly speaking

Penrose thinks of the Klein quadric as space-time (after complexification) and the original 3-space (also complexified) is then a fundamental auxiliary space (called twistor space) which is supposed to be more basic in some ways than space-time (e.g. it has fewer dimensions).

2.7. Function Spaces

If instead of a rigid rod we now have a finite piece of string then its possible positions in 3-space will require infinitely many parameters. We can describe its position by three functions $x(t)$, $y(t)$, $z(t)$ where t is a parameter on the string (e.g. distance from one end). The 'space' of all positions of the string is therefore an infinite-dimensional space. Such function-spaces occur frequently in calculus of variations problems (when we are trying to minimise some quantity depending on a function). Geometrical ideas have proved very useful in these contexts, notably in connection with 'fixed-point theorems'.

3. Conclusions

The preceding examples were designed to illustrate the fact that spaces, frequently of high dimension, arise quite naturally out of realistic situations. I chose examples from 3-dimensional mechanics to emphasise this reality. Of course a sophisticated mathematician will quite happily start with n real variables $(x_1, \ldots, x_n)$ and regard these as coordinates of a point in n-dimensional space. Such an abstract starting point may not however be entirely convincing to the sceptic who may have serious doubts about the 'meaning' of higher-dimensional geometry.

Are we now any further forward towards answering the initial question 'What is geometry?'? If geometry is not just the study of physical space but of any abstract kind of space does this not make geometry coincide with the whole of mathematics? If I can always think of n real variables as giving a point in n-space what distinguishes geometry from algebra or analysis?

To get to grips with this question we have to appreciate that mathematics is a human activity and that it reflects the nature of human understanding. Now the commonest way of indicating that you have understood an explanation is to say 'I see'. This indicates the enormous power of vision in mental processes, the way in which the brain can analyse and sift what the eye sees. Of course, the eye can sometimes deceive and there are optical illusions for the unwary but the ability of the brain to decode two- and three-dimensional patterns is quite remarkable.

Sight is not however identical with thought. We have trains of thought which take place in sequential form, as when we check an argument step by step. Such logical or sequential thought is associated more with time than with space and can be carried out literally in the dark. It is processes of this kind which can be formalised in symbolic form and ultimately put on a computer.

Broadly speaking I want to suggest that geometry is that part of mathematics in which visual thought is dominant whereas algebra is that part in which sequential thought is dominant. This dichotomy is perhaps better conveyed by the words 'insight' versus 'rigour' and both play an essential role in real mathematical problems.

The educational implications of this are clear. We should aim to cultivate and develop both modes of thought. It is a mistake to overemphasise one at the expense of the other and I suspect that geometry has been suffering in recent years. The exact balance is naturally a subject for detailed debate and must depend on the level and ability of the students involved. The main point I have tried to get across is that geometry is not so much a branch of mathematics as a way of thinking that permeates all branches.

Desert Island Theorems

Group A: Greek Geometry

A1

Pythagoras' Theorem

CHRIS DENLEY'S CHOICE

Pythagoras' Theorem: In any right-angled triangle, the square on the hypotenuse is equal to the sum of the squares on the other two sides: or $a^2 = b^2 + c^2$.

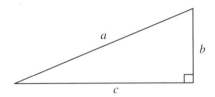

In my early days as a mathematics teacher, I once sat down to the delights of school dinner in the egalitarian dining room with the head teacher and one of my fourteen year-old pupils. The head began talking about studying mathematics, stating that he had never seen any earthly point in learning such things as Pythagoras' Theorem as there is no use for it whatsoever. So there was I doing my best to get across Pythagoras and other such fundamental results to the boy dining with us. Quite what inference he drew from the head's comments is difficult to know. The incident could have quite easily reinforced a commonly held belief in the pointlessness of mathematics, a belief probably born of the 'good old days' of education when levels of achievement were little more than a measure of a pupil's ability to regurgitate techniques. If you were lucky, as I was, the fascination for mathematics carried you through, and at least some of the relevance became self-apparent. My new head teacher it seems had not been so lucky, but that cannot excuse his unprofessional comments, which left me speechless.

Not that mathematics has to be useful to justify it as a course of study. It is a perfectly valid pursuit for its own aesthetics, elegance and fascination, in much the same way as is the study of music or art. But having worked in the aerospace

Chris Denley, Aeroelastician, BAE SYSTEMS.

industry for the last ten years I have been surprised to realise that in fact most, if not all, of the school mathematics syllabus is used either directly or indirectly in the engineering and industrial world. Had this head teacher really wanted to demonstrate mathematical irrelevance then he couldn't have picked a less appropriate example than Pythagoras' Theorem. If I'd had time to think of the perfect reply, it might have gone something like this:

'Like many basic mathematical principles, Pythagoras' Theorem, relating the lengths of the sides of a right-angled triangle, is beautifully simple and extremely powerful. Not only are there the obvious practical uses in measurement and construction, but it is involved in trigonometry and forms a crucial element for any study in the Cartesian frame of reference, such as vectors, coordinate geometry and complex numbers. With the vast range of vector quantities in the physical world (force, acceleration, etc.) the applications in physics and engineering are innumerable, for it is Pythagoras we use to combine components of vectors at right angles. Complex numbers, in which Pythagoras gives us the modulus, have wide ranging applications in engineering: electrical science, fluid mechanics, structural dynamics (vibrations), automatic control, digital signal processing, aeroelastics, to name a few. Without complex numbers there would be no oscillatory solutions to second order linear differential equations. And without these there would be no aeroservoelastic control of aircraft wings.'

This last fact is one that the head teacher would do well to bear in mind today, as he flies off in the school holidays in one of the latest airliners. He would not notice the distinct lack of wing vibration these days, but would perhaps appreciate the resulting improvement in passenger comfort and be very grateful for the reduced metal fatigue. Aerospace engineers, like everyone in the technical and practical world, not only depend daily on Pythagoras' Theorem but they too were once fourteen year-old pupils. All those years ago, the head teacher, having made his judgements about the usefulness and power of beautifully simple mathematical principles, should have reflected on this as he set off down the corridor to teach his own specialist subject, Esperanto.

A2

The Angle at the Centre of a Circle is Twice the Angle at the Circumference

CHARLIE STRIPP'S CHOICE

I was born in the mid-1960s. Like many of my generation in England, I managed to progress successfully through school and university, gaining a mathematics-based degree without ever properly encountering deductive geometry. However, when I discovered that my vocation lay in teaching mathematics and began school teaching, I realised that my older colleagues had a definite edge when it came to many 'recreational' mathematics problems. Often they also had a different and more elegant way of approaching mathematical problems in general. Their edge was that they had been taught deductive geometry. The generation I represent is a lost generation.

A colleague recommended that I seek out an old school copy of *The Elements of Euclid* – copies are not hard to find in second-hand bookshops. I found a copy, by Isaac Todhunter (1889 edition), which cost me £1 and have dipped into it frequently over the past 15 years. It started me off on the road to becoming 'geometrically literate'.

My chosen theorem states that the angle at the centre of a circle is double that of the angle at the circumference, lying on the same arc. Euclid devoted the whole of Book III to the geometry of the circle and this theorem is his Proposition 20.

It is special to me because it leads easily to other powerful results, such as the angles in the same segment are equal, the opposite angles of a cyclic quadrilateral add up to 180 degrees, the angle in a semi-circle is a right-angle and the alternate angle theorem. (Euclid has them as Propositions 21, 22, 31 and 32 respectively.) Yet its proof is readily followed by many 12-year olds with a basic knowledge of the properties of isosceles triangles. It also leads to some interesting questions such as: What happens as the point moves round the circumference?

Charlie Stripp, Mathematics Teacher, Exeter College, Devon. Vice-Chair, The Mathematical Association Teaching Committee.

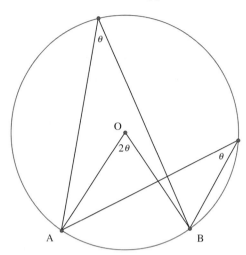

Geometrical proofs, such as the proof of this theorem, provide an excellent introduction to simple mathematical reasoning. Its development to form further theorems demonstrates, in an accessible and engaging way, how a whole set of linked results can be generated. This offers a good insight into the development of mathematics as a whole.

A3

Archimedes' Theorem on the Area
of a Parabolic Segment

TOM APOSTOL'S CHOICE

Although Newton and Leibniz are regarded as the founders of calculus, its historical development started when early Greek mathematicians used the method of exhaustion to investigate problems concerning areas and volumes. For areas of plane regions, this method employs inscribed and circumscribed polygons with an increasing number of edges. Archimedes (287–212 B.C.E.) used the method with great success. One of his many accomplishments was determining the area of a parabolic segment.

The parabolic segment in Figure A3.1(a) is the shaded region below the graph of the parabola $y = x^2$ and above the axis in the interval 0 to x. Archimedes showed that the area of this region is one-third the area of the circumscribed rectangle. This rectangle, of base x and altitude x^2, has area x^3, so the parabolic segment has area $x^3/3$. Today this result is a simple exercise in integral calculus because integration of x^2 gives $x^3/3$. But in ancient times it was an amazing achievement. In fact, it is one of the landmark discoveries that helped lay the foundations for the development of integral calculus eighteen centuries later.

Here we solve this problem without the integral calculus, by a method of Mamikon Mnatsakanian that is not only simpler than that of Archimedes but also more powerful because it applies to generalized segments in which x^2 is replaced by x^r for any positive real r.

Although the parabola has equation $y = x^2$, we shall not need this formula. We use only the fact that the tangent line above any point x cuts off a subtangent of length $x/2$ as in Figure A3.1(b). The shaded portion in Figure A3.2 is the region obtained by drawing all tangent lines to the parabola between 0 and x and cutting them off at the x-axis. We call this the tangent sweep.

In Figure A3.1(b) the parabolic segment is divided into two regions, the tangent sweep (of area S, say), and a right triangle (of area T). Note that $T = R/4$, where R

Tom M. Apostol, Teacher of Calculus, California Institute of Technology.

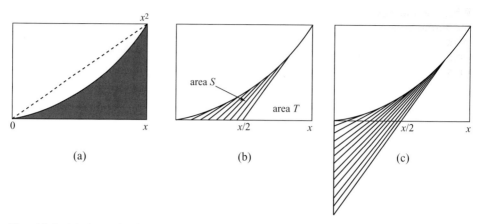

Fig. A3.1. (a) A parabolic segment. (b) The tangent sweep cut off by the x-axis. (c) The region obtained by doubling the lengths of the tangent segments in (b).

is the area of the circumscribed rectangle. We will prove that $S = T/3$, so $S + T$, the area of the parabolic segment, is $4T/3 = R/3$, as asserted.

To prove $S = T/3$, refer to Figure A3.1(c), where each tangent segment from Figure A3.1(b) is doubled in length to reach the y-axis, as shown. The area of the expanded tangent sweep is $4S$. The expanded region consists of the portion above the x-axis, with area S, and the right triangle below the x-axis, with area T. Hence $4S = S + T$, so $S = T/3$, as required.

Editor's Note

Tom Apostol gave a much fuller explanation of Mamikon's methods at a mathematical colloquium held in honour of his (Apostol's) 50 years at Caltech, and an adapted version of that talk duly appeared in the Institute's journal (Tom M. Apostol A visual approach to calculus problems, *Engineering & Science* **63**, 3 (2000), California Institute of Technology, pp. 23–31). Though the article contains a detailed description of Mamikon's treatment of the parabolic segment, it should be noted that it does differ somewhat from that presented here.

The fascinating story of Mamikon Mnatsakanian and his method is outlined in Apostol's article, in these terms:

The method was conceived in 1959 by Mamikon A. Mnatsakanian, then an undergraduate at Yerevan University in Armenia. When he showed his method to Soviet mathematicians they dismissed it out of hand and said, "It can't be right – you can't solve calculus problems that easily." He went on to get a PhD in physics, was appointed a professor of astrophysics at the University of Yerevan, and became an international expert in radiative transfer theory. He also continued to develop his powerful geometric methods. He eventually published a paper outlining them in 1981, but it seems to have escaped notice, probably because it appeared in Russian in an Armenian journal with limited circulation...

When the old Soviet system broke down, Mamikon happened to be visiting California and through the efforts of several mathematicians he was granted status as an alien of extraordinary ability. He developed further his methods for use in school mathematics. His work is now supported by *Project MATHEMATICS!*, which designs and produces computer animated mathematics videotapes and which is led by Apostol.

A4

An Isoperimetric Theorem

JOHN HERSEE'S CHOICE
(REALISED WITH THE HELP OF DOUG FRENCH)

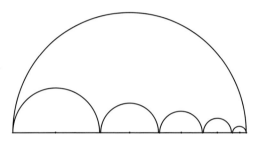

The diagram represents a large semicircle with a number of smaller semicircles of different sizes. In the diagram there are five semicircles, but there can be any number. The only restriction is that the centre of each semicircle must be on the diameter of the large semicircle and they must span the whole of that diameter.

- Prove that the total length of the perimeters of the small semicircles is equal to the perimeter of the original semicircle, regardless of their number or combination.
- Can you find other curves that can be drawn inside the large semicircle and produce a similar result? Does the large 'semicircle' have to be a semicircle?

Let D be the diameter of the large semicircle and $d_1, d_2, d_3, \ldots, d_n$ be the diameters of the smaller ones. The sum of the lengths of the arcs of the smaller semicircles is readily shown to be equal to that of the larger semicircle as follows:

$$\frac{1}{2}\pi d_1 + \frac{1}{2}\pi d_2 + \frac{1}{2}\pi d_3 + \cdots + \frac{1}{2}\pi d_n = \frac{1}{2}\pi(d_1 + d_2 + d_3 + \cdots + d_n) = \frac{1}{2}\pi D.$$

This refers to the arc length, but it immediately follows that the result holds for the perimeter by including the diameter as well. It is true for other curves, provided the

John Hersee, Executive Director of the School Mathematics Project (SMP), 1976–85. President, The Mathematical Association, 1992–93.

curves are similar, because it is only necessary to replace the constant $\frac{1}{2}\pi$ in the proof above by the appropriate constant of proportionality and to take as diameter whatever base line the curves stand on.

The diagram below shows the graph of $y = \sin x$ together with the graph of $y = \frac{1}{2}\sin 2x$ between $x = 0$ and $x = \pi$. Since the two sections of the second curve are similar to the first curve, the total lengths of the two curves are the same.

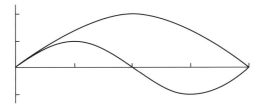

Indeed, the result can be applied to much more exotic curves like that shown in the next diagram. We do not need either to represent the curve algebraically or to determine the length of the curve to be sure that the lengths are the same.

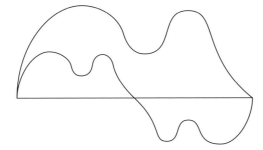

Furthermore the result is not restricted to smooth curves – it applies to all similar shapes. Similar triangles are particularly simple and striking, because the parallel lines needed to construct the diagram provide an immediate visual proof.

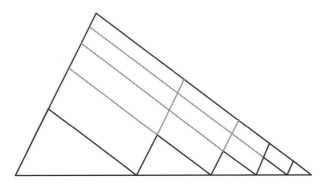

A5

Ptolemy's Theorem, Its Parent and Offspring

TONY CRILLY'S AND COLIN FLETCHER'S CHOICE

What makes Ptolemy's Theorem such a great result is its plasticity. It can be pushed and pulled to give all sorts of classical results and surprising connections. The 'no frills' version of the theorem applies to a *cyclic* quadrilateral:

Ptolemy's Theorem: If the vertices of a quadrilateral $ABCD$ lie on a circle then

$$AB \cdot CD + BC \cdot DA = AC \cdot BD,$$

where AB and CD, and BC and DA are the lengths of opposite pairs of sides and AC and BD the lengths of its diagonals.

Claudius Ptolemy (c.100–c.175 A.D.) lived and worked in Alexandria [1]. Today he is most widely known as the originator of the astronomical theory which places the Earth at the centre of the Universe, the model subsequently overthrown by Galileo and Copernicus. Not so well known is his contribution to optics and the *Geography* which contained 'the measurement of the world' so esteemed by the Renaissance map-makers [2]. Ptolemy's *Syntaxis Mathematica* ('the great compilation') was written in about 150 A.D. It became known in the medieval period as the *Almagest*, a corruption of the Arabic *almegiste* ('the greatest') and it ran to thirteen books [3, p. 77]. Ptolemy's celebrated geometrical theorem appears in Book One. It continues to be given a full measure of attention by mathematicians and historians [4–7].

Proofs The proof given by Ptolemy is undoubtedly elegant [8, p. 36]. The first shock is that the proof may not be due to Ptolemy at all. Tobias Dantzig has said the 'nimble virtuosity' of the proof suggests the theorem 'did not spring from the brain of an astronomer, but was the discovery of some brilliant geometer, most likely

Tony Crilly and Colin Fletcher, Mathematicians, respectively at Middlesex University and University of Wales, Aberystwyth. Originators of the concept of 'Desert Island Theorems'.

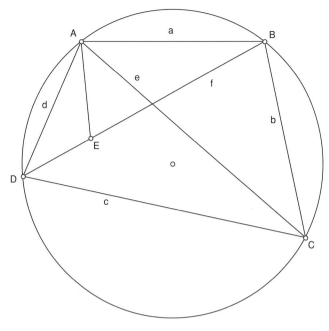

Fig. A5.1.

Apollonius' [9, p. 173]. Notwithstanding, we should be glad that Ptolemy recorded it for posterity.

In Figure A5.1, the construction of the point E holds the key: we choose E on BD so that $D\hat{A}E = B\hat{A}C$. Since $\triangle DAE$ is similar to $\triangle BAC$ we have

$$\frac{ED}{BC} = \frac{DA}{AC}.$$

Using $D\hat{A}E = B\hat{A}C$ again, we have $D\hat{A}C = B\hat{A}E$. Thus $\triangle DAC$ is similar to $\triangle BAE$ and so

$$\frac{BE}{AB} = \frac{CD}{AC}.$$

Therefore

$$BE + ED = \frac{AB \cdot CD}{AC} + \frac{BC \cdot DA}{AC}$$

and since $BE + ED = BD$, the theorem follows.

The theorem can also be proved using trigonometric formulae, and, in another way using the Cosine Rule [10, 11]. However, we believe that purely geometrical arguments are the most appealing in the context of basic propositions. Here the theorem can be proved using geometric inversion, mapping a point P to its inverse P' with respect to a point A and the relation $AP \cdot AP' = k^2$. Using this approach,

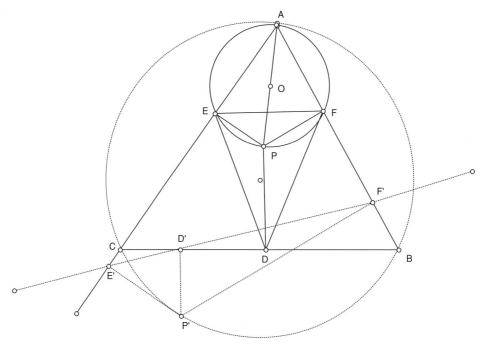

Fig. A5.2.

the cross-ratio

$$[A, B; C, D] = \frac{AC/CB}{BD/DB} = \frac{AC \cdot DB}{CB \cdot BD}$$

(which is invariant under inversion) can be used to express the theorem [12]:

$$[B, C; A, D] + [B, A; C, D] = 1.$$

A surprising and quite different proof of Ptolemy's Theorem can be obtained from the properties of the pedal triangle [13, p. 42].

Given $\triangle ABC$, the small triangle $\triangle DEF$ in Figure A5.2, the one subtended by the feet of the perpendiculars from P, forms the *pedal triangle*. We now find relations between the sides of the pedal triangle and those of $\triangle ABC$. First EF. Applying the Sine Rule in $\triangle ABC$, we obtain

$$\frac{BC}{\sin A} = 2R$$

where R is the radius of the circle which circumscribes $\triangle ABC$. Applying it to $\triangle AEF$ circumscribed by the circle $AEPF$ gives

$$\frac{EF}{\sin A} = 2\left(\frac{AP}{2}\right)$$

and hence

$$EF = \frac{BC \cdot AP}{2R}.$$

Similarly, $DE = \frac{AB \cdot CP}{2R}$ and $DF = \frac{AC \cdot BP}{2R}$.

Although Figure A5.2 shows the point P in 'normal position' (*inside* $\triangle ABC$), it can be placed anywhere in the plane and, if necessary, the sides of $\triangle ABC$ extrapolated for the purposes of constructing perpendiculars. In the case where P lies on the circumcircle of $\triangle ABC$, the pedal triangle degenerates to a straight line (the Simson line, or the would-be 'Wallace-Simson line' illustrated with 'dashed' symbols in Figure A5.2). On this line

$$DE + DF = EF$$

and therefore

$$AB \cdot PC + AC \cdot PB = BC \cdot PA,$$

the statement of the theorem when P, B, A, C, form a cyclic quadrilateral $PBAC$.

Ptolemy's Theorem is occasionally given in an extended form, a form which also provides a converse to the plain version we stated at the beginning:

1. Ptolemy's Theorem (Parent)

For a convex quadrilateral $ABCD$, $AB \cdot CD + BC \cdot DA \geq AC \cdot BD$, with equality if and only if the quadrilateral is cyclic.

2. Offspring

Applying the theorem to a rectangle (which is a cyclic quadrilateral) Pythagoras' Theorem follows immediately! In the case where the diagonals of the cyclic quadrilateral are orthogonal there is an immediate relationship between the lengths of the sides and the area A of the quadrilateral, namely

$$ac + bd = ef = 2A.$$

Without requiring the diagonals to be orthogonal, McNab [14] shows that

$$16R^2 = (ab + cd)(ac + bd)(ad + bc)/A^2$$

where R is the radius of the circumscribing circle and A is the area of the cyclic quadrilateral; this is reworked in [15]. In terms of the ratios of diagonals, the result

$$\frac{e}{f} = \frac{ad + bc}{ab + cd}$$

is given in [16] and also by Cayley a hundred years previously who derived it from an algebraic point of view [17].

Ptolemy's Theorem and the application of the full Sine Rule

$$\frac{a}{\sin A} = \frac{b}{\sin B} = \frac{c}{\sin C} = 2R$$

in the circumscribed $\triangle ABC$ leads to the familiar trigonometric identities of the type

$$\sin(A + B) = \sin A \cos B + \cos A \sin B$$

and for this reason the expressions for $\sin(A \pm B)$, $\cos(A \pm B)$ are known in some quarters as *Ptolemy's formulae* [18, pp. 183–184].

The theorem has tempted number theorists to find integer-sided cyclic quadrilaterals. Combining triangles formed with Pythagorean triples such as (3,4,5) and (8,15,17), for example, Barisien (around 1913) produced a cyclic quadrilateral with (in the terms of Figure A5.1), $a = 75$, $b = 68$, $c = 40$, $d = 51$, $e = 77$, $f = 84$. The diameter of the circumscribing circle in this example is $2R = 85$ and the diagonals are orthogonal [19, 2: pp. 216–221]. Many notable mathematicians have been attracted to finding such quadrilaterals from Bascara in the Middle Ages to Euler in the eighteenth century with a flowering of this art-form in the nineteenth.

The application of Ptolemy's theorem to the following optimization problem is nothing short of ingenious; but as with many proofs in elementary geometry, it is almost impossible to know who gave it first. The problem itself has been referred to as Steiner's problem [20, pp. 354–359]. It is required to find the Fermat point of a triangle, the location of the point P which minimizes the sum $PA + PB + PC$.

We shall illustrate it here in the case where none of the angles of the given $\triangle ABC$ exceeds $120°$. On a side, say BC, of the given $\triangle ABC$ construct an *equilateral* triangle $\triangle BCD$, and apply the extended Ptolemy's Theorem to the quadrilateral $PBDC$ and we have $PB \cdot CD + PC \cdot BD \geq PD \cdot BC$. Since $\triangle BCD$ is equilateral, $PB + PC \geq PD$ and so $PA + PB + PC \geq PA + PD$.

Referring to Figure A5.3, for the left-hand-side to be a minimum, P should be at F which is on the line AD *and* on the circle. As a corollary this shows that the Fermat point F is the intersection of the three circles so constructed from the three sides of the given $\triangle ABC$. Notice here the link with *Napoleon's Theorem* which states that centres of the three circles form an equilateral triangle.

3. The Extended Family

Arthur Cayley (1821–1895) investigated problems of the 'Ptolemy Theorem' type at various times. For instance, he solved the problem where the four points are not

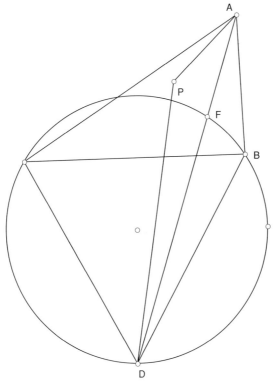

Fig. A5.3.

required to lie in the plane [21]. In his very first paper, written while an undergraduate, he considered the geometry of the mutual distances of five points in space, and as a side result obtained Ptolemy's Theorem by way of determinants [22]. His solution (Figure A5.1) was written in terms of the symmetric determinant:

$$D = \begin{vmatrix} 0 & a^2 & e^2 & d^2 \\ a^2 & 0 & b^2 & f^2 \\ e^2 & b^2 & 0 & c^2 \\ d^2 & f^2 & c^2 & 0 \end{vmatrix}.$$

He noted that $D = 0$ yields Ptolemy's Theorem. Using DERIVE we obtain the complete and symmetric factorization of D:

$$(ac + bd + ef)(ac + bd - ef)(ac - bd + ef)(ac - bd - ef).$$

The study of squared distances forms a link with modern invariant theory and the study of Gröbner bases of the 1990s [23].

Perhaps the most startling generalization of the basic Ptolemy's Theorem is the one offered by the Irish mathematician John Casey (1820–1891). A self-taught mathematician, he first established himself in the world of elementary schools around his home town of Kilkenny, County Cork and in Tipperary. In between his teaching duties, he had the temerity to solve a version of Poncelet's polygon problem and the geometrical proof brought him to the notice of geometers in Dublin's Trinity College. Induced to join them as a student he obtained his degree at the age of forty-two. A stream of geometrical papers later, he gained membership of the Royal Irish Academy in May 1866. A specialist in the geometry of circles and conics he wrote influential textbooks [24].

One form of Casey's Theorem states that if four circles touch a fifth circle externally at four points then $d_{12}d_{34} + d_{23}d_{41} = d_{13}d_{24}$, where $d_{i,j}$ denote the lengths of the common tangents. If the four circles reduce to four points then Ptolemy's Theorem is obtained [25, p. 95].

The theorems which have been discussed here are essentially products of the period to the end of the nineteenth century. In the twentieth Ptolemy's Theorem entered the mathematical literature in an abstract axiomatic formulation [26]. The charm of the old geometrical theorem was pressed into the service of the strict deductive canon.

'**Definition:** A metric space is said to be *ptolemaic* if . . .' but that, as they say, is another story.

References

1. G. J. Toomer Ptolemy, in C. C. Gillispie (Ed.), *Dictionary of Scientific Biography* (Charles Scribner's Sons, 1975), 11, 186–206.
2. J. Lennart Berggren & Alexander Jones, *Ptolemy's Geography* (Princeton University Press, 2000).
3. Ivor Grattan-Guinness, *The Fontana History of the Mathematical Sciences* (Fontana, 1997).
4. G. J. Toomer (Trans.), *Ptolemy's Almagest* (Duckworth, 1984).
5. Lucs N. H. Blunt, Philip S. Jones & Jack D. Bedient, *The Historical Roots of Elementary Mathematics* (Dover, 1988).
6. Carl B. Boyer, *A History of Mathematics* (John Wiley, 1968).
7. Thomas L. Heath, *History of Greek Mathematics*, 2 vols. (OUP, 1931).
8. J. L. Heiberg (Ed.), *Claudii Ptolemaei Opera Quae Exstant Omnia* (Teubner, 1898, 1957).
9. Tobias Dantzig, *The Bequest of the Greeks* (Greenwood Press, 1954).
10. Ho-Joo Lee, A trigonometric proof of Ptolemy's theorem, *Math. Gaz.* **85** (2001), 479–480.
11. A. G. Sillitto, Ptolemy's theorem, *Math. Gaz.* **50** (1966), 388.
12. www.maths.gla.ac.uk/~wws/cabipages/klein/ptolemy.html
13. H. S. M. Coxeter & S. L. Greitzer, *Geometry Revisited* (MAA, 1967).
14. D. S. MacNab, Cyclic polygons and related questions, *Math. Gaz.* **65** (1981), 22–28.
15. Larry Hoehn, Circumradius of a cyclic quadrilateral, *Math. Gaz.* **84** (2000), 69–70.

16. L. E. Ellis, The diagonals of a cyclic quadrilateral, *Math. Gaz.* **71** (1987), 228–229.

17. Arthur Cayley, Note on the two relations connecting the distances of four points on a circle, *Coll. Math. Papers* **12**, 576–577.

18. Morris Kline, *Mathematical Thought from Ancient to Modern Times* (OUP, 1972).

19. L. E. Dickson, *History of the Theory of Number*, 3 vols. (1919–1923); reprint (Chelsea, 1971).

20. Richard Courant & Herbert Robbins, *What is Mathematics?* (OUP, 1941).

21. Arthur Cayley, Problem 2466, *Educational Times*, vol. 8, 86–87. (*Coll. Math. Papers*, **7**, 585–587).

22. Arthur Cayley, On a theorem in the Geometry of Position, *Cambridge Mathematical Journal* **2** (1841), 267–271 (*Coll. Math. Papers* **1**, 1–4).

23. Bernd Sturmfels, *Algorithms in Invariant Theory* (Springer, 1993).

24. John Casey, *A Sequel to the First Six Books of the Elements of Euclid* (Hodges Figgis, 1881).

25. D. M. Y. Sommerville, *Analytical Conics* (George Bell, 1945).

26. Leonard M. Blumenthal, *Theory and Applications of Distance Geometry* (Oxford: Clarendon, 1953).

Editor's Note

For the authors' invocation of desert island theorems see:

Tony Crilly & Colin Fletcher, Desert island theorems, *Math. Gaz.* **82** (1998), 2–7.

Tony Crilly, Desert island theorems: My magnificent seven, *Math. Gaz.* **85** (2001), 2–9.

II

The History of Geometry

2.1

Introductory Essay: A Concise and Selective History of Geometry from Ur to Erlangen

CHRIS PRITCHARD

Mankind has come a long way in its understanding of shape, position and movement. It has done so with a singular lack of uniformity of 'progress' and rarely for more than four centuries at a time in any particular part of the world. In their early 'intuitive stage' of geometrical appreciation, no doubt humans were awed by the movements of the heavenly bodies, and conscious of the variety of speeds attained by different animals in pursuit of food or safety. But they were perhaps barely aware of the fixed 'shape' that a plumb line makes with the surface of still water or of the forms found in rocks and minerals. This was to change with the emergence of societal existence when their rather loose feel for geometry grew into an empirical understanding. For the first time, there was a need to solve problems arising out the possession of land – geometry means land measurement – and the construction of homes, irrigation systems and monuments to the gods. To a greater or lesser extent, such needs emerged independently amongst the great riverine civilizations of the Tigris and Euphrates, the Nile, the Indus, and the Yangtze and Huang Ho [1, 2, 3].

The earliest geometry belongs to the Sumerian period of the Babylonians, say 3000 BC. The Babylonians were interested in the areas of some plane figures. They mistakenly calculated the area of a general quadrilateral by taking the product of the average lengths of opposite sides. This propensity to averaging was a feature of the early geometers here and elsewhere [4, p. 7]. It can be seen again in the Babylonian rule for the volume of a basket in the shape of the frustum of a right cone, $V = \frac{1}{2}h(B_1 + B_2)$, where B_1, B_2 are the areas of the two bases. Of their successes, a familiarity with the Pythagorean relationship as early as 2600 BC is noteworthy, the context being that of finding the length of a chord given the radius and segment depth, as is the consideration of regular polygons in the Susa tablets [2, pp. 46–48].

Even these achievements pale in comparison with those of Egyptian geometry. Four thousand years ago, and perhaps even before that, the peoples of the Nile successfully resolved a number of complex practical problems using empirical formulas. Of course, evidence of their skill in geometry is to be found in every pyramid and every monumental stone the Egyptians cut and assembled. (And the same can be said of the Maya and the Inca.) But they also left us a small number of mathematical papyri of which the Moscow Papyrus (1850 BC) and the Rhind Papyrus (1650 BC) are the most important. The recipe-book approach to geometrical demonstration adopted by the scribes consisted of working through a problem and then remarking that all similar problems are tackled in the same manner. If there was any generality or proof then it was no more than implied. Yet among the exemplar solutions to the twenty-six geometrical problems considered in the two papyri there are some impressive results, including the procedures in the Moscow papyrus for finding the area of the curved surface of a hemisphere in Problem 10 and for the volume of the frustum of a square pyramid in Problem 14. Florian Cajori wrote [5, p. 15] that 'Not only in mathematics, but also in mythology and art, Hellas owes a debt to other countries. To Egypt Greece is indebted, among other things, for its elementary geometry'. Egypt was not the only cradle of geometry but, as we shall see, only Egypt, albeit under Greek influence, went on to provide a nursery as well. To this extent, is there validity to the hotly debated 'Black Athena' thesis of Martin Bernal that Greek civilization was profoundly influenced by the Egyptians and Phœnicians and that nineteenth century historians were successful in obfuscating that influence [thesis in 6, counter-argument in 7].

The earliest Chinese geometry is found in the *Zhōubì suànjīng*. This text was composed during the Western Hàn Dynasty in the first century BC [2, pp. 25–32]. It is notable for invoking Pythagoras' Theorem, under the name of Gōugŭ, in respect of right-angled triangles arising from the shadows cast by vertical sticks. The theorem is also the subject of the last chapter of the *Jiŭzhāng suànshù* (*Nine Chapters of the Mathematical Art*), a collection of mathematical methods brought together two centuries later during the Eastern Hàn Dynasty. The first chapter on 'field measurements' contains numerous formulas (some accurate, some approximate) for the areas and volumes of everyday objects, π taken to be 3. The calculation of volumes of truncated solids is particularly well treated. For example, the volume of the right tetrahedron formed by slicing obliquely through a cuboid of sides x, y, z is correctly given as $V = \frac{1}{6}xyz$. Later in this chapter, Kiang discusses an ingenious fifth century dissection of a cube not into tetrahedra but into *yang mă* or square-based pyramids.

The geometrical methods employed by these civilizations lacked structure and system; they were *ad hoc*. Joseph Needham summed it up nicely when he wrote [3, Foreword]: 'if China developed no Euclidean deductive geometry, there was plenty of empirical geometry there'. He added that 'perhaps like the ancient Babylonians, the Chinese always preferred algebraic methods', even to the extent that 'by the thirteenth century AD, they were the best algebraists in the world'.

None of the riverine civilizations proceeded from an inductive, practical geometry to deductive, demonstrative geometry. This step was taken by the Greeks in their many colonies around the coasts of the Mediterranean.

Thales of Miletus is said to have demonstrated some of the features of congruent triangles and proved that the angle in a semicircle is right. In what is now southern Italy, Pythagoras (or perhaps one of his followers) used and may have proved the theorem named after him. But the critical moves towards proof based on a logical sequence of true statements were taken between 600 BC, the time of Thales, and Euclid's compiling of the *Elements* around 300 BC [8, p. 70]. They may have been taken by Hippocrates in his now lost compendium of theorems established from axioms and postulates. In his *Manual of Greek Mathematics* [9, p. 112], Thomas Heath quotes Proclus' words:

Plato, who came next . . . caused mathematics in general and geometry in particular to make a very great advance, owing to his own zeal for these studies; indeed every one knows that he filled his writings with mathematical discourses and strove on every occasion to arouse enthusiasm for mathematics in those who took up philosophy.

And there could be no clearer indication of Plato's zeal than that he had inscribed over the Academy's entrance 'Let no man enter who is ignorant of geometry'. Greek demonstrative geometry was a great product, brilliantly promoted among the contemplating classes of the Mediterranean littoral. Plato's pupil, Aristotle, though no geometer as such, soon provided the clearest distinctions to date among the basic geometrical entities and helped to make geometry a more exacting discipline. Axioms are universal self-evident truths. Postulates are assumptions, not at all self-evident. Definitions are attached to things which exist, such as points and lines, or to those still requiring to be shown to exist, i.e. everything else.

In all likelihood, Thales and Pythagoras had learnt geometry on their travels in Egypt. The three most influential Greek geometers, Euclid, Archimedes and Apollonius, who were to follow them in the third century BC, were drawn to the city of Alexandria, built on the Nile delta by Alexander the Great in 332 BC. The arrival of Euclid to head up the school of mathematics was a major coup for Alexander's successor, Ptolemy I, Soter. Ptolemy's vision was to turn the fledgling trading port into an important centre of learning and culture. We may infer that Archimedes spent time in Egypt from the fact that his name is associated with an irrigation 'screw' for drawing the waters of the Nile to higher elevations but the suggestion that he studied at Alexandria under Euclid's pupils is without direct evidence. He certainly recognised the importance of the Alexandrian school for it was to its mathematicians, Conon and Eratosthenes, that he communicated many of his results. Apollonius did study at Euclid's school and Alexandria became his home for much of his active life. If the Nile valley was one of the main cradles of geometry then its delta provided the nursery.

Euclid's *Elements* consist of 22 definitions, five postulates (of which the last would prove the catalyst for later developments in geometry), five common notions or axioms and 465 derived propositions, systematically linked and arranged into thirteen books. The subject matter is pure geometry (both plane and solid), geometrical number theory and geometrical algebra. Bought, studied and cited on a massive scale for centuries, it is a peerless classic amongst scientific works.

A minor detour is now required, for while the material for Euclid's *Elements* was being collected and organised, there were two largely separate developments in Greek geometry. Firstly, Antiphon and Eudoxus outlined the *method of exhaustion*, in response to the challenge thrown down by Zeno's paradoxes. The method allows curvilinear areas to be approximated more and more closely by using inscribed polygons. It is really a limit avoidance argument and consequently rather unlike the latter-day calculus with which it is often compared. Secondly, there was an exploration of curves other than the straight line and the circle. To some extent this higher geometry was developed through attempts to address particular problems. However, historians have tended to give undue weight to the three classical problems – the duplication of the cube, angle trisection and the quadrature of the circle. According to Eves, the duplication problem spawned Menaechmus' conic sections and Diocles' cissoid; out of the trisection problem came Hippias' quadratrix, Nicomedes' conchoid and Archimedes' spiral and this last curve also supplied a solution to the quadrature problem [8, ff. 81]. The primacy of the Euclidean tools of ungraduated straight edge and collapsible compasses is also a relatively modern slant. Quadrature, incidentally, is the construction of a square equal in area to a given area, a classic example being Hippocrates' quadrature of the lune, dated roughly 440 BC and selected by William Dunham as the first of his dozen 'great theorems of mathematics' [10].

For sheer originality and ingenuity, there is nothing to match the geometrical works of Archimedes. Heath says of them [9, p. 281]:

The treatises themselves are, without exception, models of mathematical exposition; the gradual unfolding of the plan of attack, the masterly ordering of the propositions, the stern elimination of everything not immediately relevant, the perfect finish of the whole, combine to produce a deep impression, almost a feeling of awe, in the mind of the reader.

In all, Thomas Heath produced ten scholarly books on the mathematical sciences of the Greeks and it was fitting that his Presidential Address to The Mathematical Association in the early 1920s should focus on the role of infinitesimals in the geometry of Archimedes in particular. It is reproduced in this chapter.

The vast majority of Archimedes' output has survived to the present day: even his book on heuristics, *Method*, was rediscovered in the early twentieth century. The results they contain are undoubtedly the crown jewels of ancient geometry but regrettably, in such a short history, we must make do with mention of a few diamonds and their settings. In *Measurement of a Circle*, Archimedes 'trapped' a circle between regular polygons of 96 sides to calculate tight bounds for the value of π. The method of exhaustion was also put to effective use in *Quadrature of the Parabola* and *Conoids and Spheroids*, in the former to calculate the area of a segment of a parabola, in the latter to find the volumes of paraboloids, hyperboloids and ellipsoids of revolution. In the *Book of Lemmas* we find the lovely theorems on the arbelos and salinon (shoemaker's knife and salt cellar respectively), shapes produced by the concatenation of semicircular arcs. The Kohinoor of theorems is to be found in *Sphere and Cylinder*, and it is this: for a sphere wrapped in a cylinder of equal height, both the volumes and the curved surface areas are in the ratio 2:3. We may assume that Archimedes himself believed this to be his favourite gem, for he requested that it be inscribed on his tombstone.

Among lost results are a method of constructing a regular heptagon in a circle and possibly the formula now known as Heron's formula for the area of a triangle. And whilst the attachment of Plato's name to the regular polyhedra does disservice to Pythagoras and Theaetetus, we tend to accept Pappus' attachment of Archimedes' name to the thirteen semi-regular polyhedra, despite the loss of the relevant treatise.

By contrast to Archimedes, as much as a half of Apollonius' geometry has been lost, though we are left with seven of the eight volumes of his masterpiece on conic sections, the first four in the original, the next three in Arabic translation. What a labour of love it must have been for the translator Taliaferro if he could express this view of the *Conics* [11, pp. 599–600]:

If on first appearance this treatise should seem to the reader a jumble of propositions, rigorous indeed, but without much rhyme or reason in their sequence, then he can be sure he has not read aright ... There are one or two hypotheses at least that can order the apparent wanderings of parabolas, hyperbolas, and ellipses through the first four books. Such hypotheses are the analogies between the three sections, and especially the development of the analogy between the hyperbola and the ellipse reaching its culmination, in the first book, with the final theorem, the construction of conjugate opposite sections ... And this analogy between the hyperbola and ellipse now stands on the threshold of a vast development. For this theorem, coming as a climax to the first book, makes possible the main theme of the second book: the asymptotes, those strange lines all but touching each opposite section and forming a single bound between each adjacent pair, so making the hyperbola an all but closed section, a puckered ellipse, a mouth turned inside out. And in the third book, the fruits of this analogy are gathered as in the especially nice case of Proposition III.15.

Plane and solid geometry provided solutions to numerous practical problems on the ground but could not be used as a framework for the mathematics of the heavens above. For that a new geometrical discipline was devised, sphærica or spherics. Notable works include the systematic account by Theodosius, built up in the manner of the *Elements* and including a consideration of great and small circles, and the *Sphærica* of Menelaus. Menelaus demonstrated that the sum of the angles of a spherical triangle exceeds two right angles. These developments took place after the golden age of Greek geometry, Theodosius living in Asia Minor around 100 BC, Menelaus, yet another Alexandrian, around 100 AD. Ostensibly astronomical in nature, these texts, together with Ptolemy's spherical trigonometry, formed bridges to Indian and Muslim geometry and trigonometry. In relation to spherical geometry, there is an open question, as has been pointed out by Jeremy Gray [12, pp. 536–537]. If spherical geometry was effectively the first non- Euclidean geometry, since it dealt with the properties of space on the surface of a sphere, then why was Euclidean geometry thought by mathematicians to be the one and only geometry right up until the middle of the nineteenth century?

1. Shafts of Sunlight Through a Cloudy Sky?

With the demise of Greek civilization, there followed several centuries of inactivity and ignorance, the so-called Dark Ages. Much of the geometry already produced was permanently or temporarily lost but some was translated into Arabic and re-covered by the West in the late Renaissance. Muslim and Hindu mathematicians acted as a little more than conduits from the ancient Greeks to the modern word. Such is the 'Eurocentric trajectory' thesis so prominent in the historiography of mathematics. For many aspects of mathematics, 'transmission with enhancement and extension' would better reflect the reality, though perhaps less so in geometry than in algebra and trigonometry. Condemnation of the 'Eurocentric trajectory' has come from George Gherveghese Joseph in his well-received *Crest of the Peacock* [13]. Building on the ideas of David W. Henderson, Joseph offers in this chapter, a study of the geometrical representation of the square root in different mathematical traditions. In so doing he demonstrates that 'a geometry did exist outside Greek mathematics which had moved beyond mere numerical relations or practical sur-veying considerations to active search for general proofs'. The theme is taken up by Lesley Jones in an article on the influence of the Indian Vedic square on Muslim geometric design.

In the *Paulisa siddhānta* we see not only transmission with enhancement but also with change of focus. This is an astronomical work, composed in India by a refugee from Alexandria and incorporating the spherical trigonometry of Ptolemy. But it is only one work among many: the *Pancasiddhāntika* of Varāhamihira (5th century AD), the *Khanda-khādyaka* of Brahmagupta (7th century AD), the astronomical treatises of al-Khwārizmi (9th century) containing numerous new results including the cosine rule for the surface of the sphere, and many more. (A comprehensive account is given by Rosenfeld [14].)

For pure geometry, we look first to Brahmagupta. In his three-volumme *Metrica*, Heron provided the definitive study of mensuration to perhaps 75 AD, thought to be the time of writing. Here he gave his formula for the area of a triangle given all three sides. In his masterpiece, *Brahma-sphuta-siddhānta* (628 AD), Brahmagupta extended the formula to cyclic quadrilaterals. He also found formulas connecting the sides and diagonals of a cyclic quadrilateral and unveiled a method of generating cyclic quadrilaterals from Pythagorean triples [15, p. 49]. For such ingenuity, Brahmagupta earned the admiration of Michel Chasles, identified by Coolidge as the first geometer to write on the history of his discipline [16]. Yet it is the inconsistencies in Brahmagupta's geometry which many commentators have noted. Julian Lowell Coolidge found it 'hard to be patient with one who mixes truth and error so freely' [4, p. 16]; a thousand years earlier, the astronomer Al-Biruni (973–1050) dismissed it as a mixture of 'pearls, shells and sour dates'.

Jamshīd al-Kāshī gave π to sixteen decimal places in 1424, an accuracy unsurpassed for almost two centuries. His work is particularly poorly known and so the publication of an extract of Van Brummelin's article in this book should go some way towards correcting the situation. Al-Kāshī's strategy was to follow Archimedes, his tactics to employ a recurrence relation to move from n-gons to $2n$-gons and thus to increase the number of sides exponentially, rapidly increasing the accuracy of π's value.

2. The 'Art and Chart' Geometry of the Renaissance

Renaissance geometers sought to solve the practical problems of the day whilst recovering and reconstructing the geometry of the Greeks. There were two practical questions to be addressed.

1. In painting, how is it possible to produce on a two-dimensional surface a realistic representation of a three-dimensional scene?
2. For purposes of navigation, how can the curved surface of the globe be represented on a flat surface?

These questions, requiring an understanding of perspective and (map) projection for their solution, invoked what might be termed the 'art and chart geometry' of the Renaissance. The best introductions to Renaissance geometry are two books by Morris Kline [17, 18]. The geometry of perspective was advanced by an ingenious experiment undertaken by Filippo Brunelleschi, the architect of the dome of Santa Maria del Fiore, Florence's great cathedral. From a point just inside the cathedral, and looking out through the doors to the Baptistery of San Giovanni, he produced a perspective painting of the scene on a small panel and into this panel he drilled a small hole at the vanishing point. The story is taken up by Ross King in *Brunelleschi's Dome* [19, p. 37]:

Standing ... on the exact spot ... where Filippo had executed the panel – the observer was to turn the painted side of the panel away from himself and peer through the small aperture. In his other hand he was to hold a mirror, the reflection of which, when the glass was held at arm's length, showed (in reverse) the painted image of the Baptistery and the Piazza San Giovanni. So lifelike was this reflection that the observer was unable to tell whether the peephole revealed the actual scene ... or only a perfect illusion of that reality.

The experiment was carried out in or before 1413, the painted panel lost probably in the looting of Florence by the French in 1494. The technicalities were published by Alberti in *Della Pittura* in the 1430s and further developments were made by Uccello, Piero della Francesca (*De Prospectiva Pingendi*, ca. 1478), Leonardo (the lost *Trattato della Pittura*) and Albrecht Dürer. The collective enterprise of the geometer-artists produced such advances that Kline could claim [18, p. 203] that the

Renaissance painters ... produced the first really new mathematics in Europe. In the fifteenth century they were the most accomplished and also the most original mathematicians.

We see that originality in Luca Pacioli's book of 1509 on the golden section. *De Divina Proportione* is notable for the magnificent drawings of solids contributed by Leonardo.

A different but related problem was how to represent the surface of a sphere on a flat surface whilst retaining some of the interrelationships between features. The first step in solving this conundrum was taken in 1569 by the Flemish cartographer, Mercator (Gerhard Kremer, 1512–94) when he devised what we still refer to as Mercator's projection to produce the map *Nova et aucta orbis terrae descriptio*. From the outset, Mercator's objective was to produce a map which sailors could use to accurately plot courses. He discovered how, with just compasses and protractor, to make a map on which angles (and hence bearings) are preserved by adopting a conformal projection. The Cambridge mathematician, Edward Wright (1558–1615), improved on the method and wrote a book which provided tables for the relevant calculations. He explained how the projection worked in the following manner. Using a spherical balloon as a surrogate for the earth, place it in a glass cylinder and begin to blow it up so that the first contact is between the equator and the glass. Now further inflate the balloon until a large part of the surface of the balloon is pressed against the cylinder. There is now a direct transfer between each of the points on the surface of the globe and the cylindrical surface, which afterwards may be unfurled to reveal the details on a flat surface.

Paul Cohen and Robert Augustyn have recently provided an account of the history of Mercator's projection in the latter part of the sixteenth century [20]. Between 1584 and 1593, the cartographer Jodocus Hondius lived in exile in London, the authorities in Holland intolerant of his religion. Here he learnt directly from Wright the mathematics underpinning Mercator's projection. Upon returning to the continent, he began work on a map of the world, apparently lifting ideas directly from the book Wright was writing, *Certaine Errors in Navigation . . .* (London, 1599). This map, which lay undiscovered until 1993, is thus identified as the first to use the Mercator projection after Mercator's own map of 1569. Previously, Wright's map of 1599 was given this distinction. This apparent theft of priority explains why Wright vented his spleen against Hondius in the pages of his book. Actually, Hondius' approach consists of an informal geometrical construction and not the calculation-laden approach of Wright. The unpleasantness did nothing to stop Mercator's projection from becoming the standard projection for maritime mapping in the 17th and 18th centuries and beyond. The next great advances in the geometry of map projections would come with the publication of equal area projections by the mathematicians Johann Heinrich Lambert in 1772 and Karl Mollweide in 1805.

3. The Projective Geometry of Desargues

Out of the work of the geometer-artists developed projective geometry, notably in the work of the architect and engineer, Girard Desargues (1591–1661). The new discipline drew on Pappus' cross-ratio of four collinear points and was anticipated to some extent in the first modern European study of conics, a book of 1522 by Johannes Werner. Desargues began to explore questions such as: if a shape drawn in one plane be projected onto another (oblique) plane what geometrical properties are shared by the original and the image or differentiate between them? Working in isolation, he wrote a book on the intersection of cone and plane, certainly 'one of the most unsuccessful great books ever produced' [21, p. 393]. It was published in 1639 with a prolix title and a 'style and nomenclature . . . weird beyond imagining' [4, p. 89]. Plagued by its terminology – the straight line is variously *rameau* (branch) or *tronc* (trunk) for example – it is no surprise that it was ignored by his contemporaries, with the exception of the young Blaise Pascal (1623–1662) and one of Desargues' students. Both Desargues' Theorem and Pascal's (Hexagram) Theorem appear amongst the Desert Island Theorems in this book and Desargues' geometry is explored further by Swinden in this section. However, the manuscript on conic sections in which the latter originally appeared is now lost and Pascal's claim to have discovered 400 corollaries from it is not open to verification.

The student was Philippe de la Hire (1640–1718). It occurred to him that it should be possible to reduce a problem about an ellipse (or other conic section) into a problem about a more elementary shape, a circle. The theorem named after him concerns the relationship between what became known as *poles* and *polars* in the *principle of duality*, (explained below) and is outlined by Jeremy Gray [22, pp. 88–89].

4. Recovering the Adumbrated Analysis of the Greeks

Synthetic and analytic geometries derive their names from two forms of inquiring after truth, found in Plato, but enunciated by Theon. Synthesis is the deduction of a valid conclusion from known axioms, Euclidean geometry being a prime example. Clearly, its theorems are only as good as its axioms. In analysis, something is assumed to be true and a result is reached that is known to be true. Though analysis appears to stand on relatively shaky foundations, it is a powerful tool for attacking problems. And problems that yield to analytic solution can then have their proofs clad in the vestiges of synthetic geometry. It is likely that many of the theorems of Greek geometry were discovered by analysis but with all traces of heuristic being removed to leave elegant watertight proofs [23, p. 30]. As Archimedes' *Method* attests, not every Greek geometer felt obliged to uphold this 'discovery-proof-obliteration of heuristic' paradigm. But it was the last of the great Greek geometers, Pappus, who realized how important it is to understand how geometers attack problems. In Book VII of his *Mathematical Collection* he identified 'heretical' treatises of Euclid, Apollonius, Aristaeus and Eratosthenes that revealed something of the process of discovery. Only in these books were attempts made to attack problems by reducing them to problems with ready solutions. Regrettably, as Mahoney noted [23, p. 31],

> except for a few examples in Euclid and Archimedes, Greek geometrical analysis shared the fate of Pappus' *Mathematical Collection* and most of the works cited in Book VII. Largely ignored or unknown by the Arabic writers, it lay hidden in Greek manuscript until the sixteenth century, when it shared in the general revival of Greek mathematics.

That revival was led by the editors Francesco Maurolico (1494–1575) and Frederigo Commandino (1509–1575). Commandino translated the extant books of Apollonius' *Conics*, and Maurolico even attempted to reconstruct the lost fifth book on maxima and minima, setting a fashion for the times. In the next generation, François Viète (1540–1603) translated Apollonius' lost *Tangencies* (1600) and became so associated with such work that he was dubbed 'Apollonius Gallus'. He was succeeded in this work by Marino Ghetaldi (1566–1626) who, in the period 1607–13, restored Apollonius' *Plane Loci*. It was this work which so influenced the inaugurators of analytic (or coordinate geometry), Pierre de Fermat (1601/07–1665) and especially René Descartes (1596–1650). Descartes, had a poor opinion of Greek geometers who 'with a sort of low cunning, deplorable indeed, suppressed this knowledge' of their analytic methods.

5. Analytic Geometry

Analysis became algebraic with the publication in 1615 of Viète's treatises on the *Recognition and Emendation of Equations*, for here we have problems analysed through a process of classification into families of equations with known methods of solution. One of the difficulties that Viète confronted was that of homogeneity: how is it possible to combine arithmetically quantities such as a square and a cube? Each of his equations had dimension and that dimension was the equation's degree. Problems reducible to determinate equations in one unknown presented little challenge but to Viète there was no clear way of framing loci questions in equation form and it was this challenge that the *Conics* posed.

In 1637 two manuscripts were circulating in Parisian mathematical circles. Both propounded the marriage of geometry and algebra: more explicitly, both made a connection between two-dimensional geometric curves and algebraic equations in two unknowns. Descartes' *Geometry*, an essay appended to his *Discourse on the Method*, was in galley proof form. Descartes had been in possession of its novel methods for up to a decade. Fermat's *Introduction to Plane and Solid Loci* had a shorter period of gestation prior to 1637 and was not widely available until 1679. For these reasons, coordinate geometry is correctly referred to as Cartesian [23, p. 73].

Descartes and Fermat had overcome the problems which so troubled Viète partly through the realization that the dimension of an equation should lie not in its degree but in the number of unknowns needed for its analysis. Henceforth, geometrical quantities would be denoted in symbolic form as powers of an unknown, their ancestry surviving only in their spoken form. As Descartes wrote [24]:

Here it must be observed that by a^2, b^3, and similar expressions, I ordinarily mean only simple lines, which however, I name squares, cubes etc., so that I may make use of the terms employed in algebra.

Descartes and Fermat found Apollonius' *Plane Loci* deficient in different ways. To Fermat there were undue restrictions on the scope on the theorems it contained; to Descartes the theorems were unnecessarily complicated. They alighted upon a radical idea, locus problems became tractable if put in the form of indeterminate equations in two unknown equations which were without meaning in pure geometry. Fermat found he was able to classify equations into seven families with known solutions and went on to show how other equations might be reduced to one or other of them. He came to see results found by analysis to be in no need of the formal proof which synthesis could supply.

6. The Age of Feuding Heroes

To the nineteenth century is given the name of the 'heroic age of geometry'. But it was also a time of bitter feuding between the supporters of synthetic geometry and its analytic counterpart.

Mathematicians have been especially fickle in their devotion to geometry, 'which, of all the branches of mathematics, has been most subject to changing tastes from age to age' [21, p. 572]. Despite standing on the 'threshold of a new era' in the seventeenth century, it lost out to algebra and analysis in the eighteenth century. It was as if reborn in the nineteenth, nurtured at the *École Polytechnique* in Paris and especially in the hands of Gaspard Monge (1746–1818) and his students. A century on, in September 1904, Jean Gaston Darboux (1842–1917), a leading French geometer and permanent secretary to the *Académie des Sciences*, reviewed that heroic age in a talk given at the Congress of Arts and Science in St Louis. We are fortunate that the *Mathematical Gazette* was given permission at that time to carry the text of the address. As a geometer, Darboux was blessed with a 'rare combination of geometrical fancy and analytical power' [25, p. 227]. As an historian, it need only be noted that Florian Cajori relied heavily on the address when revising his authoritative history of mathematics for the second edition of 1919 [5]. The St Louis address is reproduced in these pages.

If it were possible to single out a geometer from the first half of the nineteenth century it would surely be Jean-Victor Poncelet (1788–1867). Under the influence of Monge, a sponsor of projective geometry, he turned from the analytic geometry of his youth to synthetic geometry. The greater generality of the analytic, however, he attempted to extend to the synthetic. For example, he noted how the intersecting chords theorem was equally valid as an intersecting secants theorem, and then by allowing C and D to coincide found a related result about secants and tangents. In all three figures drawn below, triangles AXC and DXB are similar.

Hence,

$$\frac{AX}{XD} = \frac{CX}{XB}$$

and it follows directly that $AX \cdot XB = CX \cdot XD$.

Since, in the last figure $CX = XD$, the result is modified to $AX \cdot XB = XD^2$.

In the early 1820s, Poncelet was joined by Charles Julien Brianchon (1785–1864), another of Monge's students, to unveil the 'nine-point circle'. Described by Coolidge [26, p. 21] as possibly 'the most beautiful theorem in elementary geometry that has been discovered since the time of Euclid', it is the subject of

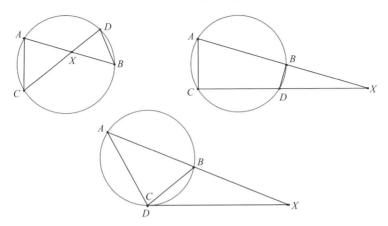

Adam McBride's Desert Island Theorem in this volume. For further developing this micrcocosm of geometry, the nine-point theorem is often associated with Karl Wilhelm Feuerbach (1800–34).

Coolidge wrote that 'I have given more attention to Poncelet than to previous writers on projective geometry because he really saw far deeper' [4, p. 95]. Poncelet's great work, *Traité des Propriétés Projectives des Figures*, was begun whilst he was incarcerated in the Russian military prison at Saratov (1813). Here he differentiated between loci which are invariant under projection from plane to plane and those which are not. Poncelet went on to allow imaginary points into his synthetic geometry, so that a line and a circle always intersect, as do any two circles.

In appraising nineteenth century contributions to synthetic geometry, Carl Boyer focussed not on Poncelet but on a more prolific and systematic geometer, Jakob Steiner (1796–1863), whom he considered a latter-day Apollonius [21, p. 574]. Such a view would have found universal approval amongst the German-speaking mathematicians of the day for whom Steiner would grace the most elaborate pedestal. Another synthetic geometer much influenced by Poncelet was Karl Georg Christian von Staudt (1798–1867). He 'cut loose from algebraic formulæ and from metrical relations . . . and then created a geometry of position, which is a complete science in itself, independent of all measurements' [5, p. 294]. Coolidge refers to von Staudt as a 'striking original and profound geometer' [27, p. 66], notable for demonstrating that all the theorems of projection and intersection hold equally in the real and complex domains.

Meanwhile Julius Plücker (1801–68), by initial inclination a synthetic geometer, 'crossed the floor of the house' in the opposite direction to Poncelet and joined the analytic geometers, Joseph Diaz Gergonne (1771–1859) and Gabriel Lamé (1795–1870). Together, they did much to simplify the algebraic representation of analytic geometry. In particular, they promoted the use of three numbers to represent a point in the plane, so-called homogeneous coordinates. Such triples, which can accommodate the point at infinity, proved to be powerful tools in projective geometry. In the 1840s, the English mathematician Arthur Cayley (1821–95) understood the possibilities of Plücker's work for the geometry of n dimensions if the homogeneous coordinates are arranged into determinants.

A most striking breakthrough during this period came with the discovery of the principle of duality. The principle concerns the interchangeability of 'points' and 'lines' in the propositions of projective geometry. We know that a line is determined uniquely by any two points through which it passes: likewise, a point is determined uniquely by any two lines which pass through it. In 1806, Brianchon had extended Pascal's geometry of the conic section, discovering an unusually satisfying dual theorem. Pascal's Hexagram Theorem and its dual, Brianchon's Theorem, may be stated thus:

Pascal's Theorem: The six vertices of a hexagon lie on a conic, if and only if the points of intersection of the three pairs of opposite sides lie on a straight line.

Brianchon's Theorem: The six sides of a hexagon are tangential to a conic if and only if the lines joining the three pairs of opposite vertices meet at a point.

Brianchon had his doubts about the applicability of the principle but its power was in no doubt. Once a theorem was proved, its dual was effectively proved too. Gergonne squabbled with Poncelet over the issue of priority of discovery. Such disputes are often unpleasant for the parties concerned but they can help to clarify issues which prove troublesome and so it was here, with regard to the domain of validity. The analogue of the principle of duality in analytic geometry was discovered by Plücker in 1829.

The Leipzig mathematician and astronomer, August Ferdinand Möbius (1790–1868), working largely in ignorance of contemporary advances in geometry, produced a short book on geometrical mechanics in 1827. The scope of *Der Barycentrische Calcul* proved rather broader than the centres of gravity of its title. The book considered problems such as how to divide a directed line segment in a given ratio, much as we now find in introductions to vectors. The centre of gravity of a triangle is defined by a triple of weights deemed to be hanging from its vertices. Möbius called these triples barycentric coordinates and enclosed them in square brackets so that they would not be confused with Cartesians. They enjoy an advantage over Cartesians in the study of projective transformations. There is no better introduction to Möbius' geometrical mechanics than Jeremy Gray's paper [22].

This is an opportune place to mention geometrical constructions. Lorenzo Mascheroni (1750–1800) showed in *Geometria del Compasso* (1797) that the Euclidean constructions which had to that point been carried out with straight edge and compasses could be drawn using just the compasses. In 1913, Ernest William Hobson, (1856–1933) the third Sadleirian Professor of Pure Mathematics at

Cambridge (following Cayley and A. R. Forsyth), chose Mascheroni as the subject of his Presidential Address to The Mathematical Association. Like Darboux' address in St Louis, it is cited by Florian Cajori in his famous history [5, p. 268] and reproduced below. After Mascheroni, interest in constructions was shown only sporadically by geometers. Poncelet and Steiner demonstrated that collapsible compasses could be dispensed with in favour of fixed compasses and a straight edge. Then in the early twentieth century, Francesco Severi (1879–1961) showed that figures could be constructed given the location of the centre of a circle and a short arc length. A science of geometrography was instituted by Émile Lemoine (1840–1912) in order to classify and compare construction methods [8, pp. 99–100].

Finally, the synthetic geometry of the triangle and the circle, thought largely exhausted by the Greeks, burst into new life in the nineteenth century. The contribution of Henri Brocard (1845–1922) is remembered by Laura Guggenbuhl in this section, whilst those of Feuerbach, Miquel and Morley are to be found amongst the Desert Island Theorems.

7. Non-Euclidean Geometry

The discovery of non-Euclidean geometries in the second quarter of the nineteenth century had consequences for geometry itself, for physics and for philosophy. We know Euclid's troublesome fifth postulate in John Playfair's form, essentially that, given a line and a point not on that line, there can be drawn through the point one and only one line which is parallel to the given line. Some of the mathematicians of ancient Greece had realised that the status of this postulate differed from that of the other four. It had appeared to them to be less than self-evident. Proclus, writing in the fifth century, noted the work of Posidonius, Geminus and Ptolemy in the field, before embarking on his own attempt to obtain the fifth postulate [28, pp. 2–7]. As we have seen, Arab mathematicians were drawn more to algebra and trigonometry than to geometry, 'but one aspect of geometry held a special fascination for them – the proof of Euclid's fifth postulate' [21, p. 266]. Almost exactly a thousand years ago, Alhazen attempted to show that if a quadrilateral has three right angles then the fourth angle is also right. And a century later, Omar Khayyam, the mathematician-poet, tried to prove that given a quadrilateral with two right angles, the remaining two angles are also right. Both assumed the fifth postulate in trying to derive it. A thirteenth century attempt by Nasir Eddin was translated into Latin by John Wallis. In turn, Wallis' work formed the starting point for investigations by Lambert and Girolamo Saccheri (1667–1733), whose *modus operandi* was proof by contradiction. However, they failed to fully appreciate the alternative geometries they were beginning to fashion. The parallel postulate generated such indignation in Jean-le-Rond D'Alembert (1717–83) that in 1759 he described its definition and properties as 'the scandal of the elements of geometry' [28, p. 52].

A major breakthrough came in the 1820s when Nicholai Lobachevski (1793–1856) produced a geometry which William Kindgom Clifford (1845–79) would later characterize as 'quite simple, merely Euclid without the vicious assumption'. He had allowed more than one line passing through the point to be parallel to the given line. Janos Bolyai (1802–60) independently produced a similar geometry. Karl Friedrich Gauss (1777–1855) appears to have explored such a geometry too, but the extent of his researches, which he never sought to publish, is still not clear. There are a small number of letters to friends describing the difficulties of the subject of 'non-Euclidean geometry' – a phrase he coined – and two short notes on the theory of parallels were discovered amongst his papers [28, pp. 64–75]. His oft-quoted dismissal of Bolyai's originality was graceless, if not disgraceful.

A geometry in which there are no parallel lines was unveiled in the 'most celebrated probationary lecture in the history of mathematics' [21, p. 588]. It was a lecture delivered by Bernhard Riemann (1826–66) at the University of Gottingen in 1854. A further 'twist' was added by Clifford in 1873. Relaxing the Euclidean definition of parallel with respect to the need for the lines to lie in the same plane, he discovered the spiralling, equidistant lines, known as 'Clifford Parallels', which exist in the elliptic space of Riemann. Michael Atiyah has selected this discovery as his Desert Island Theorem for this volume.

By the end of the nineteenth century, largely through the work of Eugenio Beltrami (1835–1900), Lobachevskian, Euclidean and Riemannian geometries were recognised as a 'trinity', each associated with a particular type of surface [9, p. 307]. Some of the main distinguishing features of these geometries, distilled from Davis & Hersh [29, p. 222], are shown in the table below:

	Lobachevskian or hyperbolic	Euclidean	Riemannian or elliptic
Associated surface	pseudosphere or tractroid (surface of constant negative curvature)	plane (surface of zero curvature)	sphere (surface of constant positive curvature)
Number of lines through a point parallel to a given line	2 or more	1	0
Parallel lines	never equidistant	equidistant	do not exist
Angle sum of triangle	less than 180 degrees	180 degrees	more than 180 degrees
Area of a triangle	proportional to the defect of its angle sum	independent of its angle sum	proportional to the excess of its angle sum

Two new models for hyperbolic geometry were discovered by Felix Christian Klein (1849–1925) and Henri Poincaré (1854–1912). Klein's geometric space is confined to the interior of a circle and there are innumerable lines through a point parallel to a given chord which do not meet that chord inside the circle. The straight lines are replaced by circular arcs in Poincaré's version and it was H. S. M. Coxeter's illustrations of this geometry which formed the template for the *Circle Limit* series of drawings by the Dutch graphic artist, Maurits Escher [30, chapters 16, 21].

8. Topology

Topology is one of the newer branches of geometry, which from Klein's perspective would be seen as the study of the properties of figures remaining invariant under continuous deformation. A topologist, it is quipped, is someone who does not know the difference between his coffee cup and the doughnut alongside. Topology features in this book only among the Desert Island Theorems, but a brief stroll over the Königsberg bridges, along the single edge of the Möbius band and into (or is it out of?) the Klein bottle should be a pleasant experience so long as we do not tie ourselves in knots.

As a distinct focus of study, topology begins with two investigations of Leonhard Euler (1707–83). The first of them began with an attempt to explain what must already have been known to the people of Königsberg, that it was not possible to take a stroll around the city crossing each of bridges across the Pregel once (and only once) before returning home. We can represent the four areas of land (two banks of the river and two islands in mid-stream) by four vertices and the seven bridges by 'edges'. This enables a graph to be drawn:

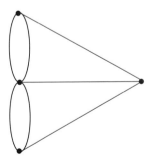

Vertices are considered to be of odd or even degree according to the number of edges meeting at them. As Euler recognized, a graph is traversible if each and every vertex is even. For the bridges problem, this means that the tour is possible if each and every area of land is served by an even number of bridges. From three areas three bridges may be taken and from the fourth five may be taken, and hence the tour is impossible. Euler published his results in a paper of 1736. The inclusion of the phrase 'geometry of position' in the title of the paper demonstrates that Euler was aware that this new geometry was one in which distance was of no consequence. A notable contribution to this combinatorial topology came in the nineteenth century with the invention by William Rowan Hamilton (1805–65) of a puzzle. The challenge it posed was to trace a route along the edges of a regular dodecahedron visiting each vertex exactly once. From two-dimensional representations of such figures there developed a theory of 'Hamiltonian circuits'. When an associated problem was posed by Frederick Guthrie in 1850 it was natural that his teacher Augustus De Morgan should raise it with Hamilton. Were four colours sufficient to colour any map? The subsequent history is given in brief by Derek Holton in his Desert Island Theorem and need not be reiterated here.

The second of Euler's investigations yielded the relationship $V - E + F = 2$, where V, E and F are, respectively, the number of vertices, edges and faces a solid shape posseses. Euler intimated the result in a letter of 1750 to Christian Goldbach (1690–1764), believing it to be true for all solid shapes. In the second of two papers he published on the subject in 1752, Euler offered a proof based on a dissection into tetrahedra. But the universality of the relationship proved illusory. For solid shapes with holes punched through them, an extra term needs to be incorporated, as was discovered by Antoine-Jean Lhuilier (1750–1840) in 1813 [31, pp. 106–107]. The 'Euler characteristic' is no longer 2 but $2 - 2g$, where g is the number of holes. These investigations of Euler and Lhuilier raised questions concerning the nature of surfaces and led to the discovery by Möbius of the twisted, single-surface band named after him. A student of Gauss, Johann Benedict Listing (1808–82) discovered the band independently and in the same year of 1858. Incidentally, it was from the title of Listing's first book, *Vorstudien zur Topologie* (1847) that 'topology' received its name. His second book (1861) featured the Möbius band as part of a detailed development of the theme inaugurated by Euler. In relatively recent times, Martin Gardner popularized the strange, elementary properties of the band in his column in *Scientific American*. These properties were investigated first by Peter Guthrie Tait (1831–1901) in his papers on knots in the 1870s and 1880s [32]. (A little of the history of the topology of knots, notably the contribution of Tait, is given by Ruth Lawrence in her Desert Island Theorem, elsewhere in this volume.) The aesthetic and intriguing qualities of the Möbius band were recognized more widely in the twentieth century with their depiction in Escher's graphic art and their adoption by modern sculptors.

9. Felix Klein

At the turn of the twentieth century Felix Klein was the most influential of geometers. In a narrow sense he is remembered because of his association with the

'Klein bottle', a topological shape having no inside or outside and formed by pasting together the edges of a Möbius band. But there was a breadth to Klein's view of geometry which such a focus disowns.

Klein had studied physics under Plücker at a time when Plücker had almost abandoned physics for geometry. But just as he gained his doctorate, aged nineteen, his mentor suddenly died and Klein was left to complete Plücker's unfinished book on analytic geometry. He could easily have gone on to promote this branch of geometry at the expense of other geometries but he had come under the influence of Sophus Lie, already researching into transformation groups and had become interested in just how geometries were interrelated. In his inaugural professorial address of 1872 at the University of Erlangen he suggested that group theory held the key because geometry could be seen as the study of those properties that remain unchanged under any particular group of transformations. Such studies constituted the so-called 'Erlanger Programm', which was quickly translated from German into six other languages. Henceforth, the geometries of ancient Greece, Euclidean and spherical, the projective geometry of Desargues and Poncelet, the non-Euclidean geometries of Lobachevski, Bolyai and Riemann and many, many others would all be studied under one umbrella.

References

1. Richard J. Gillings, *Mathematics in the Time of the Pharaohs* (Cambridge, MA: MIT Press, 1972; New York: Dover, 1982).
2. Otto Neugebauer, *The Exact Sciences in Antiquity*, 2nd Edn. (Brown University Press, 1957; New York: Dover, 1969).
3. Lǐ Yan & Dù Shíràn, *Chinese Mathematics: A Concise History*, Trans. John N. Crossley & Anthony W.-C. Lun (Oxford: Clarendon Press, 1987).
4. Julian Lowell Coolidge, *A History of Geometrical Methods* (Oxford: Clarendon Press, 1940). Coolidge offers a fourfold division of the history of geometry; synthetic, algebraic, differential, topological.
5. Florian Cajori, *A History of Mathematics*, 4th Edn. (Chelsea, 1985).
6. Martin Bernal, *Black Athena: The Afroasian Roots of Classical Civilization*, 2 vols. (New Brunswick: Rutgers University Press, 1987, 1991).
7. Mary Lefkowitz, *Not Out of Africa: How Afrocentrism Became an Excuse to Teach Myth as History* (New York, 1995).
8. Howard Eves, *An Introduction to the History of Mathematics*, Revised Edn. (Holt: Rinehart and Winston, 1964).
9. Thomas Heath, *A Manual of Greek Mathematics* (Oxford: Clarendon Press, 1931).
10. William Dunham, *Journey Through Genius: The Great Theorems of Mathematics* (John Wiley 1990 & Penguin, 1991).
11. R. Catesby Taliaferro, Translator's introduction to *Conics* by Apollonius of Perga, in *Great Books of the Western World*, vol. II (Encyclopædia Britannica & University of Chicago, 1952), 599–600.
12. John Fauvel & Jeremy Gray, *The History of Mathematics: A Reader* (Macmillan Education/Open University, 1987).

13. George Gherveghese Joseph, *The Crest of the Peacock: Non-European Roots of Mathematics* (Penguin, 1991).
14. B. A. Rosenfeld, *A History of Non-Euclidean Geometry: Evolution of the Concept of a Geometric Space*. Trans. Abe Shenitzer (Springer-Verlag, 1988).
15. Chris Pritchard & Brahmagupta, *Mathematical Spectrum* **28** (1995/96), 49–51.
16. Michel Chasles, *Aperçu historique sur l'origine et le développement des methods en géométrie*, Paris (1837). (Chasles was a contemporary of Steiner and so the geometry of that period is missing. Having little knowledge of German, his coverage of German contributions was also inadequate.)
17. Morris Kline, *Mathematics in Western Culture* (Oxford University Press, 1953; Pelican, 1972).
18. Morris Kline, *Mathematics: A Cultural Approach* (Addison-Wesley, 1962).
19. Ross King, *Brunelleschi's Dome: The Story of the Great Cathedral in Florence* (Chatto & Windus, 2000; Pimlico, 2001).
20. Paul E. Cohen & Robert T. Augustyn, *The Magazine: Antiques* (January, 1999), 214–217.
21. Carl B. Boyer, *A History of Mathematics* (John Wiley, 1968).
22. Jeremy Gray, Möbius's geometrical mechanics, in John Fauvel, Raymond Flood & Robin Wilson (Eds.), *Möbius and his Band* (Oxford University Press, 1993), 78–103.
23. Michael Sean Mahoney, *The Mathematical Career of Pierre de Fermat, 1601–1665* (Princeton University Press, 1973) (Second edition, 1994).
24. René Descartes, *The Geometry of René Descartes*, Trans. David Eugene Smith & Maria L. Latham (Chicago/London: Open Court, 1925).
25. L. P. Eisenhart, Darboux's contribution to geometry, *Bulletin of the American Mathematical Society* **24** (1918).
26. Julian Lovell Coolidge, The heroic age of geometry, *Bulletin of the American Mathematical Society* **35** (1929), 19–37.
27. Julian Lowell Coolidge, *A History of Conic Sections and Quadric Surfaces* (Oxford: Clarendon Press, 1945).
28. Roberto Bonola, *Non-Euclidean Geometry: A Critical and Historical Study of its Development*, Trans. H. S. Carslaw (Dover, 1955).
29. Philip J. Davis & Reuben Hersch, *The Mathematical Experience* (Boston: Birkäuser, 1981; Pelican, 1983).
30. Eli Maor, *To Infinity and Beyond: A Cultural History of the Infinite* (Boston: Birkäuser, 1987).
31. Norman Biggs, The development of topology, in John Fauvel, Raymond Flood & Robin Wilson (Eds.) *Möbius and his Band* (Oxford University Press, 1993), 104–119.
32. Chris Pritchard, Aspects of the life and work of Peter Guthrie Tait, in *James Clerk Maxwell Commemorative Booklet* (Edinburgh: International Congress on Industrial and Applied Mathematics, 1999). The booklet is not paginated.

2.2

Greek Geometry with Special Reference to Infinitesimals

SIR T. L. HEATH

It may be convenient that, before approaching the special subject of my address, I should indicate very briefly the stages through which Greek geometry passed from its inception to the time of Archimedes and Apollonius, when it reached its highest development.

The story begins with Thales, who lived approximately from 624 to 547 B.C., and whose work therefore belongs roughly to the first half of the sixth century. Thales travelled in Egypt and (so we are told) brought geometry from thence into Greece. So far as we can judge from the available records, Egyptian geometry consisted almost entirely of practical rules for the mensuration of plane figures, such as squares, triangles, trapezia, and of the solid content of measures of corn, etc., of different shapes. They were also able to construct pyramids of a certain slope by means of a particular arithmetical ratio, which is in fact the cotangent of the angle of slope. They knew that a triangle with sides in the ratios of the numbers 3, 4, 5 is right-angled, and they used the fact as a means of drawing right angles. But we look in vain in Egyptian records for any general theorem or for any vestige of a proof of such. What is remarkable about Thales is not the few elementary propositions attributed to him, namely that a circle is bisected by any diameter, that the base angles of an isosceles triangle are equal, that if two straight lines cut one another the vertically opposite angles are respectively equal, that the triangles of Euclid I. 26 are congruent, and that the angle in a semicircle is a right angle; the vital fact is that Thales was the first person to think of *proving* such things, and thereby originating the idea of geometry as a science in and for itself. It was here that the Greek instinct for science emerged; the Greek was not satisfied with facts, he wanted to know the why and wherefore, and he was not content till he could get a *proof* or an explanation which commended itself to his reason. It was by this new point of view, this inspiration, that the Greeks came to create the science

First published in *Mathematical Gazette* **11** (March 1923), pp. 248–259.

of mathematics and, with it, scientific method. Kant calls it nothing less than an intellectual revolution. A light, he says, broke on the first man who demonstrated the property of the isosceles triangle whether his name was Thales or what you will; since from that point onward the road that must be taken could no longer be missed, and the safe way of a science was struck and traced out for all time.

After Thales come Pythagoras and the Pythagoreans. Pythagoras lived from about 572 to 497 B.C., so that his work may be taken to belong to the second half of the sixth century, while his successors cover a large part of the fifth century. Of Pythagoras we are told that he transformed geometry into a subject of liberal education, investigating the principles of the science from the beginning. That is to say, he was the first to make geometry a connected system beginning with definitions and the other necessary preliminary assumptions. From the story that he bribed a promising pupil to learn geometry by offering him sixpence for each proposition that he mastered, we may infer that the subject was developed in a series of propositions. Pythagoras himself is credited with the theorem of Eucl. I. 47, with the discovery of a theory of proportion (numerical in character) and with the construction of the cosmic figures, the five regular solids. The construction of the dodecahedron involves that of a regular pentagon, and that again the division of a straight line in extreme and mean ratio, which is a particular case of the method known as the *application of areas*. The simplest case of this method is that of Euclid I. 44, 45, which corresponds to simple division in arithmetic. The Pythagoreans extended the method to cases of application with a certain excess or defect as in Euclid VI. 27–29, which propositions amount to the geometrical solution of the most general form of quadratic equation, provided that it has real roots. Application pure and simple, and application with excess or defect, is the form in which Apollonius expresses the fundamental properties of the three conics, and this is why he gave the three conics for the first time the names parabola, hyperbola and ellipse.

The Pythagoreans then devised and applied extensively the two powerful methods of proportion and application of areas. They also knew the properties of parallel lines, and proved the theorem that the three angles of a triangle are together equal to two right angles. In short, the Pythagoreans developed the bulk of the elementary parts of geometry corresponding to Books I., II., IV., VI., and probably III., of Euclid's *Elements*. They also discovered the incommensurable, at all events in the case of the diagonal of a square in relation to its side. This had the disconcerting result of revealing the defect in the Pythagorean theory of proportion, which applied only to commensurable magnitudes.

The geometry of which we have so far spoken belongs to the *Elements*. But, before the body of the *Elements* was complete, the Greeks had advanced beyond the *Elements*. By the second half of the fifth century B.C. they had investigated three famous problems in higher geometry, (1) the squaring of the circle, (2) the trisection of any angle, (3) the duplication of the cube. The great names belonging to this period are Hippias of Elis, Hippocrates of Chios, and Democritus.

Hippias of Elis invented a certain curve described by combining two uniform movements (one angular and the other rectilinear) taking the same time to complete. Hippias himself used his curve for the trisection of any angle or the division of it in any ratio; but it was afterwards employed by Dinostratus, a brother of Eudoxus's pupil Menaechmus, and by Nicomedes for squaring the circle, whence it got the name *quadratrix*.

Hippocrates of Chios is mentioned by Aristotle as an instance to prove that a man may be a distinguished geometer and, at the same time, a fool in the ordinary affairs of life. He occupies an important place both in elementary geometry and in relation to two of the higher problems above mentioned. He was, so far as is known, the first compiler of a book of Elements; and he was the first to prove the important theorem of Eucl. XII. 2, that circles are to one another as the squares on their diameters, from which he further deduced that similar segments of circles are to one another as the squares on their bases. These propositions were used by him in his tract on the squaring of *lunes*, which was intended to lead up to the squaring of the circle.

Hippocrates also attacked the problem of doubling the cube. He did not indeed solve it, but he reduced it to another, namely that of finding two mean proportionals in continued proportion between two given straight lines; and the problem was ever afterwards solved in that form.

Democritus wrote a large number of mathematical treatises, the titles only of which are preserved. From the title "On irrational lines and solids" we gather that he wrote on irrationals. As we shall see later, Democritus realised as fully as Zeno the difficulty connected with the continuous and the infinitesimal. Democritus, again, was the first to state that the volume of a pyramid or a cone is one third of that of the prism or cylinder respectively on the same base and of equal height, though as regards the cone at least he could not give a rigorous proof; this was reserved for Eudoxus.

We come now to the first half of the fourth century B.C., the time of Plato, and here the important names are Archytas, Theodorus of Cyrene, Theaetetus and Eudoxus.

Archytas was the first to solve the problem of the two mean proportionals; he used a wonderful construction in three dimensions, determining a certain point as the intersection of three surfaces, a cone, a half-cylinder and an anchor-ring or torus with inner diameter *nil*.

Theodorus, Plato's teacher in mathematics, extended the theory of the irrational by proving that $\sqrt{3}, \sqrt{5}, \ldots, \sqrt{17}$ are all incommensurable with 1. Theodorus's proof was evidently not general; and it was reserved for Theaetetus to comprehend all such irrationals in one definition, and to prove the property generally as in

Eucl. X. 9. Much of the content of the rest of Euclid's Book X. (dealing with compound irrationals), as of Book XIII. on the five regular solids, was also due to Theaetetus.

Eudoxus, an original genius second to none (unless it be Archimedes) in the history of our subject, made two discoveries of supreme importance for the further development of Greek geometry.

(1) As we have seen, the discovery of the incommensurable rendered inadequate the Pythagorean theory of proportion, which applied to commensurable magnitudes only. It would no doubt be possible, in most cases, to replace proofs depending on proportions by others; but this involved great inconvenience, and a slur was cast on geometry generally. The trouble was remedied once for all by Eudoxus's discovery of the great theory of proportion, applicable to commensurable and incommensurable magnitudes alike, which is expounded in Euclid's Book V. Well might Barrow say of this theory that "there is nothing in the whole body of the Elements of a more subtile invention, nothing more solidly established." The keystone of the structure is the definition of equal ratios (Eucl. V., Def. 5); and twenty-three centuries have not abated a jot from its value, as is plain from the facts that Weierstrass repeats it word for word as his definition of equal numbers, and it corresponds almost to the point of coincidence with the modern treatment of irrationals due to Dedekind.

(2) Eudoxus discovered the method of exhaustion for measuring curvilinear areas and solids, to which, with the extensions given to it by Archimedes, Greek geometry owes its greatest triumphs. The method is seen in operation in Euclid XII. 1–2, 3–7 Cor., 10, 16–18. Props. 3–7 Cor. and 10 prove the theorems about the volume of the pyramid and cone first stated by Democritus.

Menaechmus, a pupil of Eudoxus, was the discoverer of the conic sections, two of which, the parabola and the hyperbola, he used for solving the problem of the two mean proportionals. If $a : x = x : y = y : b$, then $x^2 = ay$, $y^2 = bx$ and $xy = ab$. These equations represent, in Cartesian co-ordinates, and with rectangular axes, the conics by the intersection of which two and two Menaechmus solved the problem; in the case of the rectangular hyperbola it was the asymptote-property which he used.

We pass to Euclid's times. A little older than Euclid, Autolycus of Pitane wrote two books, *On the Moving Sphere*, a work on Sphaeric for use in astronomy, and *On Risings and Settings*. The former work is the earliest Greek text-book which has reached us intact.

Euclid flourished about 300 B.C. or a little earlier. Besides his great work, the *Elements*, he wrote other books on both elementary and higher geometry, and on the other mathematical subjects known in his day, namely astronomy, optics and music. To elementary geometry belong the *Data*, and *On Divisions* (*of figures*); also the *Pseudaria*, now lost, which was a sort of guide to fallacies in geometrical reasoning. The treatises on higher geometry are all lost; they include (1) the *Conics*, which covered almost the same ground as the first three books of Apollonius's *Conics*; (2) the *Porisms*, in three books, the importance and difficulty of which can be inferred from Pappus's account of it and the lemmas he gives for use with it: (3) the *Surface-Loci*, for which also Pappus gives lemmas; one of these implies that Euclid assumed as known the focus-directrix property of the three conics, which is absent from Apollonius's *Conics*.

In the period between Euclid and Archimedes comes Aristarchus of Samos, famous for having anticipated Copernicus. Accepting Heraclides's view that the earth rotates about its own axis, Aristarchus went further and put forward the hypothesis that the sun itself is at rest, and that the earth, as well as Mercury, Venus, and the other planets, revolve in circles about the sun.

One work of Aristarchus, *On the sizes and distances of the Sun and Moon*, is extant in Greek. Thoroughly classical in form, it lays down certain hypotheses, and then deduces therefrom, by rigorous geometry, limits to the sizes and distances of the sun and moon. Though the book contains no word of the heliocentric hypothesis, it is highly interesting in itself, because we here find geometry used for the first time with a *trigonometrical* purpose. In effect Aristarchus finds arithmetical limits to the values of certain trigonometrical ratios of small angles, namely,

$$\frac{1}{18} > \sin 3° > \frac{1}{20}, \quad \frac{1}{45} > \sin 1° > \frac{1}{60}, \quad 1 > \cos 1° > \frac{89}{90}.$$

We now come to the third century B.C., which marks the culmination of Greek geometry in the works of Archimedes and Apollonius. Archimedes belongs to the second part of my subject. Of Apollonius it need only be said that he was called the "Great Geometer" out of admiration for the *Conics*, a treatise of the greatest originality and power. He produces his conics in the most general way from any oblique cone, and develops their properties with reference to axes which are in general oblique, namely a diameter and the tangent at its extremity, the principal axes only appearing as a particular case. The most remarkable portion of the work is perhaps Book V., in which Apollonius treats of normals to the conics as maximum and minimum straight lines drawn from different points to the curve; in particular he works out some intricate propositions, from which we can without difficulty deduce the Cartesian equations of the evolutes of the three conics respectively.

I can now turn to the second part of my task and endeavour to trace the evolution of the infinitesimal calculus in Greek geometry. The Pythagoreans, as we said, discovered the incommensurable. Closely connected with the idea of the incommensurable is the fact that mathematical magnitudes are divisible *ad infinitum*. Applying the principle to, say, a straight line, we have to consider to what we ultimately come, assuming (what is actually impossible) that we can carry the division to an end. Is it an indivisible line, or a point, or what? The Pythagoreans could not admit the possibility of an indivisible line: and, on the other hand, Pythagorean writers like Nicomachus are clear that you cannot make up a line out of points. What then is the ultimate product of the infinite division of a line, assuming it to be carried out? If it is neither an indivisible line nor a point, is it something which is different from both, in short, an infinitesimal? Zeno denied this, his argument having apparently taken this form. If magnitudes are divisible *ad infinitum*, the division must end, not in the infinitely small, because that is by hypothesis still divisible, but in *nothing*; if, however, you reverse the process and add nothing to itself for ever and ever, you will never produce any finite magnitude at all. The argument in the first of Zeno's paradoxes (the *Dichotomy*) is much the same. Before a moving thing can get any distance it must have travelled half that distance; before it has travelled the half it must have travelled the half of that, and so on, to the smallest distance; how then can it start at all? Democritus stated a similar dilemma about continuous variation in his question about consecutive sections of a cone parallel to the base. Suppose that we have a section of a cone parallel to its base and "indefinitely near" to it (as the phrase is), what are we to say of this section? Is it, said Democritus, equal, or not equal, to the base? If it is equal, so will the next consecutive section be, and so on; thus the cone will really be not a cone at all but a cylinder. If, on the other hand, it is not equal to the base, and in fact less, the surface of the cone will be jagged, like steps, which is very absurd.

There is no doubt that the paradoxes of Zeno, and the difficulty of answering them, profoundly affected the formal development of mathematics from his time onward. Antiphon the Sophist, a contemporary of Socrates, had asserted, in connexion with attempts to square the circle, that if in a circle we inscribe successive regular polygons, beginning from a triangle or a square and continually doubling the number of sides, we shall sometime arrive at a polygon the sides of which will coincide with the circumference of the circle. Warned by the unanswerable arguments of Zeno, mathematicians henceforth substituted for this the statement that, by continuing the construction, we can inscribe a polygon approaching equality with the circle *as nearly as we please*. The method of exhaustion used, for the purpose of proof by *reductio ad absurdum*, the lemma proved in Eucl. X. 1 (to the effect that, if from any magnitude we subtract not less than its half, then from the remainder not less than half, and so on continually, there will sometime be left a magnitude less than any assigned magnitude of the same kind, however small); and this again depends on an assumption which is practically contained in Eucl. V. Def. 4, but is generally known as the Axiom of Archimedes, stating that, if we have two unequal magnitudes, their difference (however small) can, if continually added to itself, be made to exceed any given magnitude of the same kind (however great).

It was Eudoxus who proved by the method of exhaustion the propositions about the volumes of a cone and a pyramid; and it is probably to Eudoxus that Euclid owed the proofs of the theorems that circles are to one another as the squares on their diameters, and that spheres are to one another as the cubes on their diameters.

The notion of *exhausting* the circle was, however, that of Antiphon, and he is entitled to credit for an idea which proved so fruitful. Geometers continued to use the method of exhausting an area indicated by Antiphon, while they barred themselves, in theory at least, from carrying the process to the point of taking a limit. They would certainly have denied that an inference of this kind could constitute a valid proof. There is a passage in the recently discovered tract of Archimedes, the *Method*, which bears this out. The only form of proof which was considered to be conclusive was the double *reductio ad absurdum* which was part of the method of exhaustion.

Some remarks about this method of proof are necessary at this stage. The process will be familiar to you from Euclid XII. 2. That proposition and those about the volume of the cone and pyramid suggest that Eudoxus approximated to the figure to be measured by inscribed figures only, the result of which is that the second part of the proof has to be inverted. Archimedes avoided this by approximating both from above and from below, so to speak, *i.e.* by circumscribing figures as well as inscribing them, so that both parts of the proof are direct.

Writers have sometimes observed that, before we can apply the method of exhaustion, we must already know the result to be proved, implying that the method is of no use for discovering such results. This is scarcely true, because the method of exhaustion does not consist exclusively of the proof by *reductio ad absurdum*; it includes the process of exhausting the area or volume to be measured, and this in itself often indicates the result quite clearly.

But, in view of the criticism, it is desirable to consider exactly how the Greeks, and Archimedes in particular, actually obtained the various results which were then to be established by the rigorous proof. We can distinguish three classes of cases.

In the first the result was divined in a way which, however careful the Greeks were to avoid mentioning such a thing, amounted to passing to the limit. For example, there can be no doubt that the theorem that circles are to one another as the squares on their diameters was inferred on Antiphon's lines: since similar inscribed polygons are in the ratio in question, the same must be true of similar regular polygons inscribed in the circles when the number of sides is indefinitely increased, and their length correspondingly diminished, and therefore of the circles themselves from which the polygons will then be indistinguishable. Another case is the proposition in Archimedes' *Measurement of a Circle*, to the effect that the area of a circle is equal to a triangle with height equal to the radius and base equal to the circumference of the circle: the very form of this enunciation suggests that it was arrived at by regarding the circle as the sum of an indefinitely large number of isosceles triangles with the centre as common vertex and with equal indefinitely small chords of the circle as bases. Again, when Democritus concluded that the volume of a cone is one third of that of the cylinder with the same base and height, he must have inferred this from the consideration that, the corresponding proposition being true of a pyramid with a polygon of any number of sides as base, it is still true when the number of sides of a regular polygon forming the base is indefinitely increased, and the sides correspondingly diminished, so that the polygon tends to become indistinguishable from a circle and the pyramid from a cone. But perhaps the clearest case of all is in the *Method* of Archimedes, where he says that from the volume of a sphere (already known to be what we write as $\frac{4}{3}\pi\rho^3$) he inferred that the surface of the sphere is $4\pi\rho^2$; his words are, "Starting from the fact that any circle is equal to a triangle the base of which is the circumference, and the height the radius, of the circle, it occurred to me that, just in the same way, any sphere is equal to a cone the base of which is the surface, and the height the radius, of the sphere."

The second class of case is that in which the process of exhausting the area or content itself indicates the result. In this class of case what happens is that we get a series of terms that can be summed; and there are two varieties according to the nature of the series that has to be summed. One case is simple, depending on the summation of the geometrical progression

$$1 + \frac{1}{4} + \left(\frac{1}{4}\right)^2 + \cdots \text{ad inf.}$$

This is all that is required in Archimedes' second method of finding the area of any parabolic segment. He inscribes in the segment a triangle with the same base and vertex, then in the two segments left over two triangles with the same vertex and base respectively, in the four segments left over four more triangles, and so on. If A is the area of the first triangle, the sum of the next two triangles is $\frac{1}{4}A$, the sum of the next four triangles $(\frac{1}{4})^2A$, and so on. Adding the series of triangles, we have

$$A\left\{1 + \frac{1}{4} + \left(\frac{1}{4}\right)^2 + \cdots\right\}.$$

Archimedes sums n terms of the series, stating the result in the form

$$A\left\{1 + \frac{1}{4} + \left(\frac{1}{4}\right)^2 + \cdots + \left(\frac{1}{4}\right)^{n-1}\right\} + \frac{A}{3}\left(\frac{1}{4}\right)^{n-1} = \frac{4}{3}A.$$

To find the sum of the series continued *ad infinitum* we should simply observe that $(\frac{1}{4})^{n-1}$ can be neglected when n is indefinitely increased, so that the sum of the series to infinity is $\frac{4}{3}$ and the area of the segment is $\frac{4}{3}A$. Although Archimedes probably inferred the result in this way, he does not say so, but simply declares that the area is $\frac{4}{3}A$, and then proves it by *reductio ad absurdum* with the help of the summation to n terms only.

The second kind of series the summation of which gives the desired result is more important, the series taking such forms that their summation is really equivalent to an *integration*. In the works of Archimedes there are some six cases where the investigation gives the equivalent of an integration. Three of the actual integrals are $\int x\,dx$, $\int x^2\,dx$ and a combination of the two. The volume of any segment of a paraboloid of revolution is obtained in a form equivalent to $C\int_0^c x\,dx$, where C is a constant. In three cases (those of the volume of half a spheroid, the area cut off by a spiral between two radii vectores or one radius vector and the initial line, and the first investigation of the area of a parabolic segment) the procedure is equivalent to finding the integral $\int_0^b x^2\,dx$, and in the case of the spiral the integral $\int_b^c x^2\,dx$ also. The investigation of the volume of any segment of a hyperboloid of revolution amounts to finding the integral $\int_0^b (ax + x^2)\,dx$, which, by means of a certain device, is made to serve for finding the volume of any segment of a spheroid also.

I will illustrate by the simplest case, the paraboloid of revolution. Taking an oblique segment of it and a section of the segment by a plane through the axis of revolution, we have a parabola referred in general to oblique axes, PV and PE the tangent at its extremity. Let the equation of the parabola be $y^2 = px$. PV being the axis of the paraboloidal segment to be measured, let PV be divided into n equal small parts of length h. Figures are inscribed and circumscribed to the segment made up of short oblique cylinders as indicated in the figure. To avoid taking account of the constant due to the inclination of the axes, Archimedes compares each small cylinder with the corresponding portion of the whole cylinder EQ'.

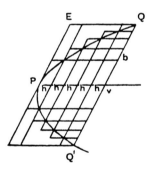

Fig. 2.2.1.

If y is the ordinate corresponding to the abscissa rh, then the small cylinder having as base the section of the paraboloid through y parallel to the tangent plane at P is to the corresponding portion of the cylinder EQ' as y^2 to QV^2, while $y^2 = p \cdot rh$ and $QV^2 = p \cdot nh$. We have then

$$\frac{\text{(circumscribed figure)}}{\text{cylinder } EQ'} = \frac{p(h + 2h + \cdots + nh)h}{p \cdot nh \cdot nh}$$

and

$$\frac{\text{(inscribed figure)}}{\text{cylinder } E\,Q'} = \frac{p(h + 2h + \cdots + (n-1)h)h}{p \cdot nh \cdot nh};$$

and Archimedes simply says that the first of the two ratios is $>$ and the second $< \frac{1}{2}$.

Now, if we sum the series, the ratios work out to

$$\frac{1}{n^2} \cdot \frac{1}{2}n(n+1) \quad \text{or} \quad \frac{1}{2}\left(\frac{n+1}{n}\right)$$

and

$$\frac{1}{n^2} \cdot \frac{1}{2}n(n-1) \quad \text{or} \quad \frac{1}{2}\left(\frac{n-1}{n}\right).$$

When, therefore, Archimedes declares that the volume of the segment is one-half that of the cylinder, he is practically taking the limit of the summation when h is indefinitely small and n indefinitely large, but so that $nh = c$, and saying that

$$\text{Limit of } h(h + 2h + \cdots + nh) = \frac{1}{2}c^2, \quad \text{or} \quad \int_0^c x\,dx = \frac{1}{2}c^2.$$

The corresponding inequality which Archimedes gives in the case of the area cut off by the spiral $r = a\theta$ between the initial line and the radius vector to any point on the first turn of the spiral works out to

$$h^2 + (2h)^2 + \cdots + (nh)^2 > \frac{1}{3}n(nh)^2 > h^2 + (2h)^2 + \cdots + \{(n-1)h\}^2.$$

He has in Prop. 10, *On Spirals*, summed the series

$$1^2 + 2^2 + \cdots + n^2,$$

his result being equivalent to $\frac{n(n+1)(2n+1)}{6}$, from which the above inequalities readily follow.

If P be any point on the first turn, r the radius vector OP bounding the required area, θ the angle through which the generating line has turned from the initial line to OP, we suppose r to be divided into n parts h and draw the radii vectores $h, 2h, 3h, \ldots, (n-1)h$, which of course divide θ into equal parts θ/n. Drawing circular arcs with O as centre through the extremities of $h, 2h, 3h, \ldots, (n-1)h$ and $nh(OP)$, we have figures circumscribed and inscribed to the spiral which are made up of sectors of circles with vertical angle θ/n and radii $h, 2h, 3h, \ldots$.

The area of the circumscribed figure is

$$\frac{\theta}{2n}\{h^2 + (2h)^2 + \cdots + (nh)^2\} \quad \text{or} \quad \frac{\theta}{2nh} \cdot h\{h^2 + (2h)^2 + \cdots + (nh)^2\},$$

and that of the inscribed figure is

$$\frac{\theta}{2nh} \cdot h\{h^2 + (2h)^2 + \cdots + ((n-1)h)^2\}.$$

Archimedes then practically takes the limit of $h^3 \cdot \frac{n(n+1)(2n+1)}{6}$ when h is indefinitely small and n indefinitely large, while $nh = a\theta$, to be $\frac{1}{3}(nh)^3$, and the resulting area is

$$\frac{\theta}{2nh} \cdot \int_0^{nh} x^2 dx \quad \text{or} \quad \frac{\theta}{2a\theta} \cdot \frac{1}{3}(a\theta)^3 = \frac{\theta}{2} \cdot \frac{1}{3}(a\theta)^2.$$

In the case of a sphere and a segment of a sphere Archimedes does not proceed, as in the case of the spheroid, by means of one of the integrals above mentioned, probably because he has to find the surface as well as the volume of the sphere and segment, and to find the surface requires a different integration. In the case of the sphere and segment, therefore, he inscribes and circumscribes polygons to the generating circle and uses the figures generated by the revolution of these polygons about the axis, which figures are of course inscribed and circumscribed to the sphere

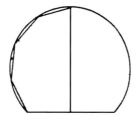

Fig. 2.2.2.

or segment. The inscribed polygon is obtained by dividing the circumference of the section of the sphere or segment into $2n$ equal parts and joining the successive points of division; the circumscribed polygon has its sides parallel to those of the inscribed polygon. Each of the equal sides of the polygons subtends at the centre an angle a/n in the case of the segment (where $2a$ is the angle subtended by the arc at the centre), and π/n in the case of the whole circle.

To take the case of the segment, Archimedes proves by geometry the equivalent of the formula

$$\frac{2\left\{\sin \frac{a}{n} + \sin \frac{2a}{n} + \cdots + \sin(n-1)\frac{a}{n}\right\} + \sin a}{1 - \cos a} = \cot \frac{a}{2n},$$

and deduces that the surface of the inscribed figure is

$$\pi a^2 \cdot 2 \sin \frac{a}{2n} \left[2\left\{ \sin \frac{a}{n} + \sin \frac{2a}{n} + \cdots + \sin(n-1)\frac{a}{n} \right\} + \sin a \right],$$

that is,

$$\pi a^2 \cdot 2 \cos \frac{a}{2n}(1 - \cos a).$$

The effect of summing the series within brackets and taking the limit, which is practically what Archimedes does, is to show that

$$\pi a^2 \int_0^a 2\sin\theta\, d\theta = 2\pi a^2 (1 - \cos a).$$

Enough has been said to show the remarkable power in the hands of an Archimedes of the method of exhaustion even with the purely geometrical means available. The disadvantages of it were that it had, so far as the proof by *reductio ad absurdum* is concerned, to be applied separately to each case, and that there were only a limited number of real integrations which pure geometry could compass. The Greeks were thus limited because they had not made the discovery that differentiation and integration are the inverse of one another (a fact which was first fully proved by Barrow), and they were without the advantage, from the point of view of both, arising from algebraical notation and the modern discovery of the development of various functions in the form of series.

Of considerations corresponding to the differential calculus there are very few traces in Greek geometry. One case, however, seems certain, though nothing of the kind is expressed in the text. There can be no doubt that Archimedes obtained the

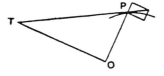

Fig. 2.2.3.

sub-tangent property of the spiral $r = a\theta$ by a consideration of the instantaneous direction of the motion of the point describing the spiral at any point P of the curve. With our notation we have, by similar triangles,

$$\frac{OT}{r} = \frac{r\,d\theta}{dr} = \frac{r}{a},$$

and $OT = r^2/a$ or $r\theta$, which is the equivalent of Archimedes's statement of the property. We know next to nothing of the content of Apollonius's lost treatise on the *cochlias* or cylindrical helix; but if, as is probable, he gave the properties of the tangent to the curve at any point, he would no doubt similarly determine the direction of the tangent as the instantaneous direction of the motion of the describing point.

To return to cases normally demanding integration or its equivalent. The Greeks were by no means limited to what they could accomplish by direct integration. An alternative to direct integration was the reduction of the actual problem to another integration, the result of which was already known. A large proportion of

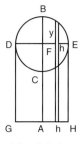

Fig. 2.2.4.

the propositions in the *Method* of Archimedes had this object. But there are other cases on record where this must have been the means of solution. Dionysodorus, we are told, found the volume of the torus or anchor-ring described by the revolution of a circle of radius a about a straight line in its plane at a distance c from the centre of the circle ($c > a$). Dionysodorus stated the result in this form:

$$\frac{\text{circle } BDCE}{\frac{1}{2} \text{ (parallelogram } DH)} = \frac{\text{(volume of torus)}}{\text{(cylinder rad} \cdot EH \text{ and height } GH)}.$$

This gives

$$\text{(volume of torus)} = \frac{\pi a^2 \cdot 2\pi c^2 a}{ca}$$
$$= 2\pi c \cdot \pi a^2.$$

We can imagine how Dionysodorus arrived at this result. Dividing FE into equal small parts h and taking the corresponding ordinates y of the points on the circle $BDCE$, we find that the section of the torus formed by the revolution of the double ordinate $2y$ about the axis GH is $\pi(c + y)^2 - \pi(c - y)^2$ or πcy. We have therefore to find

$$\sum 2\pi cyh \quad \text{or} \quad 2\pi c \cdot \sum h\sqrt{a^2 - (rh)^2}$$

within the proper limits when h is indefinitely diminished. It would be obvious that $\sum h\sqrt{a^2 - (rh)^2}$ is then the area of the circle $BCDE$, which is known, so that there is no need for an integration. The result is therefore, as stated, $2\pi c \cdot \pi a^2$.

The cubature of the torus is otherwise historically interesting. Kepler, about 1615, solved it by means of infinitesimal sections. Any plane through the axis of revolution cuts the torus in two circular sections. Hence, said Kepler, we may regard the torus as made up of an infinite number of very thin discs. These discs are thinnest on the side towards the centre of the torus and thickest on the outside, and the mean thickness is the thickness at the centre of the discs. Adding the discs, we find that the volume of the torus is the same as if the circle passing through the centres of all the circular sections were straightened out and the torus turned into a cylinder of the same section. That is, the volume is $2\pi c \cdot \pi a^2$. Professor Loria has remarked that Kepler's conception was bold and original. But the Greeks had in effect anticipated Kepler therein; for Heron of Alexandria, after mentioning Dionysodorus's solution, adds that the measurement can also be done in another way "if the torus is straightened out and so becomes a cylinder," and works out the figures on this basis.

The case of the torus investigated by Dionysodorus throws, I think, some light on the evolution of the more general proposition, claimed by Pappus as his own, which is in effect an anticipation of the theorem attributed to Guldin, that the volume of the solid formed by the revolution of any plane figure about an axis in its plane is the product of the area of the figure and the length of the path described by its centre of gravity. Divide the figure into narrow strips all of breadth h and all perpendicular to the axis of revolution.

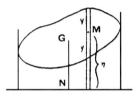

Fig. 2.2.5.

Let M be the centre of gravity of the strip of length $2y$, and η its distance from the axis of revolution. Then $2yh$ by its revolution about the axis produces a solid flat ring of content $h \cdot \pi\{(\eta + y)^2 - (\eta - y)^2\} = 2\pi \cdot 2yh\eta$. Now $\sum 2yh\eta$, when h is indefinitely diminished, is what we call the *moment* of the figure about the axis. What was wanted therefore to establish Pappus's theorem was simply the proposition that the moment of the whole figure placed where it is is equal to that of the whole figure supposed concentrated at its centre of gravity. The Greeks of course did not speak of "moments." They would have supposed the perpendicular GN from the centre of gravity G of the whole figure to the axis to be produced to H, so that $GN = NH$, and would have said that the figure placed where it is would balance about N a like figure supposed concentrated and placed at the point H. This is the phraseology of Archimedes's *Method*. Now we find that in the *Method* it is tacitly assumed, for each figure that is there treated, that when a figure is balanced where it is about a certain fulcrum against another figure on the other side, the former figure may be replaced by an equal weight placed at its centre of gravity. Archimedes in fact uses this proposition for the purpose of finding the centre of gravity of certain figures in a way corresponding to

$$\bar{\eta} \cdot \lim \sum 2yh = \lim \sum 2y\eta h \quad \text{or} \quad \bar{\eta} \cdot \int 2y dx = \int 2y\eta dx.$$

The theorem must therefore have been generally accepted at the time. No doubt it would be proved by some sort of integration.

I may perhaps suitably conclude with some remarks about Archimedes's treatise, the *Method*. I have mentioned that it illustrates the ingenuity with which Archimedes here obtains the results of certain integrations, not directly, but by reduction to other integrations the results of which are already known. But this treatise, found at Constantinople in 1906, has the supreme interest that it shows us for the first time how Archimedes originally discovered some of his results, namely by a peculiar method of his own depending on mechanical considerations. It is also a most important document in the history of infinitesimals, for here we find infinitesimals used almost without disguise.

The method is to weigh elements of the figure to be measured (X) against those of another (B) in such a way that the elements of B act at different points, namely where they are, but the elements of X act at one point. The latter point is a point on the diameter or axis of the figure to be measured produced in the direction away from the figure. The diameter or axis with the produced portion is imagined to be the bar or lever of a balance; the point of suspension is the extremity of the diameter or axis of the figure X. The content of B, as well as its centre of gravity, being known, its weight can be supposed to act as one mass at its centre of gravity. The corresponding elements of B and X, which are weighed against each other, are sections of B and X respectively by straight lines or planes parallel to the tangent line or plane at the extremity of the diameter or axis.

But the interesting fact is that the elements are spoken of as *straight lines* in the case of plane figures and *plane areas* in the case of solids. Herein Archimedes anticipated Cavalieri, who similarly speaks of figures being made up of linear or plane sections respectively, although in fact the elements are in the first case indefinitely narrow strips (areas), and in the second case indefinitely thin plane laminae (solids). The essential thing, as with Cavalieri, is that, though the number of the elements in each figure is infinite, the number is the same in both figures because the figures have the same height.

I will illustrate by one simple case, that of a right segment of a paraboloid of revolution.

BAC is a section through the axis of the paraboloid, AD the axis, ABC a section by the same plane of the cone with A as vertex and the circular section through BC as base, $EBCF$ the section by the plane through BAC of a cylinder on the same base as the cone.

PNQ is any double ordinate in the parabola BAC, and PQ produced both ways meets EB, FC in L, M.

Produce the axis DA to H, making HA equal to DA, and imagine HAD to be the bar of a balance, A being the point of suspension.

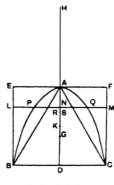

Fig. 2.2.6.

Now, by the property of the parabola,

$$DA : AN = BD^2 : PN^2,$$

or

$$HA : AN = LN^2 : PN^2$$
$$= (\text{circle of rad} \cdot LN) : (\text{circle of rad} \cdot PN)$$
$$= (\text{circle in cylinder}) : (\text{circle in paraboloid}).$$

Consequently the circle in the paraboloid, if placed with its centre of gravity at H, balances the circle in the cylinder placed where it is and acting at N. Similarly for the respective circles in the two figures made by *all* plane sections at right angles to AD. Therefore the paraboloid acting as one mass at H balances the cylinder placed where it is, which again can be supposed to act as one mass concentrated at its centre of gravity, *i.e.* at K where K bisects AD.

Since then

$$(\text{cylinder}) : (\text{paraboloid}) = HA : AK$$
$$= 2 : 1,$$

the content of the paraboloid is half that of the cylinder and $\frac{3}{2}$ that of the cone ABC.

In the next proposition Archimedes uses the same procedure for finding the centre of gravity of the paraboloidal segment. This time he weighs the paraboloid where it is against the cone ABC supposed concentrated at H, thus:

Since

$$PN^2 : BD^2 = AN : AD, \quad \text{and}$$
$$BD^2 : RN^2 = DA^2 : AN^2, \quad \text{it follows that}$$
$$PN^2 : RN^2 = DA : AN = HA : AN.$$

Now PN is a radius of the section of the paraboloid, and RN a radius of the section of the cone ABC, by the plane section through LM and perpendicular to AD. Therefore the circle in the paraboloid where it is balances the circle in the cone when placed with its centre of gravity at H. So for the sections by all other planes at right angles to AD.

Therefore the paraboloid where it is balances the cone ABC supposed concentrated at H.

If now G is the centre of gravity of the paraboloid, we may suppose its weight to act at G.

Therefore

$$(\text{paraboloid}) : (\text{cone}) = HA : AG.$$

And the paraboloid is $1\frac{1}{2}$ times the cone; therefore

$$HA = \frac{3}{2}AG,$$

or

$$AG = \frac{2}{3}AD.$$

2.3

A Straight Line is the Shortest Distance
Between Two Points

J. H. WEBB

The popular definition of a "straight line" as "the shortest distance between two points" is one of the few pieces of mathematical jargon in general circulation in the English language. This is unfortunate, for it is not a good definition, and one may well wonder who originated it. Euclid is not to blame, for he defined a straight line as "that which lies evenly between its points". (This is indeed cryptic, and Sir Thomas Heath, in a lengthy analysis of the original Greek, had to confess that "the language is . . . hopelessly obscure".)

Heath attributes the "shortest distance" definition to Legendre, whose *Éléments de géométrie* ran into many editions in the early nineteenth century. Yet to blame Legendre for our bad definition would not be quite fair, for Legendre originally wrote: "La ligne droite est le plus court chemin d'un point à un autre." Let us pass over the translation of "droite" as "straight" with the observation that "right line" was often used synonymously with "straight line", to a serious mistranslation: "chemin" is rendered as "distance" instead of "path" or "route". To say that a straight line is a special sort of path makes sense, but to define a straight line as a distance does not.

The first English translation of Legendre's work was brought out by David Brewster, Professor of Physics and later Principal of the University of Edinburgh, and it is here that our "shortest distance" definition first appeared. However, if one reads the title page carefully, it is clear that Brewster was not the actual translator, but only the editor of the translation. The translator's name is not mentioned, though he had written an eight-page "Introduction on proportion".

An American version of Brewster's translation was published soon after. Though it was freely "revised and adapted" by Charles Davies, of the Military Academy, West Point, the definition of a straight line is faithfully copied from Brewster's edition.

First published in *Mathematical Gazette* **58** (June 1974), pp. 137–138.

Who was the anonymous translator whose translation slip became part of the English language on both sides of the Atlantic? He was a young schoolmaster of Kirkcaldy, near Edinburgh, who was engaged by Brewster to do the work for the handsome fee of £50. He had been a promising mathematician in his student days at Edinburgh University, but was later to become famous as an essayist and historian. His name was Thomas Carlyle.

Carlyle's early prowess is mentioned by his biographers, and his published letters contain several references to mathematical problems, and to his work on translating Legendre. It is evident from these letters that the work soon palled, for Carlyle eventually persuaded his brother John to take over "this thrice-wearisome Legendre" and finish the job. Carlyle never again returned to mathematics.

Editor's Note

For a detailed discussion of the translating of Legendre's geometry, see Alex D. Craik 'Geometry versus analysis in early 19th century Scotland: John Leslie, William Wallace and Thomas Carlyle. *Historia Mathematica* **27** (2000), 133–163.

2.4

On Geometrical Constructions By Means
of the Compass

E. W. HOBSON

The subject that I have chosen for my address is a special point connected with the theory of the constructions of Euclidean Geometry. I propose to shew, in a simple manner, that the essential elements of all the constructions made in this part of Geometry can be obtained by the employment of the compass alone, without the aid of the ruler. Before proceeding, however, to this main point of my address, I desire to make a few remarks upon the general theory of the constructions of Euclidean Geometry.

The usual statement made, when it is wished to assign the scope of the Euclidean constructions, is that the constructions are such as can be made by employing two instruments only, the ruler and the compass. This statement suffers from the serious defect that it takes no account of a certain fundamental distinction. The neglect of this distinction not only leads to confusion of thought in relation to the science of Geometry, but it has also been a fruitful source of misunderstandings as regards the real nature of some well known special problems of Geometry that have attracted the interest of a great number of men throughout many centuries. The distinction to which I allude is that between abstract or theoretical Geometry on the one hand, and practical or physical Geometry on the other hand. On the practical side, Geometry is a physical science in which the objects dealt with – points, straight lines, circles, etc. – are physical objects, the relations between which are to be ascertained and described, with a view to dealing with actual spatial relations with an accuracy sufficient for all practical purposes. On the theoretical side, Geometry is an abstract science concerned with the relations between objects, which, although they are called by the same names – points, straight lines, circles, etc. – as before, are no longer physical objects. The properties and relations of these ideal objects are assigned by means of a scheme of definitions and postulations suggested by the observations made in physical Geometry. These objects are idealisations, obtained

First published in *Mathematical Gazette* **7** (March 1913), pp. 49–54.

by abstraction, of those physical objects that bear the same names. The properties and relations in theoretical Geometry are absolutely precise, whereas the corresponding properties and relations are realised only in some greater or less degree in any physical objects which we may take as representative of the ideal objects. In fact, there is an inevitable element of inexactness in all our actual spatial perceptions which is eliminated when we pass from physical to abstract Geometry. The obliteration of the distinction between abstract and physical Geometry is furthered by the fact that we all of us, habitually and necessarily, consider both aspects of the subject at one and the same time.

We may be thinking out a chain of reasoning in abstract Geometry, but if we draw a figure, as we must do in order to fix our ideas, it is excusable if we do not always remember that we are not in reality reasoning about the objects in the figure, but about objects which are their idealisations and of which the objects in the figure are only an imperfect representation. Even if we only visualise, we see the images of physical objects which are only approximative representations of those objects about which we are reasoning. It would take me too far into the philosophy of the subject if I were to attempt to give the grounds upon which the necessity rests, of rising from a practical to a theoretical treatment of the subject of spatial relations, as an essential condition for a really fruitful scientific development of the subject of Geometry.

Rulers are physical objects by means of which physical straight lines, of approximate straightness and of small but uncertain thickness, can be constructed. A compass is a physical object by means of which physical circles, of small thickness, can be constructed. When we say then that Euclid's constructions are such as can be made with a ruler and a compass it is clear that the statement is primarily one relating to the physical constructions which lead to the counterparts and the practical representations of the ideal objects with the determination of which Euclid, in his abstract Geometry, is concerned. The assertion is only a completely accurate one if it be taken to mean that the Euclidean postulates as to the existence of straight lines and circles imply that the corresponding practical constructions can be made with an indefinite degree of accuracy, subject only to the limitations of the instruments employed, the ruler and the compass. The unconscious assumption that the converse of this is true, namely that any construction that can be made practically by means of the ruler and compass, to an unlimited degree of approximation, necessarily corresponds to an ideal Euclidean construction or rather determination, allowable in accordance with the Euclidean postulates, is an error which, I think, accounts in a large measure for the aberrations of the circle squarer and of the trisector of angles. These two problems, that of squaring the circle and that of trisecting an angle, are soluble as problems of practical Geometry, by employing the ruler and compass, even with such accuracy that the errors would be imperceptible; but the corresponding ideal problems are, as we now know, incapable of solution by Euclidean modes of determination. Thus it is not true that any problem of construction which can in practical Geometry be solved by means which require only the use of the ruler and compass has, corresponding to it in abstract Geometry, a problem that can be solved by methods restricted in the Euclidean mode.

Let us consider what the Euclidean postulations in this matter amount to. In the first place it is postulated that any two assigned points A, B determine uniquely a straight line (A, B); the whole straight line, not merely the segment between A and B; the points A and B being incident on this straight line. It is further postulated that a unique circle $A(B)$ exists, of which A is the centre, and on which B is incident; similarly there exists the circle $B(A)$. To say that these postulations of existence amount to allowing the use of the ruler and compass is to leave the abstract standpoint and to pass over to that of practical Geometry. In the abstract, as long as we are assured of the existence of the straight line (AB), or of the circle $A(B)$, as determinate objects of known nature, we have all we want; we can use them and reason about them and about their relations to other entities; the notion of drawing or constructing them is a notion appropriate to the physical, not the abstract, side of the subject. But we have not yet made clear what the effect of the complete Euclidean scheme of postulations is, in relation to problems of what is usually called construction, but would be better called determination. Every such problem is in its essence reducible to the determination of a certain number of points which shall satisfy certain conditions prescribed in the problem. How then are we to be allowed to consider a point as determined? When this question has been answered we then know the exact scope, and the limitations, of the possibility of Euclidean constructions. In Euclidean Geometry a point is determined in three ways only, in accordance with the following rules:

If A, B, C, D are four assigned points, then –

(1) There exists in general one point P, regarded as determinate, which is incident on both the straight lines (A, B) and (C, D), called their intersection. In case the four points are all incident on one straight line this determination fails, as also if (A, B) and (C, D) are parallels.
(2) A point P which is incident both on the straight line (A, B) and on the circle $C(D)$, when such a point exists, is regarded as determinate.
(3) A point P which is incident both on the circle $A(B)$ and on the circle $C(D)$, when such a point exists, is regarded as determinate.

Whenever, in a Euclidean problem of determination, a point is to be determined, it must be given by the use of the rules (1), (2), (3), these rules being repeatedly used any finite number of times. In each case in which one of these rules is employed it must be shewn that it does not fail to determine the point. In this last requisite

Euclid's own treatment of problems is sometimes defective, as for example in the first proposition of the first book, where (3) is employed without proof that the two circles intersect one another.

The practical constructions corresponding to (1) involve the use of the ruler only, those corresponding to (3) require the use of the compass only, and those corresponding to (2) require the use of both ruler and compass.

It is an interesting question whether all three modes of determining a point are really necessary for the problems of construction, or whether any restrictions on their use may be made without diminishing the range of the problems that can be solved. In fact, more than one such restriction can be made. I propose to shew you in detail that all the Euclidean constructions or determinations can be made by means of (3) alone, without the use of (1) or (2); that is to say, I shall shew that any point that can be determined by (1) or by (2) can also be determined by (3). In other words, any point required for the purposes of a Euclidean construction can be determined as an intersection of two circles. In practical Geometry this means that the compass alone suffices for all the constructions that can be made by means of ruler and compass. All the essential points of a required figure may be obtained by using the compass alone; the ruler may be completely discarded, unless, after the cardinal points of the required figure have been obtained, we desire to fill in the straight lines of the figure, by joining pairs of the cardinal points, for which purpose we must naturally have recourse to the ruler.

In order to shew that the rules (1) and (2) are reducible to (3), it is necessary to solve a short chain of four problems by means of (3), as follows:

(α) Figure (2.4.1) If P, A, B are three given points, it is required to find the image of P in the straight line (A, B). This image P' is given as the second point of intersection of the two circles $A(P)$, $B(P)$, and is thus, in accordance with (3), determinate.

(β) Figure (2.4.2) If A, B are two given points, it is required to find by means of (3) a point B' on the straight line (A, B) such that $AB' = 2AB$, or such that AB' is a given integral multiple of AB.

Determine A_1 as an intersection of $A(B)$ with $B(A)$, then determine A_2 as the intersection of $A_1(B)$ with $B(A)$. Lastly A', the other extremity of the diameter of the circle $B(A)$ through A, is determined as the intersection of $A_2(B)$ with $B(A)$. The point A' is the point required, such that $AA' = 2AB$. By continued repetition of this procedure it is casy to see that a point $A^{(n)}$ on the straight line (A, B) can be determined, so that $AA^{(n)} = n \cdot AB$, where n is any assigned integer.

(γ) Figure (2.4.3) Having given three points A, B, P, it is required to determine by means of (3) the image, or inverse point of P, with respect to the circle $A(B)$.

Determine Q and Q' as the intersections of $A(B)$ with $P(A)$. Determine P' as the second point of intersection of $Q(A)$ and $Q'(A)$.

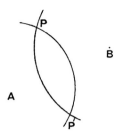

Fig. 2.4.1.

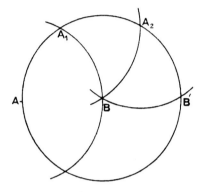

Fig. 2.4.2.

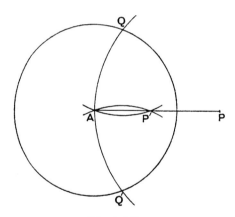

Fig. 2.4.3.

Triangles $AP'Q$, AQP are similar, and thus $\frac{AP'}{AQ} = \frac{AQ}{AP}$, or $AP \cdot AP' = AQ^2$, hence P' is the required inverse point. This method fails in case AP is less than half the radius of the circle. In that case we employ (β) to determine the point $P^{(n)}$ on AP, where $AP^{(n)} = n \cdot AP$, n being so chosen that $AP^{(n)}$ is greater than half the radius of the circle. We then as before determine the inverse point $P'^{(n)}$ of

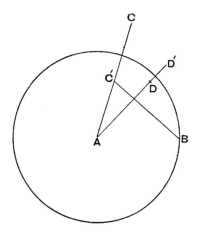

Fig. 2.4.4.

$P^{(n)}$ with respect to the circle.* Lastly, we determine P' by means of (β), so that $AP = n \cdot AP'^{(n)}$.

(δ) Having three points A, B, C assigned, it is required to determine by means of (3), the centre of the circle on which they are incident.

Let C', Figure (2.4.4), the inverse point of C with respect to the circle $A(B)$, be determined by (γ). Let D', the image of A in the straight line (B, C') be determined by (α). Let D, the inverse point of D' with respect to the circle $A(B)$, be determined by (γ); then D is the centre of the circle on which A, B, C are incident. This can be proved in a simple manner, if it be remembered that the inverse, with respect to $A(B)$, of the circle on which A, B, C are incident is the straight line BC'.

We are now in a position to shew that the methods (1) and (2), of determining a point, are reducible to the method (3). First, as regards (1), in which the point is to be determined by the condition that it shall be incident on each of the straight lines $(A, B), (C, D)$.

Take any circle whose centre O is not incident on (A, B) or on (C, D). Determine, by means of (γ), the points A', B', C', D' which are the inverse points of A, B, C, D with respect to this circle. Determine, by means of (δ), the centre H of the circle on which O, A', B' are incident, and also the centre K of the circle on which O, C', D' are incident. Employing (3), the second point of intersection P' of the two circles $H(O), K(O)$ is determinate. The required point P is the inverse point of P' with respect to the circle of centre O, and can therefore be determined by means of (γ). The proof is simply carried out by employing the known properties of inverse figures.

Next consider (2), in which the point is to be determined as one which is incident on the straight line (A, B) and on the circle $C(D)$. Determine, by means of (β), the

---

* The construction here given may be employed to bisect the straight line (AB). Determine by means of (β) the point B' incident on (AB) such that $AB' = 2AB$. The point required can then be determined by (γ) as the inverse point of B' with respect to the circle $A(B)$.

points A', B' which are the inverse points of A, B with respect to the circle $C(D)$. Then, by means of (δ), determine the centre O of the circle on which C, A', B' are incident. Determine, by (3), the intersections, P', Q' of the circles $C(D)$ and $O(C)$. The required points P and Q, the intersections of (A, B) and $C(D)$ are now determined by means of (β) as the inverse points of P' and Q' with respect to the circle $C(D)$. This method fails in case the C is incident on (A, B). In that case we must have recourse to an inversion with respect to some other circle of which the centre is not on (A, B).

It has now been completely established that any required point that can be determined as the intersection of two straight lines, or as an intersection of a straight line and a circle, may be determined as an intersection of two circles. In practical Geometry, the use of the compass alone is therefore sufficient for the determination of the cardinal points of a required figure.

The first treatment of Euclidean constructions by means of the compass without the ruler was given by L. Mascheroni in a work entitled *Geometria del compasso* that was published in 1797 at Pavia. This work contains the solution of a large number of problems by the employment of the compass alone, and is interesting on account of the ingenuity displayed in many of the solutions. Various other treatises and memoirs on the same lines have been since published. Among these I may mention a paper by L. Gérard (*Math. Ann.*, vol. 48, 1897), in which a construction of the kind we have been considering is given for the division of the circumference of a circle into 17 equal parts. The first writing, however, in which a systematic proof was given of the possibility of making all the Euclidean constructions by means of the compass alone is contained in a paper by A. Adler (*Zur Theorie der Mascheronischen Konstruktionen*, Wiener Sitzungsberichte, vol. 99, 1890). In this paper the method of inversion, which I have made the basis of the treatment in the present address, is systematically employed.

In the case of any particular problem of construction the figure would be of quite unwieldy complication if the determination of each of the required points were carried out by working through all the steps in accordance with the procedure I have indicated. This will, however, in any particular case be unnecessary; by the exercise of some geometrical dexterity a comparatively simple construction for attaining the required end will be discoverable. My object, in indicating the solutions of the short chain of problems by the method described, was to provide a proof of the possibility of solving all problems of Euclidean construction by means of the compass, but I did not intend to lay down a process which should be carried out in each particular problem of construction.

As regards other restrictions upon the unlimited use of the ruler and compass which may be introduced without diminution of the range of soluble problems, I may mention that, as was shewn by Geometers in the sixteenth century, all the

Euclidean constructions can be made by means of a ruler and a compass with its arms making a fixed angle with one another. It was shewn by Brianchon and his followers that many constructions in which ruler and compass are usually employed can be made by the use of the ruler alone; more recently the limits of the possibilities in this direction have been carefully examined. The remarkable fact was established by Poncelet and Steiner that if we have given, once for all, a single fixed circle, with its centre known, then all the Euclidean constructions can be carried out by employing the ruler alone. Although not all the Euclidean constructions can be carried out by the use of the ruler with a single straight edge alone, it can be shewn that all these constructions can be carried out by employing a ruler with two parallel straight edges, or even by employing a ruler with two straight edges inclined to one another at any angle (not 180°).

Time does not allow me to prove the truth of these last statements, but I have indicated a field in which the working out of detailed problems affords unlimited scope for the exercise of the ingenuity of those persons, of which the number is much larger than is often supposed, who are endowed by nature with an interest in the special problems of elementary Geometry.

Even for purely didactical purposes, some use, I think, might be made of ideas of this order. Many of the recent changes in the methods and in the subject matter of mathematical teaching have been such as to leave less scope than formerly for the display, on the part of students, of any ingenuity they may possess. No doubt, ingenuity is not one of the faculties possessed in any considerable degree by the Intellectual Democracy. Our methods of teaching and the matter taught are very properly chosen chiefly with a view to meet the needs of the average student, and have regard, in the main, to average capabilities. There are, however, individual students in every institution who do rise above the average in respect of natural interest in, and capacity for, a particular study such as Geometry. For these, a strictly moderate amount of time spent on problems which afford scope for some ingenuity will be well spent, and will at least perform the important function of stimulating in them a real interest in matters which are apt otherwise to strike them as being so purely mechanical in their nature as to afford little or no scope for initiative.

2.5

What is a Square Root?

GEORGE GHEVERGHESE JOSEPH

A widely accepted view among historians of mathematics is: Mathematics outside the sphere of Greek influence (such as India and China) was algebraic in inclination and empirical in practice which provided a marked contrast to Greek mathematics which was geometric and anti-empirical. A simple illustration of the difference in approach between the Greeks and others was in tackling a problem such as: Solve for x the equation: $x^2 = N$.

The Greeks would seek a geometric solution which involved taking the side of a square of area N while the Indian and Chinese approach would be similar to the algebraic procedure we adopt today of taking the square root of N.

Like a number of such generalisations there is more than a grain of truth, though this is not to argue that all mathematical traditions not influenced by the Greeks were essentially algebraic without any analytical tradition in geometry. The main argument of this paper is that a geometry did exist outside Greek mathematics which had moved beyond mere numerical relations or practical surveying considerations to active search for general proofs. The reasoning in these proofs was mostly based on one basic premise: figures of dissimilar shape can have the same area and that they can be dissected and then reassembled for purposes of proving this fact.

1. The Concept of a Square Root

A student first introduced to the square root of a number N is told that it is the number which gives N when multiplied by itself. Examples such as:

The square root of 4 is 2 since $2 \times 2 = 4$

The square root of 9 is 3 since $3 \times 3 = 9$

First published in *Mathematics in School* **26**, 3 (May 1997), pp. 4–9.

are used to illustrate the point. And a geometrical interpretation shows a square of 9 square units with side (root) of 3 units. Or, in a general form, where the square root is the solution to the quadratic $x^2 = N$. This reinforces the view that the square root exists irrespective of the value of N.

The problem really starts when the student has to find the square root of N where $N = 2, 7, 14, \ldots$. The definition of the square root given earlier is not very helpful for it is not possible to find a whole number multiplied by itself which will give either 2 or 7. Neither is a geometrical interpretation sufficiently general to be helpful. No doubt the square root of 2 is the diagonal of a unit square, by applying the so-called Pythagorean theorem. But how would one interpret the square root of, say, 14 geometrically?

Faced with this problem of incommensurability, the reactions of different mathematical traditions are interesting. The Pythagoreans came across lengths which were incommensurable when determining the mean proportion of the two sides of a rectangle needed for 'squaring the rectangle'. The discovery of the diagonal of a square of side one unit, $\sqrt{2}$, caused such a scandal that a pupil of Pythagoras, Hippasus, who compounded the scandal with public disclosure, was supposed to have perished at sea. The memory of this scandal still remains in the terminology of modern mathematics. Numbers expressible as a ratio of two integers are called *rational* numbers, whereas numbers such as the length of a unit square or the value of π not expressible as a ratio are known as *irrational* ('un-ratio-able'). It is interesting in this context that the etymology of the word 'rationalism' comes from the Latin word *ratio*, which is a translation of the Greek word *logo* meaning mathematical ratio, symbolising reason itself.

Thus Greek difficulties with incommensurability arose from the attempt to establish a close correspondence between geometric and arithmetic quantities, the result being a heavy emphasis on a geometric interpretation of irrationality of numbers. Because of this geometric bias, the Greeks were not at ease with irrational numbers and consequently operations with numbers were reduced to a narrow geometric realm robbing them of considerable potency in arithmetic.

The stress in other traditions on *operations* with numbers rather than the numbers themselves meant that their mathematics steered clear of any problem with incommensurability. For example in India, surds, known as *karani*, were accepted as 'proper' numbers from early times and rules for handling them were developed. Though the rational-irrational classification did not exist in the Indian tradition, the notion of exact-inexact numbers was developed. This is reminiscent of the Babylonian distinction between 'regular' and 'irregular' numbers. In both traditions, procedures were developed for calculations with these sets of numbers.

2. The Babylonian Square Root Algorithm

The earliest version of an approximation procedure for evaluating square roots of 'irregular' numbers is from Babylonia and dates back about four thousand years. In a clay tablet, held at the University of Yale, the diagram shown in Figure 2.5.1(a) appears, with the transliteration of the Babylonian numerals into the Neugebauer notation for sexagesimal (base 60) number system given in Figure 2.5.1(b).

The number 30 in Babylonian notation is marked along the length of one side of the square. There are also two other numbers given on and below one of the diagonals. The number on the diagonal converted from the Babylonian notation to ours gives:

$$1 + 60^{-1}(24) + 60^{-2}(51) + 60^{-3}(10)$$
$$= 1 + 0.4 + 0.0146667 + 0.0000463$$
$$= 1.41421297$$

What we have here is a well-known approximation to the square root of 2, with the estimate being correct to five places of decimals! The number *below* the diagonal is the product of 30 (the side of the square) and the estimate of the square root of 2,

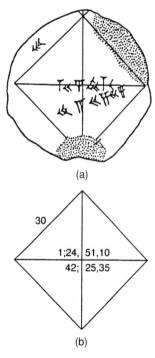

(a)

(b)

Fig. 2.5.1.

the number given on the diagonal. This product in decimal notation is 42.426389.
What does this quantity represent?

Let d be the diagonal of the square. Applying the Right Angled Triangle theorem,

$$d^2 = 30^2 + 30^2, \quad \text{so} \quad d = 30\sqrt{2} = 42.426407$$

which is the number below the diagonal expressed in our notation. Two aspects
of Babylonian mathematics are highlighted in this example. First, over a thousand
years before Pythagoras, the Babylonians knew and used this result. Second, there
is the intriguing question as to how the Babylonians arrived at their remarkable
estimate of the square root of 2. We may well find the answer in a method known
as Heron's procedure, named after an Alexandrian mathematician who lived two
thousand years later. Represented in modern symbolic algebra, Heron's procedure is:

Let N be the number whose square root is sought and the positive number a be
a 'guess-estimate' of the answer. Then

$$N = a^2 + e$$

where the difference (or 'error') e can be positive or negative. We try next to find
a better approximation for the square root of N which we denote as $(a + c)$.

It is obvious that the smaller the 'error', e, the smaller is c relative to e. Thus we
impose the following condition on c:

$$N = (a + c)^2 = a^2 + e \quad \text{or} \quad 2ac + c^2 = e \tag{1}$$

If we made a sensible guess for a in the first place, then c^2 will be very small
relative to $2ac$ and may therefore be ignored. So (1) becomes:

$$c = \frac{e}{2a} \tag{2}$$

Hence, from (1) and (2), an approximation for the square root of N is

$$(a + c) = a + \frac{e}{2a} = a_1$$

Now take

$$a_1 = \left(a + \frac{e}{2a}\right)$$

as the new 'guess-estimate' and repeat this process to get $a_2, a_3, \ldots$, which are better
and better approximations. Implied in this iterative procedure is the assumption (The
Completeness Axiom) that the sequence of approximations converges to some real
number.

To illustrate this approximation procedure consider the question we started with – how did the Babylonians obtain their estimate for square root of 2 as 1.41421297?

Step 1: If $a = 1$, then $c = \frac{e}{2a} = 0.5$ and $a_1 = 1.5$
Step 2: If $a_1 = 1.5$, then $\frac{e_1}{2a_1} = \frac{-0.25}{3}$ and $a_2 = 1.41667$
Step 3: If $a_2 = 1.41667$, then $c_2 = \frac{e_2}{2a_2} = -0.00246$
and $a_3 = 1.41667 - 0.00246 = 1.41421$.

This is very close to the value for the square root of 2, expressed in decimal notation, shown on the diagonal of Figure 2.5.1(b). This procedure for calculating square roots seemed to have been a standard procedure in Hellenistic mathematics, showing the Babylonian influence on the mathematics of that period. A more in-triguing question concerns the appearance of another approximation in the earliest extant mathematical writings of the Indians, known as the Sulbasutras (800 BC–500 BC).

3. The Indian Square Root Algorithm

An important source of early Indian mathematics derives from a class of ritual literature dealing with the measurement and construction of various sacrificial altars. The Sulbasutras provided such instructions for two types of rituals, one for worship at home and the other for communal worship. Square and circular altars were sufficient for household rituals while more elaborate altars involving combinations of rectangles, triangles and trapeziums were required for public worship. One of the most elaborate of the public altars was shaped like a falcon, or rather like the shadow of a falcon, just about to take flight (Figure 2.5.2). It was believed that offering a sacrifice on such an altar would enable the soul of the supplicant to be conveyed by a falcon straight to heaven. (Joseph, 1996).

Fig. 2.5.2. The first layer of a Vakrapaksa-syena altar.

In Figure 2.5.2: the wings are each made from 60 bricks of type a, and the body, tail and head from 50 of type b, 6 type c, and 24 type d bricks. Each subsequent layer was laid out using different patterns of bricks with the total number of bricks equalling 200.

The procedure for evaluating $\sqrt{2}$ arose from an attempt to construct a square altar twice the area of a given square altar, a basic design requirement for a number of constructions. The problem reduces to one of constructing a square twice the area of a given square A of side 1 unit. It is clear that for the larger square C to have *twice* the area of square A, it should have side $\sqrt{2}$ units. Also, we are given a third square B of side 1 which needs to be dissected and reassembled so that by fitting cut-up sections of square C on square A, it is possible to make up a square close to the size of square C. Figure 2.5.3(a) shows what needs to be done. The instructions in the Sulbasutras may be translated as:

Increase the measure by its third and this third by its own fourth less the thirty-fourth part of that fourth. This is the value with a special quantity in excess.

If we take 1 unit as the dimension of the side of a square, the above formula gives the approximate length of its diagonal as follows:

$$\sqrt{2} = 1 + \frac{1}{3} + \frac{1}{3 \times 4} - \frac{1}{3 \times 4 \times 34} = 1.4142157$$

A commentator on the Sulbasutras, Rama, who lived in the middle of the fifteenth century gave an improved approximation by adding a fifth and sixth term to the right-hand side of the equation, consisting of:

$$\frac{-1}{3 \times 4 \times 34 \times 33} + \frac{1}{3 \times 4 \times 34 \times 34}$$

which then gave the first seven places of decimals correctly.

It can be shown that the Sulbasutra formula, when used in evaluating $\sqrt{2}$, produces more or less the same result as one obtained from the iterated use of the Babylonian procedure discussed earlier. This has raised the possibility that this method of

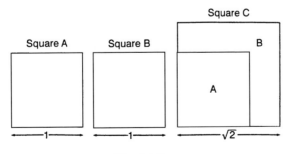

Fig. 2.5.3a.

calculating the value of the square root of 2 may have been derived from the Babylonian procedure. In considering this issue, the following points may be of relevance:

(i) During the Sulbasutra period, there is no evidence from any other source of knowledge of the use of the sexagesimal system of number reckoning.
(ii) There is no evidence that the Babylonians were aware of this precise algorithm and their intellectual heirs, the Greeks, had a number of approximations to the value of the square root of 2 but this precise method did not occur in any of their literature.
(iii) It is quite likely that the basic mode of approach was different in the two cultures: the Babylonian approach being *algebraic* and the Indian being *geometric*.

The Sulbasutras contain no clue as to the manner in which this accurate approximation was arrived at. A number of theories or possible explanations have been proposed. Of these, a plausible one is that of Datta (1932).

In Figure 2.5.3(b), two squares ABCD and PQRS of unit sides are taken. PQRS is divided into three equal rectangular strips, of which the first two are marked 1 and 2. The third strip is subdivided into three squares of which the first is marked 3. The remaining two squares are each divided into four equal strips marked as 4, 5, 6, 7 and 8, 9, 10, 11. These eleven strips are added to the other square ABCD in the manner shown in Figure 2.5.3(b) to obtain a large square *less* the small shaded square at the corner. The side of the augmented square AEFG equals $1 + \frac{1}{3} + \frac{1}{3 \times 4}$. The area of the shaded square is $(\frac{1}{3 \times 4})^2$ so that the area of the augmented square AEFG is greater than the sum of the area of the original squares ABCD and PQRS by $(\frac{1}{3 \times 4})^2$.

In order to get the area of the square AEFG to be approximately equal to the sum of the areas of squares ABCD and PQRS, cut off two tiny strips from either side of the square AEFG of width x so that:

$$2x \left(1 + \frac{1}{3} + \frac{1}{3 \times 4}\right) - x^2 = \left(\frac{1}{3 \times 4}\right)^2$$

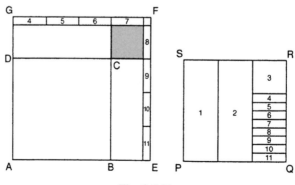

Fig. 2.5.3b.

Simplifying the above expression and ignoring x^2, an insignificant quantity, gives: $x = \frac{1}{3 \times 4 \times 34}$.

Thus the side of the square whose area equals the sum of the areas of two original squares, **ABCD** and **PQRS** or the diagonal of each of the original square is:

$$\sqrt{2} = 1 + \frac{1}{3} + \frac{1}{3 \times 4} - \frac{1}{3 \times 4 \times 34}.$$

What is particularly interesting about this line of argument is its visual mode, a form of argument also found in Chinese geometry. Described in Chinese texts as the 'out-in complementarity' principle (or what is more familiarly known as the principle of 'dissection and reassembly'), it follows from two common sense assumptions:

(i) The area of a plane or a solid figure remains the same when the figure is rigidly shifted to another place on the plane or in the space

(ii) If a plane or solid figure is cut into several sections, the sum of the areas or volumes of the sections is equal to the area or volume of the original figure.

This mode of demonstration is neither dependent on the existence of well packaged symbolic algebra nor the Euclidean mode of axiomatic deductive inference. A modern variant of this approach is found in a topic in geometry known as Dissection Theory. Let us examine how this theory helps us to understand the concept of a square root.

4. The Dissection Theory

In Dissection Theory, there is the result that every polygonal region in a plane can be cut up into a finite number of pieces and then rearranged to form a square. Consider the special case of a rectangle.

Every rectangle is equivalent by dissection to a square

The proof of this theorem is well known. In Figure 2.5.4, a rectangle ABCD (of sides a and b where $a > b$ and square AEFG (of side s) are equivalent with $s = \sqrt{ab}$. The square is placed on the rectangle as shown in the Figure. Draw GB to cut DC at H and FE at K. Let DC cut FE at L. From similar triangles KBE and GBA, we have the following result:

$$\frac{KE}{GA} = \frac{EB}{AB} \qquad \frac{KE}{s} = \frac{(a-s)}{a}$$

so:

$$KE = s\frac{(a-s)}{a}$$
$$= s - \frac{s^2}{a}$$

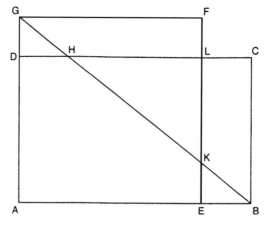

Fig. 2.5.4.

$$= s - \frac{ab}{a}$$
$$= s - b$$

So $\triangle GFK \equiv \triangle HCB$ and $\triangle GHD \equiv \triangle KBE$.

Cut out triangle HCB and trapezium DHBA from rectangle ABCD and assemble square AEFG by inserting $\triangle GFK (\equiv \triangle HCB)$ and trapezium GKEA ($\equiv$ DHBA). This completes the proof by dissection.

The method outlined suffers from two drawbacks. The proof is based on properties of similar triangles and there is often some unease about how satisfactory is the use of such facts in establishing a concrete theory of areas of polygon. The other problem is that there is no attempt at a visual explanation of $\sqrt{ab}$, which is what we are ultimately seeking.

However, the Dissection Theory is perfectly consistent with demonstrations derived from other geometric traditions which have the added advantage of avoiding the assumption that $\sqrt{ab}$ needs to exist uniquely. For purposes of illustration, let us consider two different traditions – the Indian and the Chinese.

5. The Dissection Theory in the Sulbasutras

In the writings of Baudhyana, the oldest of the Sulbasutras, appears a set of instructions for constructing a square altar whose base has the same area as the base of a rectangular altar:

In order to turn an oblong (i.e., rectangle) into a square, take the width of the oblong as the side of the square; divide the rest of the oblong into two parts and by suitable rotation, join these two parts to the two sides of the square. Fill the empty place with an added piece.

In Figure 2.5.5(a) ABCD is the given rectangle in which AB = *a* and BC = *b*. Take points E on AB and H on DC such that AD = AE = DH. Join EH. Now take points F on EB and G on HC such that EF = FB and HG = GC. Join FG. Label square AEHD as I, rectangle EFGH as II and rectangle FBCG as III. Move III with F and G now located at H and D respectively. Construct the smaller square IV to obtain the larger square AFJC as the sum of the sections I, II, III and IV. So the original rectangle ABCD has been transformed into a large square AFJC from which the small square HGJB needs to be removed. Another result from the Baudhyana's Sulbasutra states:

If you wish to remove one square from another, cut off from the larger one an oblong with the side of the smaller one, draw one of the sides of that oblong to the other side; where it touches the other side, that piece should be cut off. By this method the removal is effected.

Figure 2.5.5(c) shows the large square AFJC in Figure 2.5.5(b) from which the small square (i.e., the square HGJB or IV in Figure 2.5.5(b)) is to be removed. With B as the centre and BE the radius, construct a circle which cuts AC at K. Then

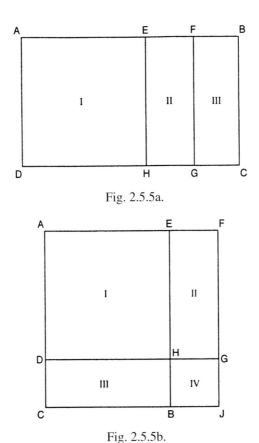

Fig. 2.5.5a.

Fig. 2.5.5b.

the square on KC is the required square whose area is equal both to the difference in the areas of the larger square AFJC and the small square HGJB and equal to the area of the original rectangle ABCD. The proof follows from the Right Angled Triangle theorem, whose knowledge is evident from a number of references in the Sulbasutras:

$$JM^2 = BM^2 - BJ^2$$
$$= BE^2 - BJ^2$$
$$= \text{Square } AFJC - \text{Square } HGJB$$

Figure 2.5.6 shows how the side of a square of length $\sqrt{ab}$ can be directly constructed from the original rectangle given Figure 2.5.5(a).

Given the length and width of the rectangle ABCD are a and b respectively where b is also the side of the square AEHD, then it is easy to deduce that

$$MG = DG + \frac{1}{2}(a - b) + b$$
$$HG = \frac{1}{2}(a - b)$$

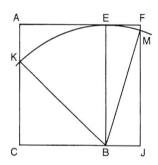

Fig. 2.5.5c.

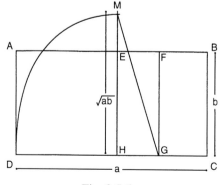

Fig. 2.5.6.

Applying the Right Angled Triangle theorem to triangle MHG, will give

$$MH^2 = MG^2 - HG^2$$
$$= \left[\frac{1}{2}(a-b)+B\right]^2 - \left[\frac{1}{2}(a-b)\right]^2$$
$$= ab$$

or $\quad MH = \sqrt{ab}$

Note that the Sulbasutra approach to square root just outlined has clear conceptual advantages over the Dissection Theory approach discussed earlier. The existence of $\sqrt{ab}$ is established without having to resort to the 'Completeness Axiom'. No use is made of any facts about similar triangles. There is no need for the area or the sides of the rectangle (or square) to be expressed in numbers. The concept of square root derives directly from the construction of the square and a simple demonstration that its area is the same as the area of the rectangle.

6. The Dissection Theory in the Chiu Chang Suan Shu

The fourth chapter of the premier Chinese mathematical text, the Chiu Chang Suan Shu (c. 200 BC) contains twenty four problems on land surveying. An important objective was to parcel out land given the area and one of the sides. Consider the following problem from the text:

> There is a (square) field of area 71824 (square) pu (or paces).
> What is the side of the square?
> Answer: 268 pu

The algebraic rationale underlying the Chinese approach may be expressed with the following symbolic notation: N is a number whose square root is a three-digit integer. α, β and γ are digits representing 'hundreds', 'tens', and 'units' place value positions respectively. So that if the square of N is a three digit number, abc, then $\alpha = 100a$, $\beta = 10b$ and $\gamma = c$. Therefore,

$$N = (100a + 10b + c)^2$$
$$= (\alpha + \beta + \gamma)^2$$
$$= \alpha^2 + (2\alpha + \beta)\beta + [2(\alpha + \beta) + \gamma]\gamma \tag{1}$$

It is simple to extend this formula to include more than three digits by expanding $(\alpha + \beta + \gamma + \delta \cdots)^2$. The Chinese used the above relationship but reversed the procedure and the ensuing calculations that resulted. The procedure is initiated by finding an appropriate value for α by 'inspection'. It is, for example, easily deduced that $\alpha = 200$ (i.e., $100a$ where $a = 2$) if we are seeking the square root of $N = 71824$. The procedure continues with calculating α^2. This quantity is then subtracted

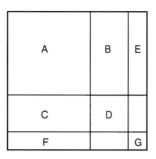

Fig. 2.5.7.

from N. We next estimate the second place of the square root (i.e., $\beta = 10b$), and then form $(2\alpha + \beta)$. We can now work out

$$N - \alpha^2 - (2\alpha + \beta)\beta = N - (\alpha + \beta)^2 \qquad (2)$$

The procedure continues along similar lines until the third component on the right hand side of (1) is calculated. If N is a perfect square, the final subtraction of this component from (2) would leave a remainder of 0. The geometric rationale for the algorithmic approach just discussed is found in Figure 2.5.7.

We begin by constructing a square $(A = \alpha^2)$ with side 200 *pu*. Two rectangular sections, $B = \alpha\beta$ and $C = \alpha\beta$, each of dimensions 200 by 60, are added together to give a total area of 24000 square *pu*. To complete the larger square shown in Figure 2.5.7, one needs to add a further square section, $D = \beta^2$, whose side is 60 *pu* and area is 3600 square *pu*. The area of the larger square is

$$A + B + C + D = 40000 + 24000 + 600 = 67600 \ pu^2.$$

The shortfall that has to be made up is $71824 - 67600 = 4224 \ pu^2$. It is seen that this is equal to the area of two rectangular strips of dimensions 260 by 8, $E = (\alpha + \beta)\gamma$ and $F = (\alpha + \beta)\gamma$, and a small square $G = \gamma^2$ of side 8, i.e., $2(260 \times 8) + 8^2 = 4224$ square *pu*. Thus the geometric representation of the procedure for extracting the square root of 71824 is equivalent to finding the length of the side of a square of area 71824 square *pu*. Figure 2.5.7 indicates clearly that the side required is $200 + 60 + 8 = 268 \ pu$. It is noteworthy that this method of extracting square roots was eventually extended to the solution of quadratic equations. Indeed, the clear connection established between extraction of roots of any degree with solution of the same degree is a feature of Chinese mathematics not present in any other early mathematical tradition.

7. Conclusion

Implicit in the discussion of the square root is the need to pay more attention to the intuitive elements in mathematics. Often in our haste to get to the more powerful analytic tools of mathematics, we ignore the 'concrete' meanings and images that are already present. Sometimes it is hard even to recognise that some meaning is missing until a student (or more usually an adult learner) asks in some bewilderment 'What does that mean?' or, given a formalist demonstration of something says: 'I know it, to a degree I understand it, but I don't *feel* it'. Such students can often only make progress, are only satisfied by, a procedure which accepts their psychological state, and works from that to an understanding which fuses, or at least deals with disharmony between, that emotional belief and their intellectual beliefs. Proofs, methods and reasoning should be rather like old fashioned demonstrations, in that they should reflect, not necessarily a chain of deductive reasoning, but rather how the human brain arrived at its current thought. It is clear that in the mathematical traditions we have examined the 'geometric' concept of a square root was seen as being important. In spite of a tendency to neglect the geometric mode of argument in favour of analytic ones, based on the relatively recent notions of Cauchy sequences and the Axiom of Completeness, for many students and even some teachers, an intuitive understanding of real numbers and operations with real numbers requires geometry. Compare the geometric imagery of the product of real numbers a and b with the multiplication of two infinite, non-repeating, decimal fractions. Try explaining the product of $\sqrt{2}$ and π to someone with limited mathematical background without the help of geometry. The early traditions that we have examined would not have attempted such a task. Instead they would have concentrated on the visual and intuitive features found in geometrical explanations.

References

B. Datta, *The Science of the Sulbas: A Study in Early Hindu Geometry* (Calcutta: Calcutta University Press, 1929).

G. G. Joseph, *The Crest of the Peacock: Non-European Roots of Mathematics* (London: Penguin, 1996).

G. G. Joseph, *The Geometry of Vedic Altars in Nexus: Mathematics and Architecture* (K. William, Ed.) (Fucecchio: Edizione Dell'erba, 1995).

R. P. Kulkarni, *Geometry According to the Sulba Sutra* (Pune: Vaidika Samsodhana Mandala, 1983).

Yong Lam Lay, The geometrical basis of the ancient Chinese square-root method, *ISIS* **61** (1970), 96–102.

D. Nelson, G. G. Joseph & J. Williams, *Multicultural Mathematics* (Oxford: Oxford University Press, 1993).

Notes

As an illustration of an intuitively convincing proof, consider the following demonstration of the result that a negative number has no square root, given by Krishna Daivajna (c. AD 1600), a commentator of the mathematical texts of Bhaskaracharya (c. AD 1100).

A negative number is not a square. Hence, how can we evaluate its square root? It may be argued that 'why cannot a negative number be a square? Surely it is not a royal command' . . . Agreed. Let it be stated by you who claim that a negative number is a square as to whose square it is; surely not of a positive number, for the square of a positive number is always positive. Not also a negative number because then also the square will be positive by the same rule. This being the case, we cannot see how the square of a number becomes negative.

2.6

An Old Chinese Way of Finding the Volume of a Sphere*

T. KIANG

On the reverse side of the Moon, just south of the Sea of Moscow, is a crater named Tsu Ch'ung-Chih.[†] Who is Tsu Ch'ung-Chih? What did he do to be thus admitted to the honoured company of Maxwell and Hertz, of Mendele'ev and the Curies, of Lomonosov and Tsiolkovsky?

Tsu Ch'ung-Chih was a Chinese mathematician-astronomer who flourished in the fifth century A.D. He made notable contributions to the calendar calculation and determined several constants with remarkable accuracy. For example, he gave a value of 27·21223 days for the length of the nodical month, the modern value being 27·21222 days. As another example, he found that the planet Jupiter completes seven and one-twelfth circuits of the heavens in every seven cycles of 12 years; this corresponds to a sidereal period of Jupiter of 11·859 years, which differs from the modern value by only 1 part in 4000.

However, it is in mathematics that Tsu Ch'ung-Chih is chiefly remembered. His best-known achievement here is his evaluation of π. He gave

$$3 \cdot 1415926 < \pi < 3 \cdot 1415927.$$

This represents a veritable *tour de force* in calculation when we remember that he had only "counting chips" to help him. Tsu Ch'ung-Chih gave also two approximating ratios for π: a coarse one of 22/7 and a fine one of 355/113. As is well-known, the former ratio was anticipated by Archimedes in the West, but the latter ratio was not known in Europe until the times of Valentinus Otto and Adriaan Anthoniszoon,

First published in *Mathematical Gazette* **56** (May 1972), pp. 88–91.

* Text of a talk first given as a Lunch-hour Lecture at University College London on February 27, 1964. Since then it was given to the Irish Astronomical Society, Dublin and Armagh Centres, and to University College Dublin Mathematical Society. Messrs. C. R. Spratt and P. Murphy, respectively of University of London and Dunsink Observatories, made solid models for illustration.

† In the Hànyǔ Pīnyīn system now in use in China, the name is spelt Zǔ Chōng-Jī. The Chinese characters are given in Figure 2.6.2.

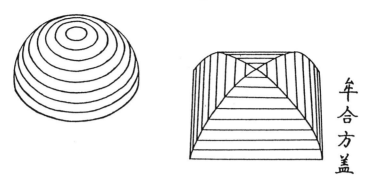

Fig. 2.6.1. Regard a sphere as a pile of circles. Replace each circle by its circumscribing square, and we have the solid figure called by ancient Chinese mathematicians "móuhéfānggài". (The Chinese characters are shown on the right.) For clarity only the upper halves of a sphere and its móuhéfānggài are drawn. Since the *areas* of a circle and its circumscribing square are in the ratio $\pi : 4$, this must also be the ratio between the *volumes* of a sphere and its móuhéfānggài.

i.e. over 1000 years later. It may be noted that this value is also correct to seven significant figures.

We now turn to the main topic of the present article, namely, Tsu Ch'ung-Chih's evaluation of the volume of a sphere. Generally speaking, the Chinese approach to this problem is quite different from Archimedes'. Whereas Archimedes approximated a sphere by a solid of revolution with a regular polygonal section, the Chinese mathematicians directed their attention on the solid figure which they called móuhéfānggài. To obtain this figure we regard a sphere as a pile of circles of varying sizes, and we replace each circle by its circumscribing square (see Figure 2.6.1). Thus, by construction, the ratio between the *volumes* of the sphere and its móuhéfānggài is the same as that between the *areas* of a circle and its circumscribing square. That is,

$$\text{sphere: móuhéfānggài} = \pi : 4. \tag{1}$$

In mathematical language, móuhéfānggài is the space common to two identical cylinders intersecting at right angles, but we could picture it as a pair of pyramids with *curved* faces fitted base to base – móuhéfānggài literally means "close-fitted square lids". The credit of being the first to conceive correctly this figure goes to Liú Huī, who lived some 200 years before Tsu Ch'ung-Chih, but it was left to Tsu Ch'ung-Chih, or, according to some historical records, to his son, to accomplish the much more difficult task of evaluating its volume.

Tsu Ch'ung-Chih divides the móuhéfānggài into eight identical *octants* (see Figure 2.6.2). Each *octant* has a square base, two upright faces and two curved faces. If the radius of the original sphere is r, then the square base has sides r, and each

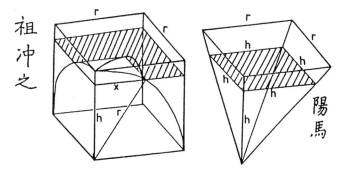

Fig. 2.6.2. Left figure shows an *octant* of a móuhéfānggài placed inside a cube having the same base. The shaded area in the same figure is the section at height *h* of the "difference-solid" between the cube and the *octant*. By an ingenious application of what is known in the West as Pythagoras' Theorem, Tsu Ch'ung-Chih (whose name in Chinese characters appears along the left margin) proved this area to be equal to h^2, and thus provided the key to evaluating the volume of a sphere. It then follows that the difference-solid has the same volume as the figure called "yángmǎ" shown on the right.

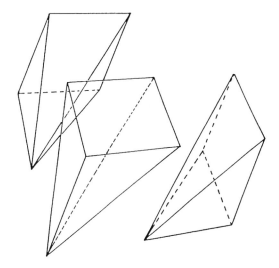

Fig. 2.6.3. Illustrating the proposition "a cube can actually be cut up into 3 yángmǎ".

upright face is a quadrant of a circle of radius *r*. Let us now, so Tsu Ch'ung-Chih says, compare the *octant* with the cube with sides *r*, or, rather, let us put the *octant* inside the cube and find the volume of their difference-solid. At height *h*, the section of the *octant* is a square whose area is, say, x^2; while that of the difference-solid is an L, whose area is then $r^2 - x^2$. But look at an upright face of the *octant* and see that *h*, *x*, *r* are the three sides of a right-angled triangle with *r* as the hypotenuse. Hence we have $r^2 - x^2 = h^2$, so that the area of the L is equal to h^2. The difference-solid therefore has the same volume as that solid whose section at height *h* is a square of sides h. But *that* is a familiar object known as yángmǎ. Thus Tsu Ch'ung-Chih arrives at the key equation

$$r^3 - \frac{1}{8} \text{ móuhéfānggài} = \text{yángmǎ}. \tag{2}$$

The rest is easy. For, in the vivid language of the Chinese mathematical classic, Jiǔ-zhāng Suànshù (The Nine Books of Calculation Art), "a cube can actually be cut up into 3 yángmǎ" (see Figure 2.6.3). Hence,

$$\text{yángmǎ} = \frac{1}{3} r^3. \tag{3}$$

From (1), (2), and (3), it then follows that

$$\text{sphere} = \frac{4}{3} \pi r^3,$$

and the problem is solved.

2.7

Mathematics and Islamic Art

LESLEY JONES

1. The Influence of the Vedic Square

The most exciting experiences in mathematical development are the discovery of unifying ideas which are central to many different topics. One such idea is the Vedic square. Not that I am claiming to have discovered the Vedic square! Its properties were known to the people of Northern India many centuries ago and it was the basis of a whole mathematical system. In AD 770 the Muslims incorporated it into their system of mathematical knowledge. It is from them that our knowledge has developed. Some of the properties of the square led to the discovery of systems which formed the basis of the intricate patterns and designs which are now familiar to us as examples of Islamic Art.

The Qur'an affects all aspects of life and the ban on representational art led to an artistic tradition which contrasted with the developments in European Art. In mosques and in the illuminated writing of Qur'ans no decoration could depict people or animals so there developed a tradition in which geometric and rhythmic patterns predominated. The people had a nomadic lifestyle so many portable belongings were decorated in a similar way. Their clothes, textiles, carpets, metalwork and pottery were adorned with intricate patterns.

There are many good reasons for exploring Islamic patterns in school. They provide a genuine cross-curricular focus, offering scope for co-ordinated work in mathematics and art. The examples are not the usual Euro-centric ones. Here is an opportunity to show that we value contributions from other cultures, both in terms of artistic expression and mathematical knowledge. Before starting the project it would be useful to familiarise all of the pupils with some of the designs. Posters and postcards can be obtained and displayed around the room. Muslim pupils may be able to lend fabrics and artefacts with patterns on them.

First published in *Mathematics in School* **18**, 4 (September 1989), pp. 32–35.

The Vedic square is created in a similar way to the familiar multiplication square. Using the digits 1 to 9 across the top row and down the first column they are then multiplied together and the results placed in the appropriate slot in the matrix. Where the result is more than a one digit number the digit sum is used in the square, thus $4 \times 4 = 16, 1 + 6 = 7$.

The pattern for multiples of 9 produces the rather strange result with 9 in every position in the final row and column. Once the square is complete the search for

1	2	3	4	5	6	7	8	9
2	4	6	8	1	3	5	7	9
3	6	9	3	6	9	3	6	9
4	8	3	7	2	6	1	5	9
5	1	6	2	7	3	8	4	9
6	3	9	6	3	9	6	3	9
7	5	3	1	8	6	4	2	9
8	7	6	5	4	3	2	1	9
9	9	9	9	9	9	9	9	9

Fig. 2.7.1.

patterns can commence. At this stage I am reluctant to continue as I would not wish to spoil the reader's fun. If you have not met the square before do pause here and try your hand at finding patterns in the square. It doesn't take very long to find out why "9" was seen as such an important number in the Vedic system. The complements in 9 have strong links in the square (e.g. compare the "8" row with the "1" row and the "2" row with the "7" row).

Consider the pattern which shows where each number occurs. Not only does each example show rotational symmetry, but each complementary pair produces a reflection.

For each pattern there is a choice of ways to join the numbers. How many ways could this be done, producing a symmetrical result?

Combined patterns produced by traced overlays or dotty paper drawings begin to be reminiscent of the Islamic patterns.

The digit pattern from each row can be used to create a "spiral". Right angles and 120° turns produce interesting results, but there is no need to be restricted to these. (The figures for this article were produced using LOGO, so the interior 120° angle is denoted 60°.) Spirals can be produced on dotty paper, or with the help of a

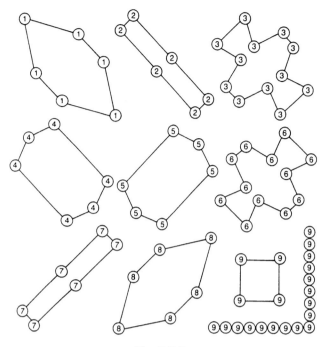

Fig. 2.7.2.

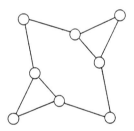

Fig. 2.7.3.

microcomputer. Alan Brine and Derek Bunyard (1) describe the "Vedic" program on Micromath Disc 3 which enables children to get straight into the patterns.

2. Using the Fibonacci Sequence as a Starting Point

In 1228 Fibonacci introduced to Europe the number sequence which was named after him. He had studied mathematics with the Arabs and it seems likely that this is where he met the sequence.

The sequence in its usual form gives us;

$$0\ 1\ 1\ 2\ 3\ 5\ 8\ 13\ 21\ 34\ 55\ 89 \ldots$$

By finding digit sums we reduce the sequence to;

$$0\ 1\ 1\ 2\ 3\ 5\ 8\ 4\ 3\ 7\ 1\ 8\ 9\ 8\ 8\ 7\ 6\ 4\ 1\ 5\ 6\ 2\ 8\ 1\ 9\ 1\ 1\ 2\ 3 \ldots$$

(The digit sums can be used to extend the pattern. You do not need to revert to the original numbers.)

Clearly the pattern will repeat itself as indicated by the shaded section above.

If we select alternate terms we note that there are two different sequences intertwined.

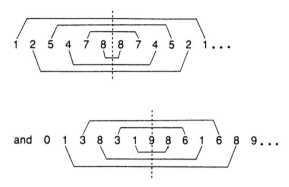

The first sequence is symmetrical. The symmetry of the second is apparent through pairing the complements in 9, but we can also note that the first and last 5 terms of the second sequence have a symmetry of their own.

Using these patterns to produce spirals we begin to see the creation of familiar motifs and patterns.

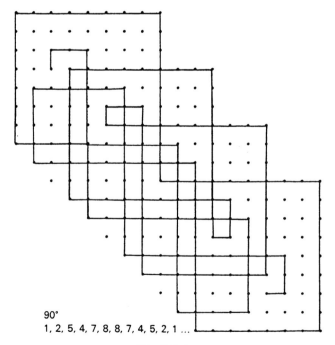

90°
1, 2, 5, 4, 7, 8, 8, 7, 4, 5, 2, 1 ...

Fig. 2.7.4.

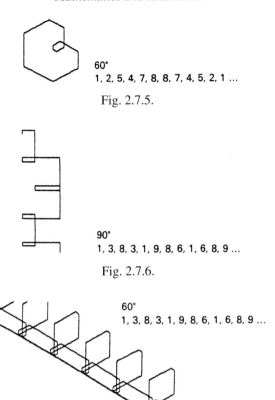

60°
1, 2, 5, 4, 7, 8, 8, 7, 4, 5, 2, 1 ...

Fig. 2.7.5.

90°
1, 3, 8, 3, 1, 9, 8, 6, 1, 6, 8, 9 ...

Fig. 2.7.6.

60°
1, 3, 8, 3, 1, 9, 8, 6, 1, 6, 8, 9 ...

Fig. 2.7.7.

Starting the sequence with different numbers and following the same routine we obtain

0 2 2 4 6 1 7 8 6 5 2 7 9 7 7 5 3 8 2 1 3 4 7 2 9 2 2 4 6...

which separates into;

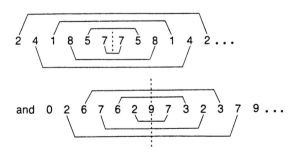

2 4 1 8 5 7   7 5 8 1 4 2 . . .

and 0 2 6 7 6 2 9 7 3 2 3 7 9 . . .

The symmetry patterns are the same, but the spiral patterns appear as in Figures 2.7.8, 2.7.9, 2.7.10 and 2.7.11.

When we explore the pattern starting with 0 3 3 we obtain a pattern with quite a different look;

0 3 3 6 9 6 6 3 9 3 3 6 9 6 6 3 9 3 3 6 9 6 6 3 9

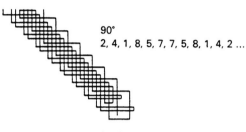

90°
2, 4, 1, 8, 5, 7, 7, 5, 8, 1, 4, 2 ...

Fig. 2.7.8.

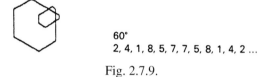

60°
2, 4, 1, 8, 5, 7, 7, 5, 8, 1, 4, 2 ...

Fig. 2.7.9.

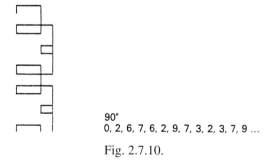

90°
0, 2, 6, 7, 6, 2, 9, 7, 3, 2, 3, 7, 9 ...

Fig. 2.7.10.

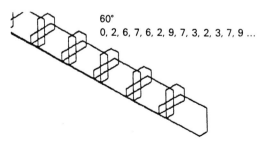

60°
0, 2, 6, 7, 6, 2, 9, 7, 3, 2, 3, 7, 9 ...

Fig. 2.7.11.

Details of a Koran frontispiece based on a decagon grid. Egypt, 14th century.
© *The British Library*

which separates into;

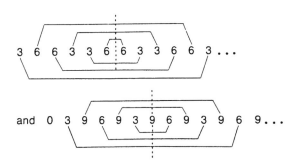

But the symmetry in the pattern remains the same (accepting that $9 + 9 = 1 + 8 = 9$).

The spiral patterns have a very different look, but provide promising and recognisable motifs for our patterns.

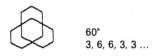

60°
3, 6, 6, 3, 3 ...

Fig. 2.7.12.

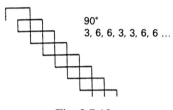

90°
3, 6, 6, 3, 3, 6, 6 ...

Fig. 2.7.13.

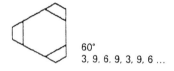

60°
3, 9, 6, 9, 3, 9, 6 ...

Fig. 2.7.14.

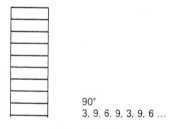

90°
3, 9, 6, 9, 3, 9, 6 ...

Fig. 2.7.15.

The pattern starting with 0 4 4 gives us

0 4 4 8 3 2 5 7 3 1 4 5 9 5 5 1 6 7 4 2 6 8 5 4 9 4 4 ...

which separates into;

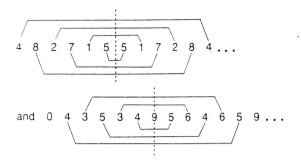

4   8   2   7   1   5   5   1   7   2   8   4 . . .

and   0   4   3   5   3   4   9   5   6   4   6   5   9 . . .

with the spiral pattern shown below.

60°
4, 8, 2, 7, 1, 5, 5, 1, 7, 2, 8, 4 ...

Fig. 2.7.16.

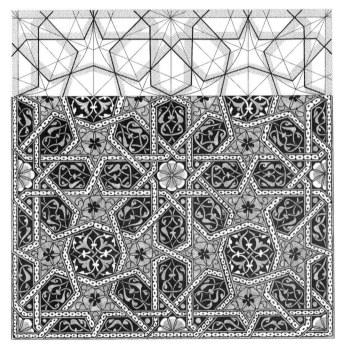

Details from a Koran illumination based on the grid shown. The slimmer shape of the star's rays and the smaller central octagon are produced by positioning the interlacing ribbon over the grid as shown TOP, i.e. with the grid line either in the centre or to one side of the ribbon, a device which allows for further variations in designs of this kind. Egypt, 15th century.

Detail from a Koran frontispiece with a design laid out on a grid similar, though not identical, to that shown on the previous page. Egypt. 1356.

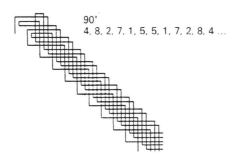

Fig. 2.7.17.

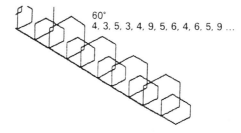

Fig. 2.7.18.

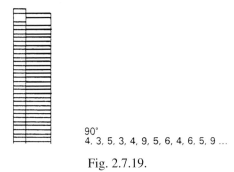

90°
4, 3, 5, 3, 4, 9, 5, 6, 4, 6, 5, 9 ...

Fig. 2.7.19.

At this point a glance at the patterns will show that the rest are contained within those already created. Once again the pairing of complements in 9 are apparent.

The 0 4 4 pattern contains the one which starts 0 5 5.
The 0 3 3 pattern contains the one which starts 0 6 6.
The 0 2 2 pattern contains the one which starts 0 7 7.
The 0 1 1 pattern contains the one which starts 0 8 8.

The reader is left to ascertain what happened to the 0 9 9 pattern.

Reference

1. D. Bunyard & A. Brine, Islamic Art, *Micromath* (Spring, 1988).

2.8

Jamshīd al-Kāshī – Calculating Genius

GLEN VAN BRUMMELEN

'What's the value of π to ten places?'

No problem, you say, as you bring out your handy pocket calculator and type the right keys. But suppose the batteries are dead, or worse yet, that calculators haven't been invented. What will you do now?

Calculating machines really only got their start in the last century. For almost all of history, people have had to do everything by hand. That meant one of three things: spend sleepless nights with pencil and paper working your way through the drudgery of arithmetic, look up somebody else's calculations and 'borrow' those, or think of some extremely clever method to get better results, faster than anyone else.

It takes real genius to take this last choice, and that term clearly applies to Jamshīd al-Kāshī. The hero of our story, an early 15th-century Iranian astronomer, probably spent many long nights with his numbers. But his insights into patterns of calculation, which led to a value of π accurate to more than twice as many places as any of the others of the time, were both remarkable and beautiful.

1. The Wandering Scholar

Our first record of al-Kāshī is in his home town of Kāshān in AD 1406, and his earliest dated treatise is from one year later. It is clear that he was interested in computation from the beginning. His *Khāqānī Zīj*, a comprehensive astronomical handbook written in 1414, is a revision of the *Īlkhānī Zīj*, the standard written some two centuries earlier by Nasīr al-Dīn al-Tūsī. In the introduction, al-Kāshī praises al-Tūsī's work highly, then lists some 70 problems that are corrected, many of them computational.

First published in *Mathematics in School* **27**, 4 (September 1998), pp. 40–44; extract

Al-Kāshī was not well off at the time, and spent a number of years wandering from place to place attempting to make a living, partly as a physician. Eventually, around 1418, he obtained a position at the school established by Sultan Ulūgh Beg, at Samarkand (now in Uzbekistan). Ulūgh Beg was a tremendous supporter of the sciences and a great astronomer in his own right. Al-Kāshī's financial troubles were over.

2. A Slice of π

Ulūgh Beg's investment more than paid off. In 1424 al-Kāshī wrote 'A Treatise on the Circumference', in which he found a value of π accurate to 16 decimal places. The next nearest anyone had come was six, and it took until 1615 before Ludolf van Ceulen bettered al-Kāshī's record.

Our explanation below won't follow al-Kāshī word for word. For one, al-Kāshī used the sexagesimal (base-60) arithmetic standard among astronomers at the time, even though he was responsible for introducing a complete base-10 system of arithmetic in another work. For another, he used a circle of radius 60, rather than our unit circle. Nevertheless, we will stay close to al-Kāshī's spirit.

The basic idea, which goes back at least to Archimedes, is to approximate a circle of radius 1 with an inscribed polygon of many sides (Figure 2.8.1). Now, we know that the circumference of the circle is 2π times the radius. If we can find a way to measure the distance around the polygon, and we've chosen a polygon with enough sides, then we will have an estimate for the value of 2π.

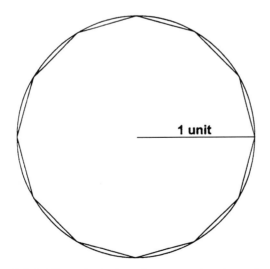

Fig. 2.8.1. The unit circle and an approximating polygon

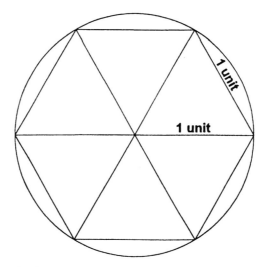

Fig. 2.8.2. Our first approximation, the hexagon

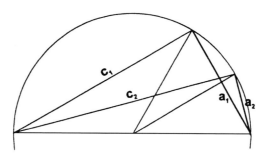

Fig. 2.8.3. From the side of a hexagon (a_1) to the side of a dodecagon (a_2)

We can get a start with a hexagon. It is easy to measure the distance around the hexagon, because it can be broken into a series of equilateral triangles (Figure 2.8.2). So, the length of one side of the hexagon, which (for reasons we will see later) we call a_1, is just 1 unit. Multiply by the six sides to get our approximation to 2π and divide by 2. We get the rather unimpressive value $\pi \approx 3$.

Our next step is to double the number of sides in our polygon from 6 to 12, a dodecagon. But how can we compute the length of a side of a dodecagon (which we shall call a_2)?

Here al-Kāshī had one of his many moments of brilliance. Draw both the hexagon and the dodecagon in the circle, as in Figure 2.8.3, but consider only the top half. Connect the point at the left extreme of the diameter of the circle to the first vertex of each polygon above the rightmost point on the diameter.

The next bit of reasoning is not for the faint of heart; casual readers may believe the formula at the bottom of this paragraph and press on. The rest of us can begin

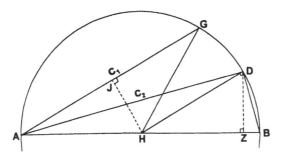

Fig. 2.8.4. The proof of the relation $c_2^2 = 2 + c_1$

by dropping perpendiculars DZ onto AB, and HJ onto AG (Figure 2.8.4). Notice that triangles DAZ and ADB are similar, because their angles are the same. So, $AB/c_2 = c_2/AZ$, which we rearrange to say $c_2^2 = AB \cdot AZ = 2(1 + HZ)$. Our attention now shifts to HZ. A famous theorem from Euclid's *Elements* tells us that $\angle BAG = \frac{1}{2}\angle BHG = \angle BHD$. This in turn tells us that triangles AHJ and DHZ are similar since they have the same angles; but in fact they are also the same size, because they both have a radius as the hypotenuse! So $HZ = AJ = \frac{1}{2}c_1$. Finally, if we substitute this into our equation above for c_2, with a bit of simplification we produce the remarkable formula

$$c_2^2 = 2 + c_1.$$

Notice that the diameter of the circle, which has length 2, is the hypotenuse of a right-angled triangle with sides a_1 and c. So, using the Pythagorean Theorem,

$$c_1 = \sqrt{2^2 - a_1^2} = \sqrt{4 - 1} = \sqrt{3} = 1.73205\ldots,$$

and from our formula,

$$c_2 = \sqrt{2 + c_1} = 1.93185\ldots$$

From here we can easily compute our dodecagon approximation to π. Using Pythagoras again, the length of the side of the dodecagon is

$$a_2 = \sqrt{2^2 - c_2^2} = \sqrt{4 - c_2^2} = 0.51763\ldots$$

Multiply by 12 sides to get the distance around the dodecagon, and divide by 2; our value for π is now $3.10582\ldots$ A little better, but still not great.

To improve our value even further, we must double the number of sides again, producing a 24-gon. This means we must compute a_3. But now we come to the crux of the matter: our two key formulas did not assume that we were dealing only with the hexagon and dodecagon; they will work when you double the sides of any polygon!

Hence, $c_3^2 = 2 + c_2$, $c_4^2 = 2 + c_3$, and so on; also $a_3 = \sqrt{4 - c_3^2}$, $a_4 = \sqrt{4 - c_4^2}$ and so on! So, in general,

$$c_n^2 = 2 + c_{n-1} \quad \text{and} \quad a_n = \sqrt{4 - c_n^2}$$

Now it's easy to get better and better results by successively doubling the number of sides in our polygons. We don't even have to draw them! The first few results are summarized in this chart; try to extend it yourself to get better and better values of π. It's very easy if you use a calculator!

Number of sides	c_n	a_n	Estimate of π
6 ($n = 1$)	1.73205	1	3
12 ($n = 2$)	1.93185	0.57163	3.10582
24	1.98288	0.26105	3.13262
48	1.99571	0.13080	3.13935
96	1.99892	0.06543	3.14103

With a polygon of 96 sides, al-Kāshī had π correct to three places. But that wasn't enough for him. To make sure he had the ultimate value of π, he decided to keep going until he had the universe from its diameter and would be in error by less the width of a horse's hair! Given estimates of the diameter of the universe at the time, that meant al-Kāshī needed the equivalent of 16 decimal places.

This leads to a problem: how do you know when you have a value accurate to so many decimal places? Al-Kāshī knew that his method would always under-estimate π because his polygons were inside the circle, so he designed a way to use polygons *outside* the circle to over-estimate π. If his under-estimates and over-estimates agreed to 16 places, then he would know he had achieved his goal.

Al-Kāshī extended the above table further and further, calculating everything to more than 16 digits. Notice that you don't really need to calculate the values of a_n until you are approaching the time that you want to stop. Eventually, after 28 rows, he had his ultimate slice of π: 3.1415926535897932. This value comes from a polygon with 805,306,368 sides! Truly a marvellous piece of computation.

References

M. Bagheri, A newly found letter of al-Kāshī on scientific life in Samarkand, *Historia Mathematica* **24** (1997), 241–256.

J. L. Berggren, *Episodes in the Mathematics of Medieval Islam* (New York: Springer-Verlag, 1986).

E. S. Kennedy, A letter of Jamshīd al-Kāshī to his father, *Orientalia* **29** (1960), 191–213.

P. Luckey, *Der Lehrbrief über den Kreisumfang. Abhandlungen der Deutschen Akademie der Wissenschaften zu Berlin*, Jahrgang 1950 No. 6. (Berlin: Akademie Verlag, 1953).

A. P. Youschkevitch & B. A. Rosenfeld, 'Al-Kāshī', in *Dictionary of Scientific Biography*, vol. VII (New York: Charles Scribner's Sons, 1973), 255–262.

2.9

Geometry and Girard Desargues

B. A. SWINDEN

It is a far cry from the Conics of Apollonius to the theory of linear spaces and matrix algebra and many men have helped to build the road between. This article is about one of the builders whose work for a time was discarded and whose name was almost forgotten. But his name finds frequent mention now, when the latter aspect of geometry bears so little resemblance to the former, and metrical geometry has given place to projective geometry with its non-metrical homogeneous coordinates, which make it possible to publish a book on coordinate geometry which does not contain a single figure.

It was only about a century ago that projective geometry began to develop as an independent branch of mathematics. Before that period its basis was almost entirely metrical. Such ideas as harmonic section, foreshadowed by Apollonius himself (c. 260 B.C.–200 B.C.) in his fourth book, were defined essentially in terms of distance. In the *Synagoge* of Pappus (end of 3rd century A.D.) there is the important theorem that if $A_1 A_2 A_3$, $B_1 B_2 B_3$ be two triads of points lying respectively on each of two straight lines, then the cross-meets such as $(A_2 B_3, A_3 B_2)$ are collinear, a theorem later generalised by B. Pascal (1623–1662), who replaced the pair of lines by any conic. With Pappus' Theorem and the fundamental propositions of incidence in a plane, *viz.* that given two distinct coplanar lines there is a unique meet, and given two distinct points there is a unique join, we can, without any use of the idea of distance, assign to each point of the plane two (or three) coordinates which are elements of a collection satisfying the ordinary formal laws of algebraic addition and multiplication and are such that the coordinates of any point on the line joining two distinct points P and Q are linear combinations of the coordinates of P and Q, *i.e.* a straight line has a linear equation.

Pappus, too, discussed the problem *ad tres et quattuor lineas*, wherein a conic is found to be the locus resulting from the motion of a point the product of whose

First published in *Mathematical Gazette* 34 (December 1950), pp. 253–260.

distances from two straight lines is equal to its distance from a third, or to the product of its distances from a third and fourth lines: the ancients had proved this for the case of three lines, but Pappus, while stating the result for four lines, gave no proof in that instance. Newton (1642–1727) gave a purely geometrical solution, in which Descartes had failed, though his method of coordinates enabled him to prove that the locus was a curve of the second order.

G. Desargues (1593–1661), who introduced the conceptions of points at infinity, conical and cylindrical projection, involution, pole and polar and the germ of the notion of cross-ratio, and R. Descartes (1596–1650) with his development of analytical methods, were contemporaries who were in constant communication and whose ideas interacted. For reasons which I suggest later Descartes' work gained considerable celebrity while that of Desargues was neglected and would have fallen into complete oblivion but for a brief mention in Blaise Pascal's *Essay on Conics* and the researches of M. Chasles.

In the early nineteenth century interest in Projective Geometry revived with L. N. M. Carnot (1753–1823), J. V. Poncelet (1788–1867), who systematically used "imaginary" points, and J. Steiner (1796–1863), but the basis was still essentially metrical. Then in 1847, K. G. C. von Staudt (1797–1867) in his *Geometrie der Lage* expounded the view that Projective Geometry should be built on an abstract basis with no reference to metrical ideas, and F. C. Klein (1849–1925) pointed out that "Projective Geometry is the study of properties invariant under a certain group of transformations, which are in fact represented analytically by linear homogeneous transformations of coordinates".[*] Bobillier and J. Plücker (1801–1868) could not have foreseen what would spring from their "methods of abridged notation" which were the genesis of so-called "trilinear" coordinates.

Girard Desargues is known to schoolboys by reason of two theorems with which his name is connected. In the ordinary Histories of Mathematics he usually receives only a few lines commemorating his theorem on the involution of the four-point conic; certainly Rouse Ball[†] gives him credit for the invention of Projective Geometry, but he places him, wrongly I think, at the end of an epoch. Of his work on Conics I shall speak later, but his theorem of homologous triangles may be mentioned now. Pappus' Theorem together with the propositions of incidence in a plane, as I have said above, enables us to establish a one-to-one correspondence between each point of the plane and a set of homogeneous coordinates which are elements of a field; if instead of Pappus' Theorem we use Desargues' Triangle Theorem, our coordinates still satisfy the basic laws of algebra except the commutative law of multiplication. In the special case when the two triads are in perspective Pappus' Theorem can be

---

[*] J. A. Todd, *Projective and Analytical Geometry*, 1947.
[†] *Short History of Mathematics*, sixth edition, p. 257.

deduced from Desargues', but not otherwise. In projective space of three or more dimensions Desargues' Theorem follows from the basic propositions of incidence and need not be[*] adjoined to them.

Desargues was born at Lyons in 1593; his father was a man of good family, apparently a notary, who had some property near Condrieu, to which Girard is said to have retired at the end of his life. We know nothing of his early life or education, nor of how he became interested in geometry; presumably he had read Euclid, Apollonius and Ptolemy, since he quoted them in his work: indeed much of our knowledge of him comes only from sources such as Baillet's *Life of Descartes* (1691) and from the *Letters of Descartes* published by Cousin in Paris, 1824–6.[**] Desargues was certainly in Paris in 1626 when Descartes went there to settle; their community of age and interest brought them together and they formed a lasting friendship. Both appear to have been interested in applying mathematics to the easement of labour. It seems certain that Desargues was not unskilled in engineering and architecture[†] since Cardinal Richelieu employed him in this capacity at the siege of la Rochelle in 1628. At the end of the war he returned to Paris where he devoted himself to the study of geometry and its applications to architecture. He joined the circle which met weekly at le Pailleur's house to discuss mathematics and natural philosophy: these meetings led to the formation of the French Academy in 1666. There he met Gassendi, Bouillaud, Pascal the elder, Roberval, Mersenne,[‡] Mydorge: it was in 1637 that Blaise Pascal, then aged 14, was admitted to this company, at the time when Desargues was engaged on his work on Conics, a point of some importance in view of later events.

Desargues' interests seem to have been very wide. He wrote, we know, on pure geometry, mechanics, architecture, gnomonics and perspective: of his work on algebra, which Descartes asked to see,[§] we have no knowledge, nor can we say to what extent he studied metaphysics, though we know that Descartes preferred his judgment on such matters "to that of three theologians".[||]

---

[*]   For further discussion see G. de B. Robinson, *The Foundations of Geometry*. Toronto, 1946. Hilbert, *Grundlagen der Geometrie*, seventh edition, pp. 85 *et seq.* shows that D.'s theorem cannot be proved in a plane without using a third dimension or some congruence axiom, and gives an example of a two dimensional space in which the theorem is not true.

[**] M. Poudra (*Oeuvres de Desargues*, Paris, 1864) collected practically every literary reference which would help to build up a picture of the history and character of D. and I acknowledge my debt to him.

[†]   The archives of Lyons contain some correspondence which passed in 1646 between the municipal authority and "the celebrated architect Desargues, a son of the borough" relating to plans for a new town hall.

[‡]   Mersenne was very much concerned with mathematicians and exponents of other branches of learning, and seems to have acted as a "clearing-house" of knowledge: results were frequently communicated to him before publication and letters from one scientist to another were passed through him; *e.g.* Desargues and Beaugrand, the King's Secretary, who were not on speaking terms, communicated through the intermediary of Mersenne. It may be that the *Letters of Mersenne*, published by Mme. Tannery during the last fifteen years, would throw still more light on Desargues, but I have not seen the relevant volumes.

[§]   Cousin, vol. 8, p. 493, 28th February, 1641.

[||] Cousin, vol. 8. p. 433, 31st December, 1640.

In order correctly to appreciate Desargues' work it is necessary to consider the background against which he wrote. After a final century of infertile existence Alexandria fell in A.D. 641 and the long history of Greek mathematics closed. Thereafter in Europe mathematics stagnated for some centuries and geometry practically disappeared. It is true that in the Middle Ages the monastic schools and some learned clerks, particularly those connected with building operations, kept alive a faint memory of an almost legendary figure called Euclid, who was the "inventor" of geometry and who was mixed up in some curious way with Egypt and perhaps with Abraham!* but speaking generally geometry was dead: its knowledge was confined to certain practical rules of measurement and results such as Pythagoras' theorem, which were almost trade secrets of the hierarchy of professional builders. During the twelfth century copies of Arabian and Greek textbooks were introduced into western Europe through the Moors of Spain, but they remained very rare: the fall of Constantinople, too, sent a wave of refugees into Italy, bringing with them some of the traditions of Greek science. Then the invention of printing in the middle of the fifteenth century, allowing the wider dissemination of the results obtained by the Arabs and the Greeks, heralded the Renaissance. During the next two centuries mathematicians were largely concerned with the development of algebra – first syncopated, then symbolic – and trigonometry (it may be noted that Desargues' work on Conics is marked by a complete absence of algebraic symbols): only about the beginning of the seventeenth century do we find a reviving interest in mechanics and geometry. The latter revival was perhaps due to J. Kepler (1571–1630) who, in a short note on Conics in his *Paralipomena* (1604), asserted the principle of continuity, introduced the name "focus" and suggested the use of the eccentric angle in dealing with the ellipse.†

Thus at this period there was a marked resurgence of mathematical progress but interest was generally concentrated on the development of analysis. Vieta's (1540–1603) introduction of symbolic algebra, brought into general use by the writings of Harriot (1560–1621) and Descartes, Cavalieri's (1598–1647) use of indivisibles,‡ Roberval's (1602–75) discussion of the nature of tangents to curves, Fermat's (1601–65) work on infinitesimals and the use of analysis in geometry, and above all Descartes' introduction of analytical geometry, were all symptomatic of the trend of the times. True, in 1522 J. Werner of Nuremberg published a small pamphlet on Conics, and F. Maurolicus (1494–1575) of Messina translated various Latin and Greek texts and discussed the conics, as Werner had done, as sections of a cone, but the general feeling, expressed by Descartes in one of his letters

---

* *British Museum, Add. MS.*, 23.198 vv. 435–538.
† Ball, l.c. p. 256.
‡ Archimedes rejected a similar method as not rigorous, v. *The "Method" of Archimedes*, edition T. L. Heath, 1912, p. 13.

to Mersenne,* was that pure geometry had no more to say,[†] and this impression persisted for well over a century.

One other point should be mentioned. Even at this time a writer frequently published his work merely by printing a few broadsheets in microscopic characters and circulating them among his immediate contacts: this method was certainly adopted by Desargues, which may account for the rarity of his written remains. The procedure did not make for rapid diffusion of ideas and it lent itself to the unscrupulous appropriation of other men's discoveries, leading to acrimonious disputes about priority of authorship.

Such broadsheet publications were the following:

1. *Méthode universelle de mettre en perspective les objets donnés réellement, ou en devis, avec leur proportions, mesures, esloignemens, sans employer aucun point qui soit hors du champ de l'ouvrage, par G. D. L., Paris* 1636 *avec privilége (de* 1630).

This was reproduced at the end of Bosse's *Perspective* of 1648, together with a note containing a number of Desargues' short memoirs designed to elucidate the main work, *viz.*

   2. (*a*) *Proposition fondamentale de la pratique de la perspective.*
      (*b*) *Autre fondement encore du trait de la perspective, ensemble du fort et du faible de ses touches ou couleurs.*
      (*c*) *Proposition géométrique* [this is Desargues' Triangle Theorem].
      (*d*) Two other geometric propositions.

3. *Perspective adressée aux théoricians:* reproduced at the end of Bosse's *Perspective* of 1648 and said by Curabelle to have been printed in 1643.

4. *Brouillon project d'exemple d'une manière universelle du S.G.D.L. touchant la practique du trait à preuves pour la coupe des pierres en l'architecture, et de l'esclaircissement d'une manière de réduire au petit pied en perspective comme en géometral et de tracer tous quadrans plats d'heures égales au soleil. Paris en aout* 1640 *avec privilege.*

In this he speaks of Fermat's method of drawing tangents and his theory of maxima and minima and he credits Roberval with the discovery of the curve "traced by a point in the diameter of a circle which rolls on a straight line". He also names as his own pupils Bosse the engraver, de la Hire the painter[‡] and Hureau a master-mason, whom, he says, understood his method of perspective in less than two hours.

---

* Cousin, vol. 8, p. 88.
[†] J. L. Coolidge, *History of Geometrical Methods*, 1947, pp. 104–5, expresses the same view with regard to the present situation in synthetic projective geometry.
[‡] Father of Philippe de la Hire, the geometer (1640–1719).

5. *Manière universelle de poser le style aux rayons du soleil en quelque endroit possible avec la règle, l'esquerre et le plomb.* 1640.

M. Poudra reconstructed this work from the criticism *Avis charitables* mentioned below.

6. *Atteinte aux évènemens des contrarietez d'entre les actions des puissances on forces.*

This was a short memoir which is lost except for a few lines quoted by Beaugrand. It seems to have been appended to the Treatise on Conics, and makes use of the word "involution" defined for the first time in that work. In effect it states that the centre of gravity of a sphere lies in the diameter common to the earth and sphere at the point conjugate to the earth's centre. Descartes, in a letter to Mersenne dated 15th November, 1638, commenting on this and partially correcting it, gave an incorrect result but was unable completely to solve the problem.

Desargues' most important publication was his Treatise on Conics, entitled *Brouillon project d'une atteinte aux évènemens des rencontres d'un cone avec un plan, par le sieur G. Desargues Lionais, Paris* 1639.*

The treatise is not easy to read, partly because, as Desargues says in his title, it is no more than a rough sketch or outline, partly because he really lets himself go in the invention of new and unusual terms. He opens with an expansion of Kepler's principle of continuity: thus parallel lines meet in a unique point at infinity, the (infinite) points at opposite "ends" of a line being regarded as coincident; parallel planes meet in a unique line at infinity; a straight line is the special case of a circle whose centre is at infinity. Then he continues with a large number of definitions leading up to the theory of involution of six points; various special cases are treated, particularly that in which certain coincidences reduce the six points to four, giving a harmonic range. The theory of projectivities on a line revolves round the double points and Desargues was quick to notice this: he establishes that the characteristic cross-ratio of an involution is -1, and obtains most of the relations between segments of a harmonic range which now appear in school textbooks. He defines involution and harmonic pencils and deduces the important result that the characteristic property of an involution or harmonic range is projective. His main weapon in establishing these results is Menelaus' Theorem on the plane triangle cut by a transversal.

Coming now to the conic, Desargues defines pole and polar and remarks that the centre and diameters are special cases. He then obtains the theorem of the involution of the complete inscribed quadrangle by projection from the circle and goes on to the self-polar triangle, with more theory of poles and polars, the involution of conjugate points and an involution of points on a conic. In a general proposition he shows

---

* D.'s critics, Curabelle and others, spoke of a work on Conics called *Leçons de ténèbres* of which we have no trace. I suspect that this was merely an alternative title to the *Brouillon-projet*.

how the axes, conjugate diameters, tangents, asymptotes (*viz.*: tangents at infinity) and foci may be obtained by projection from a circle or from another conic. He also introduces the idea of a polar plane for a solid.

Between 1626 and 1630 Desargues gave some lectures in Paris which greatly impressed his colleagues: these were probably on perspective,* architecture and gnomonics. He himself says that he was impelled to study the geometrical basis of these subjects because having noticed that the master-craftsmen, architects, painters, masons and so on, had learnt only a number of empirical rules and obtained their results in a random and groping manner resulting in uncertainty and fatigue, he was inspired with a desire to lighten their labours by discovering shorter and more intelligible rules. The pamphlets 1, 4 and 5 clearly represent the issue of his researches.

It will be noticed that all these publications emanated between 1630 and 1643. Desargues was highly appreciated by his mathematical contemporaries as is obvious from the letters of Descartes and that passage in the younger Pascal's *Essay on Conics* in which he said: "We shall also prove the following property due to M. Desargues of Lyons, one of the great intellects of this time and amongst those most expert in mathematics, particularly in conics; his writings on this subject, though not numerous, have given ample testimony of his ability to those who have been willing to take cognisance of it. I gladly acknowledge that I owe the little I have discovered in this subject to his writings and that I have tried to imitate, so far as I could, his method of treating all the conic sections generally without using the triangle through the axis. The remarkable proposition in question is etc. –" and he went on to enunciate the involution of the inscribed quadrangle. Descartes did not think that Pascal could be the sole author of this essay and attributed it to the teaching of Desargues.[†] While Leibniz said: "I think M. Descartes was right in saying that young Pascal, aged only 16 when he composed his treatise on conics, had profited by the ideas of M. Desargues and it seems clear to me that Pascal acknowledged that himself."

But Desargues came up against professional jealousy an the part of his non-mathematical colleagues. In 1642, François l'Anglois *dit* Chartres published a book from the press of Melchior Tavernier entitled *Practical Perspective by a Parisian Jesuit.*[‡] This book contained serious errors and Desargues criticised it severely in notices placarded on the walls of Paris and in a small pamphlet accusing the author of having clumsily plagiarised his own methods. The author and publisher, interested in the sale of their production, retaliated by printing several violent attacks on Desargues. They made a collection of these libels and of a number of

---

* Coolidge, l.c. p. 109, calls Desargues "the bad boy of the whole story" of Perspective.
† Cousin, vol. 8, p. 214.
‡ Probably the pamphlet was the composition of Tavernier himself based on a MS by the Friar Dubreuil.

adverse criticisms of Desargues' works which were current and published them under the ironical title *Advis charitables sur les divers œuvres et feuilles volantes du sieur Girard Desargues, Lyonnois. Paris* 1643. This collection contains items printed in 1640 and 1641 and different copies of it do not always contain the same items. In one of these variants appears the letter from Beaugrand,* the King's Secretary, dated 20th July, 1640, criticising and depreciating Desargues' "Conics" as no more than one of the lemmas of the seventh book of Pappus together with a corollary of the seventeenth theorem of Appollonius' third book: nearly the whole of the last half of the letter is concerned with Beaugrand's own discussion of the centroid of the triangle and the criticism of Desargues' note on mechanics already mentioned.

Those attacks annoyed Desargues, who like Descartes had a difficult temperament and became restive under opposition; he appears to have decided to write no more under his own name. One of his pupils was an engraver. Abraham Bosse, who though no mathematician had sufficient penetration to understand the practical application of Desargues' theories on perspective, and became his loyal supporter and friend. Desargues entrusted to Bosse the task of interpreting and working out in detail the general theory he had evolved, a task to which Bosse devoted himself with the utmost fidelity: indeed, in 1666 he published a full statement of the events which led him to resign from his post as Lecturer at the Royal Academy of Painting and Sculpture, from which it appears that he refused to obey the Governing Body's instruction to forswear the principles he had learnt from his master. Bosse's interpretations of Desargues' ideas were embodied in a number of books on Perspective, Gnomonics etc., dated from 1643 onwards: their style is very diffuse and they are consequently very difficult to follow, but three of them bear long forewords by Desargues stating that they are in complete conformity with his own methods and notions, and the "privileges" for these were granted on Desargues' application.

The critics were clearly of opinion that these three books, at least, were the work of Desargues himself, for in 1644 François l'Anglois published another attack entitled *Examen des œuvres du sieur Desargues par J. Curabelle*, in which Bosse's books were attributed to his master. The quarrel between Curabelle and Desargues, who was not a patient man, grew bitter and the latter offered to back his opinions with a stake of 100,000 livres: Curabelle accepted in the sum of 100 pistoles only and after long and acrimonious discussions on both sides negotiations broke down over the question of arbitrators. Curabelle gave his version of the story in a pamphlet called *Foiblesse pitoyable du sieur G. Desargues contre l'examen fait de ses œuvres par J. Curabelle.*

---

* Beaugrand wrote a Commentary on the Cycloid, one or perhaps two books on "Geostatics" and a commentary on Vieta's principal work.

I have said that a community of temperament and interest produced a lasting bond between Desargues and Descartes. In 1628 the latter went to La Rochelle, curious to see the vast undertaking connected with the construction of the dyke "in the design of which his friend Desargues had some share".* Even when Descartes withdrew to Holland in 1629 they maintained a regular communication and were in the habit of exchanging their views on mathematical themes, usually through the intermediary of Mersenne. Desargues used his influence with Richelieu to induce him to offer a pension to Descartes in order to persuade the latter to return to France. Descartes refused the offer but remained profoundly grateful. Desargues also supported Descartes in his quarrel with Bourdin and his discussion with Fermat. On the other hand Descartes took Desargues' part when Beaugrand was attacking the Treatise on Conics.

Though their methods of tackling the problem were different each had the same general outlook on mathematics and each in his own way carried his researches to a high degree of excellence. They were both highly sensitive to criticism and could not brook contradiction: each, aware of the value of his own concepts, was inclined to pay scant regard to the work of others.

Descartes died in Sweden in 1650 and Desargues felt his loss keenly. He withdrew to Lyons and appears to have carried out no more serious research though he continued to produce designs for some local building construction. He returned to Paris for a short visit in 1658 on the occasion of the marriage of a nephew who became his heir. He owned a country house at Condrieu to which he used to go to cultivate his garden. He died in 1661, a disappointed man. His will, dated 5th November, 1658, contained the following clause: "M. Desargues gives and bequeaths to his obliging and good friend M. Abraham Bosse, etcher, residing in Palace Yard, and in default of him to his heirs, the sum of 2000 livres payable in four instalments, etc."

Of Descartes, Pascal and Desargues the first two have lived while the last has been almost forgotten. Was Pascal so much greater a mathematician than Desargues, or can it be that his religious and philosophical *Pensées* have helped to keep his memory alive? He himself has acknowledged his debt to Desargues and is hardly likely to have been deceived in his estimate of the latter's ability. But Desargues laboured under great difficulties, some due to the trend of the times and some to his own temperament. Descartes' work was difficult to read, but it was in greater accord with current mathematical thought and it captured the imagination of contemporary scientists. Desargues had the odds against him in pursuing the course of pure geometry; he lengthened those odds by the conciseness of his statements and the extraordinary nomenclature he adopted – he speaks of "trees", "trunks," "knots,"

---

* v. Baillet's *Life of Descartes*, quoted by M. Poudra.

"branches" and all the rest of what Beaugrand called "rustic barbarisms" – and even Descartes was moved to chide him for departing from the language of Apollonius; and the battle was finally lost through his quarrel with his critics, pursued with great bitterness by him and his champion Bosse, whose break with the Royal Academy of Painting could not fail to have some effect on the newly formed French Academy. The quaintly named *Brouillon-projet* was forgotten and lost: all that remains is a copy made by Philippe de la Hire* together with a letter certifying its authenticity and a few notes made by him. These were found in a library in Paris by M. Chasles in 1845 and are now in the Library of the Academy: meanwhile de La Hire had been acclaimed as the author of the theory of poles and polars. With no knowledge of the actual work other than that to be derived from Desargues' critics, Chasles tried[†] to rehabilitate his reputation, but the pioneer was overshadowed by men like Monge, Brianchon and Poncelet who were founding the new school of descriptive geometry.

But Desargues' name is now being mentioned more frequently by geometers, and there is today a tendency to reduce previous estimates of what Pascal accomplished.[‡]

---

* Ball, l.c. p. 317 says that P. de la Hire was Desargues' favourite pupil. I think this is wrong; anyhow Philippe was no more than ten years old when Desargues left Paris. Laurent, Philippe's father, was one of the Academicians who gave Bosse a declaration of thanks for his offer to publish his works in the name of the Academy.
† *Aperçu historique, etc.* 1837.
‡ Boutroux in Pascal's *Works* (Paris, 1923) vol. 1, pp. 245 *et seq.*

2.10

Henri Brocard and the Geometry of the Triangle

LAURA GUGGENBUHL

During the last half of the nineteenth century there was a great revival of interest in the field of geometry. A large part of this interest had its origin in a simple problem submitted to a contemporary mathematical periodical by a French army officer. The problem was to find a point O within a triangle ABC such that the angles OAB, OBC and OCA would be equal. The name of the army captain who submitted the problem was Brocard – Pierre René Jean-Baptiste Henri Brocard.

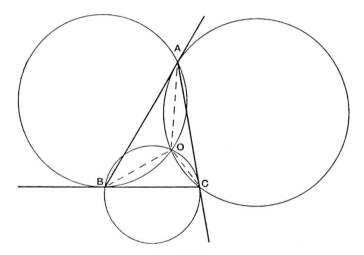

Fig. 2.10.1.

Soon thereafter many different solutions of the problem were published; and gradually a rather elaborate unit of geometric theory had developed from this source. A solution of the problem is readily accessible in a text book on Modern Geometry, and nothing more than a brief summary will be given here. The most picturesque solution is one in which circles are drawn as follows – circles tangent to the side AB

First published in *Mathematical Gazette* **37** (December 1953), pp. 241–243.

of the triangle ABC, at the vertex A, and at the vertex B respectively, and passing through the vertex C; and four other similar circles. Three of these six circles are concurrent at a point O, and the other three at a point O'. The points O and O' satisfy the conditions of the above problem, and are called the Brocard points of the triangle.

The angle OAB (angle ω) is called the Brocard angle of triangle ABC. It is a simple matter to prove from the following diagram, in which construction lines can be easily identified, that

$$\cot \omega = \cot A + \cot B + \cot C.$$

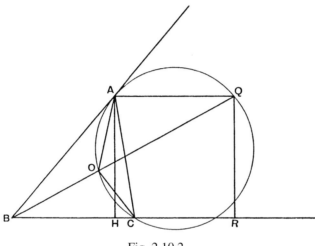

Fig. 2.10.2.

The Brocard circle of a triangle is the circle drawn on PK as diameter, where P is the circumcenter of the triangle and K is its symmedian point.[1] It passes through O and O'.

Around 1890 several books were published, which were devoted to the theory of the Brocard configuration. However, biographical details about Brocard remained relatively obscure. In English, such details were virtually non-existent. It is the purpose of this article to fill an often expressed wish to know a little more about the man.

Henri Brocard was born on May 12, 1845, at Vignot (Meuse) in the north-eastern corner of France. He was a son of Jean Sebastion and Elizabeth Auguste Liouville Brocard. No record has been found of brothers or sisters, and Brocard never married. In 1894 he published an autobiography, giving the most minute details, for the first

---

[1] For a definition and brief history of the Symmedian or Lemoine point see Florian Cajori, *A History of Mathematics*, 4th ed., New York: Chelsea, 1985, pp. 299–300.

fifty years of his life. This account tells of his education at the Lycée of Marseille, and the Lycée and Académie of Strasbourg. He attended the École Polytechnique from 1865 to 1867, and then became a member of the Engineers of the French Army. His army career was not one of active military combat. It is known that he was a prisoner of war at Sedan in 1870; but for the most part, his activities in the army were devoted to organizing courses in physics and chemistry at various military colleges, and to research, particularly in meteorology and mathematics. For several years of his army service he was assigned to North Africa, and he has been called the co-founder of the Meteorological Institute of Algiers.

Brocard became a member of the Society of Letters, Sciences and Arts of Bar-le-Duc in 1894. It is through the pages of the publications of this Society that one can follow the activities of the last twenty-six years of his life. Though he declined the honour of becoming President, he obviously found great happiness in his work as Librarian of the Society. He became interested in local affairs, and devoted himself to public service. Largely through his efforts, one of the streets of Bar-le-Duc was named in honour of Louis Joblot, a native Barrisien, who was an acknowledged but almost forgotten pioneer in the field of microscopy. When he retired from the army in 1910, Brocard was a lieutenant-colonel, and an Officer in the Legion of Honor. In the years of his retirement he spent much of his time in the pleasant garden in the rear of his house, at 75 Rue de Ducs in Bar-le-Duc, where he would use a small telescope for meteorological and astronomical observations. In spite of the fact that he lived completely alone and rarely had visitors, he always attended the quadrennial meetings of the International Congress of Mathematicians. On January 16, 1922, he was found dead at his desk, and in accordance with his specific request, he was buried in the cemetery at Vignot, next to his father and mother.

His most colourful and ambitious publication was *Notes de Bibliographie des Courbes Géométriques*. Volume I appeared in 1897 and Volume II in 1899. Probably no more than about fifty copies of this work were prepared, and it has become exceedingly rare and valuable. It was lithographed in the print-script of the author, and it was privately distributed. Although it was described as being neither a dictionary nor an encyclopaedia, it may be regarded as a source book of geometric curves. It has a painstakingly carefully prepared index of over one thousand numbered and named curves. About twenty years after the appearance of this work, virtually the same material appeared in printed form under the joint authorship of H. Brocard and T. Lemoyne.

Brocard has not yet found a place, nor even a line, in books on the history of mathematics; for his was obviously a most modest personality, and he had the humility of the true scholar. However, his influence upon his contemporaries was such that his name has been used to identify not only the Brocard points of a triangle, and the Brocard circle (which he discovered), but also many other geometric conclusions.

Perhaps some day, you too will be passing through Bar-le-Duc. If so, you might enjoy a leisurely stroll along the Quai Carnot to the municipal library. The present Librarian of the Society of Letters, Sciences and Arts is M. Rogie, and the President is M. Lucien Braye. If you are interested in such things, M. Rogie and M. Braye would surely be more than pleased to place before you with reverent pride, the Library's treasured copy of Brocard's *Notes de Bibliographie des Courbes Géométriques*.

2.11

The Development of Geometrical Methods

M. GASTON DARBOUX

In order to give an adequate account of the progress of Geometry in the century that has just come to a close, it is important to cast a rapid glance over the condition of Mathematical Science at the beginning of the nineteenth century.

We know that during the latter portion of his life, Lagrange, somewhat weary of the researches in Analysis and Mechanics, which, however, assured for him immortal glory, took up the study of Chemistry (which he described as almost as easy as Algebra), of Physics, and also devoted himself to philosophical speculation. We almost always see a similar state of mind at certain periods in the lives of the greatest men of science. From the fresh ideas of their fertile youth, ideas which they utilise to the full in the field of their labours, they have derived all that they have any right to expect; their task is done; their mental activity is impelled towards entirely new subjects. And this impulse, be it understood, had a quite peculiar force in the days of Lagrange. The field of research opened to geometers by the discovery of the Infinitesimal Calculus appeared to be almost exhausted. It really seemed that when a few more or less complicated differential equations were integrated, and a few more chapters were added to the Integral Calculus, the very limits of Science would be attained. Laplace had completed his explanation of the system of the world, and was laying the foundations of Molecular Physics. New paths were opening for the experimental sciences, and all was ready for the astonishing development they were to receive during the nineteenth century. Ampère, Poisson, Fourier, and Cauchy himself, the creator of the theory of imaginaries, were entirely taken up with the application of analytical methods to Mechanics and Molecular Physics; they seemed to believe that apart from the new domain of which they were making so rapid a survey, the limits of Theory and Science were definitely fixed.

An address by M. Gaston Darboux, of the *Institut de France*, Sept. 24, 1904, given at the Congress of Science and Art, St. Louis, U.S.A. First published in *Mathematical Gazette* **3** (December 1904), pp. 100–106; (January 1905), pp. 121–128; (March 1905), pp. 157–161; (May 1905), pp. 169–173.

From the end of the eighteenth century, the subject for which we must claim the name of modern Geometry had been in a large measure contributing to the renaissance of Mathematical Science in all its branches. It had offered to research a new and fertile path, and, in particular, it had shewn in the most striking and successful manner, that general methods are not everything in Science, and that even in the simplest subject there is much that may be done by the ingenious and inventive mind. The beautiful geometrical demonstrations of Huyghens, Newton, and Clairaut were forgotten or neglected. The fine ideas introduced by Desargues and Pascal had never been developed and seemed to have fallen on barren soil. Carnot, by his *Essai sur les Transversales* and his *Géométrie de Position*, and Monge, more especially by his beautiful theories on the generation of surfaces, had pieced together a chain the links of which had seemed to be severed. Thanks to them, the conceptions of Descartes and Fermat, the inventors of Analytical Geometry, assumed once more, in the Infinitesimal Calculus of Leibniz and Newton, the position they had lost and which they should never have ceased to occupy. As Lagrange said of Monge, "ce diable d'homme" will make himself immortal by his Geometry. And in fact, not only did Descriptive Geometry play its part in coordinating and perfecting the processes employed in all the arts, "in which precision of form is a condition of success as well as of excellence for labour and its products," but it made its appearance as the graphical translation of a general and purely rational Geometry, the fertility of which has been shewn by researches both numerous and important. At the side of his *Géométrie Descriptive* we must not forget to place his other master-piece, the *Application de l'Analyse à la Géométrie*. Nor should we forget that to Monge we owe the idea of lines of curvature, and the elegant integration of the differential equations of those lines in the case of the ellipsoid, which it was said was a matter of envy to Lagrange. We must lay stress on this character of the whole of the work of Monge. The man who reconstructed what we call modern Geometry has shewn us from the outset, and this has perhaps been forgotten by his successors, that the alliance between Analysis and Geometry was useful and fruitful, and that perhaps this alliance was a condition of success for both these branches of Mathematics.

II

To the school of Monge we owe many geometers: Hachette, Brianchon, Chappuis, Binet, Lancret, Dupin, Malus, Gaultier de Tours, Poncelet, Chasles, etc. Among these, Poncelet ranks first and foremost. He passed over everything in the works of Monge connected with Cartesian Analysis or with Infinitesimal Geometry; he devoted his attention exclusively to the development of the germs contained in the purely geometrical researches of his illustrious predecessor. He was captured by the Russians at the passage of the Dnieper, and interned at Saratoff. During the period of enforced leisure which followed he turned his attention to the proof of the principles which he developed in his *Traité des Propriétés projectives des Figures*, published in 1822, and in the great memoirs on polar reciprocals and harmonic means which date back to about this time. It may therefore be said that modern Geometry was born at Saratoff. Returning to the ideas which had been neglected since the days of Pascal and Desargues, Poncelet simultaneously introduced homology and polar reciprocals, thus at the outset throwing into relief the ideas which Science had been maturing for 50 years.

Poncelet's methods were not enthusiastically received by French analysts, being in a sense opposed to the processes of Analytical Geometry. But their importance was so great and their novelty so striking that it was not long before they were thoroughly examined. Poncelet alone had discovered the principles; on the other hand, several geometers almost simultaneously began to study them from every point of view and succeeded in deducing from them the essential results which they implicitly contained.

At this time, Gergonne was the editor of a periodical which is of inestimable value to us in the history of Geometry. The *Annales de Mathématiques*, published at Nîmes from 1810 to 1831, had been for more than fifteen years the only journal in the world exclusively devoted to mathematical research. Gergonne was in many respects a model editor for a scientific journal; but he had the defects of his qualities. He collaborated, often against their will, with the authors of the memoirs which were sent to him; he altered their papers, and sometimes made them say more or less than they had intended. However that may be, he was greatly struck with the originality and the scope of Poncelet's discoveries. A few simple ways of transforming figures were already known in Geometry; homology had even been used in the plane; no one but Poncelet had extended it to space, and no idea seemed to exist of the wealth and fertility of the method. Besides, all these transformations were *ponctuelles*, point was made to correspond to point. By the introduction of polar reciprocals, Poncelet became an inventor of the highest rank; he gave the first example of a transformation in which to a point corresponded something else than a point. Every method of transformation enables us to multiply our theorems, but the method of polar reciprocals had the advantage of bringing into correspondence two entirely different propositions. This was essentially something new. To throw this into relief Gergonne invented the plan, which has since been adopted with so much success, of writing correlative propositions in two columns. His also was the idea of substituting for Poncelet's demonstrations, which required the assistance of a curve or surface of the second order, the famous principle of duality, the significance of which, though somewhat vague at first, was sufficiently cleared up by the discussions which took place on the subject between Gergonne, Poncelet, and Plücker.

Bobillier, Chasles, Steiner, Lamé, Sturm, and many others, whose names escape me, assiduously collaborated with Poncelet and Plücker on the *Annales*. When Gergonne was appointed Rector of the Académie de Montpellier in 1831, his publication of the journal was interrupted. But his success in creating a new form of mathematical literature, and the taste for research which he had done so much to develop, had already begun to bear fruit. Quetelet had just started the Belgian *Correspondance mathématique et physique*. Crelle in 1826 had published the first sheets of the famous journal which bears his name, and in which are to be found memoirs by Abel, Jacobi, and Steiner. A considerable number of formal treatises also appeared at this time, in which the principles of modern geometry were being expounded and developed in a masterly manner.

First, there was in 1827 the *Barycentrisches Calcul* of Möbius, a really original work, remarkable for the depth of its conceptions and the lucidity and rigour of its exposition. In 1828 appeared Plücker's *Analytisch-geometrische Entwickelungen*, the second part of which was published in 1831. These were very soon followed by the same author's *System der analytischen Geometrie* (Berlin, 1835). In 1832 Steiner published his *Systematische Entwickelung der Abhängigkeit der geometrischen Gestalten von einander*, and in the next year appeared his *Geometrische Constructionen ausgeführt mittelst der geraden Linie und eines festen Kreises*, in which was proved by the most elegant examples a proposition of Poncelet's relative to the use of a single circle in geometrical constructions. Finally, in 1830, Chasles sent to the Académie de Bruxelles, which by a happy inspiration had offered a prize for the best essay on the principles of modern geometry, his celebrated *Aperçu historique sur l'origine et le développement des méthodes en Géométrie*. This was followed by the *Mémoire sur deux principes généraux de la Science: la dualité et l'homographie*, which was not published until 1837.

Time would fail me were I to endeavour to offer an adequate appreciation of these noble works, or on such an occasion as this to give an account of each of them. Such an attempt would lead us to a new verification of the general laws of the development of science. When the moment is ripe, when the fundamental principles are recognised and enunciated, nothing can check the march of ideas; the same or nearly the same discoveries are made almost simultaneously, and in all parts of the world. But although I shall not venture on a discussion of this kind, which would be of little use, and might give but little pleasure, it is important that I should try to throw some light upon a fundamental distinction in the tendencies of the great geometers who, about 1830, gave to Geometry an impulse hitherto unknown.

III

Some, like Chasles and Steiner, who devoted the whole of their lives to researches in Pure Geometry, drew a distinction between what they called *synthesis* and *analysis*; and, adopting in their general scope, if not in their detail, the tendencies of Poncelet, endeavoured to found an independent theory, a rival to the Cartesian Analysis.

Poncelet did not rest content with the inadequate resources furnished by the method of projections. He therefore imagined the famous *principle of continuity*, which gave rise to such long discussions between him and Cauchy. The principle is an excellent one when properly enunciated, and may be of the greatest service. Poncelet made the mistake of refusing to present it as a simple consequence of Analysis; Cauchy, on the other hand, would not recognise that his own objections, valid no doubt in the case of certain transcendental figures, lost their force in the applications made by the author of the *Traité des propriétés projectives*. Whatever opinion may be held on the subject of such a discussion, it at least shewed, in the clearest manner, that Poncelet's geometrical system rested on an analytical basis, and we know besides, from the unfortunate publication of the Saratoff manuscripts, that the principles which served as a foundation for the *Traité* were established by the aid of the Cartesian Analysis. Poncelet had abandoned Geometry for Mechanics, in which his work had a preponderating influence. Chasles, his junior, for whom had been created in 1847 a chair of Higher Geometry in the *Faculté des Sciences de Paris*, set to work to build up a geometrical theory entirely independent and autonomous. This he expounded in two works of the highest importance, his *Traité de Géométrie Supérieure* (1852), and the *Traité des Sections Coniques*, unfortunately left unfinished, the first part alone appearing in 1865.

In the preface to the former he very clearly indicates the three fundamental points which enable the new theory to participate in the advantages of analysis, and seem to him to mark some progress in the cultivation of the science. They are:

(1) The introduction of the principle of signs, which simplifies both enunciations and proofs, and gives to the analysis of Carnot's transversals all the scope of which they are susceptible.
(2) The introduction of imaginaries, which supplements the principle of continuity, and furnishes demonstrations as general as those of analytical geometry.
(3) The simultaneous demonstration of correlative propositions, *i.e.*, propositions which correspond in virtue of the principle of duality.

Chasles thoroughly discusses homography and correlation in his work, but in his exposition he systematically avoids the use of the transformation of figures. These, he thinks, cannot supplement direct proofs, because they mask the origin and real nature of the properties obtained by their means. There is something in this criticism, but the march of science shews us that it errs on the side of severity. Even if it often happens that transformations, when employed without discrimination, uselessly multiply the number of theorems, yet we must not forget that they often assist us to a better realisation of the propositions to which they are applied. Did not the use of Poncelet's projection, which led to the fruitful distinction between projective and metrical properties, make us recognise the great importance of this anharmonic ratio, the essential property of which is to be found in Pappus, and the fundamental rôle of which did not appear in modern geometry until fifteen centuries later?

The introduction of the principle of signs was not the novelty Chasles supposed when he wrote his *Traité de Géométrie Supérieure*. Möbius in his *Barycentrisches Calcul* had already followed up a *desideratum* of Carnot's, and had used signs in a very wide and precise manner, defining for the first time the sign of a segment and even of an area. Later on he succeeded in extending the use of signs to lengths which are not measured off on the same straight line, and to angles which are not formed around the same point. In addition, Grassmann, whose mind presents so many points of analogy with that of Möbius, was necessarily compelled to use the principle of signs in the definitions which serve as a basis for his remarkably original method of discussing the properties of extension.

The second characteristic assigned to his system of geometry by Chasles is the use of imaginaries. Here his method was really new, and he was able to illustrate it by examples of the greatest interest. Admiration will always be awarded to the beautiful theories he has left us on the confocal surfaces of the second degree, in which every known property and many new properties, as varied as they are elegant, are derived from the general principle that these surfaces are inscribed in the same developable which is circumscribed to the circle at infinity. But Chasles only introduced imaginaries by their symmetrical functions, and was therefore unable to define the anharmonic ratio of four elements when they cease to be real, in whole or in part. If only he could have established the idea of the anharmonic ratio of imaginary elements, a formula which is to be found in the *Géométrie Supérieure* (p. 118, new edition) would have immediately enabled him to give the beautiful definition of an angle as the logarithm of an anharmonic ratio – a definition which enabled Laguerre, my lamented colleague, to find the long-sought, complete solution of the problem of the transformation of those relations which contain both angles and homographic and correlated segments.

IV

While Chasles, Steiner, and later, von Staudt, were endeavouring to build up a theory that would be a rival to Analysis, and were, as it were erecting altar against altar, Gergonne, Bobillier, Sturm, and Plücker in particular, were perfecting the Cartesian

Geometry, and were constructing an analytical system in some measure adequate to the discoveries of the geometers. To Bobillier and Plücker we owe what is known as the method of *Abridged Notation*. In the last volumes of Gergonne's *Annales* are to be found a few really original pages from the pen of the former. Plücker had begun to develop the method in his first volume, very soon to be followed by a series of works in which the foundations of modern geometry are deliberately established. To the same investigator we owe tangential and trilinear coordinates, used in homogeneous equations, and also the canonical forms, the validity of which is recognised in the fruitful but sometimes deceptive method known as the *enumeration of constants*. All these timely discoveries were to infuse new blood into the Cartesian Analysis; they enabled it to give their full significance to those conceptions which the so-called *synthetic* geometry had not completely grasped. Plücker – and with his name it is only fair to couple that of Bobillier, carried off by an untimely death – must be regarded as the first to familiarise us with those methods of modern Analysis, in which the use of homogeneous co-ordinates enables us to treat simultaneously, without the reader's knowledge, as it were, a figure, and all the figures that are deduced from it by homography and correlation.

V

From this moment a brilliant period opens for geometrical research of every kind. Analysts interpret all their results, and set to work to translate them by constructions. Geometers endeavour to discover in every question some general principle – in most cases impossible to prove without the aid of Analysis – so as to deduce a host of particular consequences, closely connected with each other and with the principle from which they are derived. Jacobi's brilliant pupil, Otto Hesse, admirably develops the methods of homogeneous co-ordinates of which Plücker, perhaps, did not fully appreciate the value. Boole discovers in Bobillier's polars the first notion of a covariant; the theory of forms is created by the labours of Cayley, Sylvester, Hermite, and Brioschi. Aronhold, Clebsch, Gordan and other geometers still with us, invent the definitive notation of the theory, establish the fundamental theorem relative to the limitation of the number of covariant forms, and thus succeed in giving to the theory its fullest extension and scope.

The theory of surfaces of the second order, constructed mainly by the school of Monge, is enriched by a large number of elegant properties – mostly due to Hesse, who later finds in Paul Serret a worthy rival and an investigator who will continue his work.

The properties of the polars of algebraical curves are developed by Plücker, and in particular, by Steiner. The study of curves of the third order, which have by this been the subject of research for a considerable time, is enriched and rejuvenated by a host of new ideas. In the first place, Steiner treats by pure geometry the double tangents of curves of the fourth order, and Hesse, following in his steps, applies algebraical methods to this beautiful problem, as well as to the points of inflexion of curves of the third order.

The idea of *class*, introduced by Gergonne, the study of a paradox partially elucidated by Poncelet, relating to the respective degrees of two curves each the polar reciprocal of the other, are the immediate impulse to Plücker's researches on the so-called *ordinary* singularities of algebraical plane curves. The celebrated formulæ discovered by Plücker are later extended by Cayley and by other geometers to algebraical gauche curves, and by Cayley again and by Salmon to algebraical surfaces. Singularities of a higher order are in their turn attacked by geometers; Halphen shews that, contrary to an opinion which then widely obtained, each of these singularities may be considered as equivalent to a certain group of ordinary singularities, and his researches bring to a close for the time being this difficult and important question.

Analysis and Geometry – Steiner, Cayley, Salmon, Cremona – meet in their investigations into surfaces of the third order; and as Steiner foresaw, this theory becomes as simple and as easy as that of surfaces of the second order.

Ruled algebraical surfaces, so important in their applications, are studied by Chasles, by Cayley, the marks of whose influence are seen in every form of mathematical investigation, by Cremona, Salmon, and La Gournerie; and later by Plücker in a volume to which I shall presently recur.

The study of the general surface of the fourth order still appears to be too difficult; but that of the particular surfaces of that order with multiple points or multiple lines is begun by Plücker for wave surfaces, by Steiner, Kummer, Cayley, Moutard, Laguerre, Cremona, and many others. As for the theory of algebraical gauche curves, its elements are extended, and finally it receives through the investigations of Halphen and Noether the most notable advancement. Between their labours it is impossible on such an occasion as this to draw a distinction. A new theory with a great future before it is initiated by Chasles, Clebsch, and Cremona; it deals with all the algebraical curves which can be traced on a given surface.

Homography and correlation, the two methods of transformation which originated all the preceding investigations, receive in their turn an unexpected development; they are not the only methods which bring a single element into correspondence with a single element, as is shewn by a particular transformation briefly pointed out by Poncelet in his *Traité des propriétés projectives*. Plücker invents *transformation by reciprocal radii vectores* or *inversion*, and it was not long before Sir William Thomson and Liouville shewed its importance both in Mathematical Physics and in Geometry. Magnus, a contemporary of Möbius and Plücker, maintained that he had found the most general transformation which will bring a point into correspondence with a point. But Cremona's researches shew that Magnus transformation is only the first of a series of birational transformations which the great Italian geometer shews us how to obtain methodically, at any rate for the figures of Plane Geometry. Cremona's transformations will retain their interest for a long time to come, although later researches proved that they always reduce to a series of successive applications of the transformation we owe to Magnus.

VI

All the investigations I have mentioned, and others to which I shall presently recur, originate and find their first impulse in the conceptions of modern geometry. But I must now point out another stimulus to great progress in geometrical research. Legendre's Theory of Elliptic Functions, which had been too much neglected by French geometers, was developed and extended by Abel and Jacobi. With these great geometers, almost immediately followed by Weierstrass and Riemann, the theory of Abelian functions, which Algebra will presently attack by its own unaided resources, will bring to the Geometry of curves and surfaces a contribution the importance of which will continue to increase.

Jacobi had already used the analysis of elliptic functions to prove Poncelet's celebrated theorems on in- and circum-scribed polygons, thus opening a chapter which has been enriched by a large number of elegant results; by methods connected with geometry, he had also succeeded in integrating Abelian equations.

But Clebsch was the first to shew in a long series of investigations the whole importance of the notion of the *deficiency* of a curve, due to Abel and Riemann, by developing a vast number of results and elegant solutions, which the use of Abelian integrals appeared, so simple was it, to connect with their real point of departure. The study of the points of inflection of curves of the third order, that of double tangents of curves of the fourth order, and, in general, the theory of osculation on which the ancients and moderns had so often exercised their minds, were connected with the beautiful problem of the division of the elliptic and Abelian functions.

In one of his memoirs, Clebsch had discussed *rational* curves, or curves of deficiency zero; this led him, towards the end of a life alas too short, to consider what may be called *rational* surfaces, those which can be simply represented by a plane. Here was a vast field for research, already opened for the elementary cases by Chasles, in which Clebsch was followed by Cremona and many other savants. In this connection Cremona, generalising his researches in plane geometry, brought to light not the whole of the birational transformations of space, but some of the most interesting of these transformations. The extension of the idea of deficiency to algebraical surfaces has already begun; already also work of great value has shewn that the theory of simple or multiple integrals of algebraical differentials will find, in the study of curves and surfaces alike, an extensive field of important applications; but it is not for me to dwell any further on this topic.

VII

While the mixed methods of which we have just pointed out the principal applications were being created, pure geometers did not remain inactive. Poinsot, the creator of the theory of couples, developed by a purely geometrical method, described by him as "one in which the object of research is never for one moment out of sight," the theory of rotation of a solid body which the investigations of d'Alembert, Euler, and Lagrange seemed to have exhausted; Chasles made a valuable contribution to kinematics by his beautiful theorems on the theory of the displacement of a solid body, which have since been extended by other elegant methods to the case in which the motion has various degrees of freedom. He published his beautiful theorems in the theory of attraction which are worthy to rank with those of Green and Gauss. Chasles and Steiner found themselves on common ground in the study of the attraction of ellipsoids, and thus shewed once more that geometry has its place marked out for it in the most important questions of the integral calculus.

Steiner did not disdain to devote his attention at the same time to the elementary portion of geometry. His researches on the contacts of circles and conics, on isoperimetrical problems, on parallel surfaces, on the centre of gravity of curvature,[*] aroused the admiration of all by their simplicity and their depth.

Chasles introduced his principle of correspondence between two variable objects, which has given birth to so many applications; but here analysis assumed its rights, and studied the essentials of the principle, giving it precision, and generalising it. This was also the case with the famous theory of *characteristics* and the numerous researches of Jonquières, Chasles, Cremona, and others, who were to lay the foundation of a new branch of geometry – *Enumerative Geometry*. For several years the validity of Chasles' celebrated postulate was unchallenged; many geometers thought that it had been irrefutably established. But, as Zeuthen observed at the time, it is very difficult in proofs of this kind to recognise that there is always some weak point which the author may not have perceived; in fact, Halphen, after many fruitless attempts, definitively crowned all these researches by clearly indicating in what cases Chasles' postulate may be accepted, and when it may be rejected.

---

[*] *Krummungsschwerpunkt* [Tr.].

VIII

These are the chief investigations which restored synthetic geometry to its place of honour, and assured to it in the course of the last century its due place in mathematical research. Many illustrious workers have taken part in this great geometrical movement, but it must be recognised that their leaders and guides were Chasles and Steiner. Such was the splendour of the light cast by their wonderful discoveries that, at any rate for the moment, they threw into the shade the publications of other modest geometers, who were perhaps less desirous of discovering brilliant applications likely to kindle a love for geometry, than of building the science itself upon an absolutely solid foundation. Perhaps the labours of the latter received a more tardy reward; but their influence increases daily, and no doubt will increase still more. There is no doubt that were I to pass them by in silence, I would be neglecting one of the principal factors which will play their part in future researches. It is especially to von Staudt that I am here alluding. His geometrical work has been expounded in two works of great interest: the *Geometrie der Lage*, which appeared in 1847, and the *Beiträge zur Geometrie der Lage*, published in 1856, that is to say, four years after the *Géométrie Supérieure*.

Chasles, as we have seen, was attempting to build up a body of doctrine independent of the Cartesian analysis, and had not completely succeeded. I have already pointed out one of the reproaches that may be levelled at this system; imaginary elements are only defined in it by their symmetrical functions, which necessarily excludes them from a considerable number of investigations. On the other hand, the constant use of anharmonic ratio, transversals, and involution, requiring frequent analytical transformations, gives to his *Géométrie Supérieure* an almost exclusively metrical character, which notably differentiates it from the methods of Poncelet. Returning to these methods, von Staudt endeavoured to construct a geometry free from all metrical relations, and exclusively based upon relations of situation. It is in this spirit that his first work, the *Geometrie der Lage* (1847) was conceived. The author takes as his point of departure the harmonic properties of the complete quadrilateral and those of homologous triangles, proved solely by considerations of geometry of three dimensions, analogous to those of which the school of Monge has made so frequent a use.

In the first part of his work, von Staudt entirely omitted imaginary relations. It is only in his *Beiträge*, his second work, that, by a very original extension of Chasles's method, he geometrically defined an isolated, imaginary element and distinguished it from its conjugate. This extension, although rigorous, is laborious and very abstract. It may be in substance defined as follows: two conjugate imaginary points may always be considered as the double points of an involution on a real line; so that we pass from an imaginary to its conjugate by changing i into $-i$, and as we may distinguish the two imaginary points by making correspond to each of them one of the two different directions that may be attributed to the line. There is something rather artificial in this; the development of the theory erected on such foundations is necessarily complicated. By purely projective methods, von Staudt established a complete method for calculating the anharmonic ratios of the most general imaginary elements. As in the case of all geometry, projective geometry employs the notion of order, and order involves number. It is not therefore surprising that von Staudt was obliged to build up in detail his method of calculation, but the ingenuity which he displayed in arriving at his conclusions must be admired. In spite of the efforts of the distinguished geometers who have attempted to simplify its exposition, we fear that this part of von Staudt's geometry, like the otherwise so interesting geometry of that profound thinker Grassmann, will not prevail against the analytical methods which have now come into almost universal favour. Life is short; geometers know and also practise the principle of least action. In spite of these fears, which should discourage no one, it seems to me that, in the first form that was given to it by von Staudt, projective geometry must necessarily become the companion of descriptive geometry; that the one is called upon to revive the other in spirit, processes, and applications. This has already been felt in several countries, and notably in Italy, where the great geometer Cremona did not disdain to write for schools an elementary treatise on projective geometry.

IX

In the preceding sections I have tried to follow and to exhibit clearly the remotest results of the methods of Monge and Poncelet. By inventing tangential and homogeneous coordinates, Plücker had seemed to exhaust all that the method of projections and of polar reciprocals could supply to analysis. Towards the end of his life he returned to his early researches, and gave them an extension which was to widen in unexpected directions the domain of geometry.

Preceded by innumerable researches on systems of straight lines by Poinsot, Möbius, Chasles, Dupin, Malus, Hamilton, Kummer, Transon, and especially by Cayley, who was the first to introduce the notion of line coordinates, – researches which originated either in statics and kinematics, or in geometrical optics, – Plücker's line geometry will always be regarded as the part of his work in which we meet with the newest and most interesting of his ideas. It is important that Plücker should have been the first to build up a methodical treatment of the straight line as an element, but that is nothing compared to what he discovered. It is sometimes said that the principle of duality brings into evidence the fact that the plane as well as the point may be considered as an element of space. That is true, but by adding to the plane and the point the straight line as a possible element of space, Plücker was led to recognise that any curve or surface may also be considered as elements of space, and thus sprang into existence a new Geometry, which has already inspired a large number of treatises, and which will stimulate even more in the future. A beautiful discovery of which I shall speak further on has already connected the geometry of spheres and that of straight lines, and has suggested the introduction of the notion of the coordinates of a sphere. The theory of systems of circles has already commenced. No doubt it will be developed with the detailed investigation of the representation, which we owe to Laguerre, of an imaginary point in space by an orientated circle. But before describing the development of these new ideas which have given fresh life to the infinitesimal methods of Monge, I must retrace my steps for a moment, to touch on the history of the branches of geometry which I have so far neglected.

X

Among the works due to the school of Monge, I limit myself here to the consideration of those which are connected with *finite* geometry; but some of his pupils devoted themselves in particular to the development of the new ideas of infinitesimal geometry, which were utilised by their master in the study of curves of double curvature, lines of curvature, and the generation of surfaces – ideas which are partially expounded in the *Application de l'Analyse à la Géométrie*. Among them we may mention Lancret, the author of beautiful researches on gauche curves, and especially Charles Dupin, the only one, perhaps, who followed all the paths which were opened up by Monge.

Among other works, we owe to Dupin two volumes, the authorship of which Monge would not have been ashamed to acknowledge: the *Développements de Géométrie pure*, which appeared in 1853, and the *Applications de Géométrie et de Mécanique* which dated from 1822. In them we find the idea of the *indicatrix* which was, according to Euler and Meunier, to revolutionise the whole theory of curvature, the conceptions of conjugate tangents and of asymptotic lines which have taken so important a place in recent researches. We cannot forget the determination of the surface, the lines of curvature of which are circles; nor, in particular, the Memoir on triply orthogonal surfaces, in which is to be found, together with the discovery of the triple system formed of surfaces of the second degree, the celebrated theorem with which the name of Dupin will always be connected.

Under the influence of these works, and of the revival of synthetic methods, the geometry of the infinitely small resumed in all researches the permanent position which Lagrange had hoped to mark out for it. It is a remarkable thing that the geometrical methods thus revived were to receive their liveliest impulse on the publication of a memoir which at first, at least, seemed to be connected with analysis pure and simple. I mean Gauss's celebrated volume *Disquisitiones generales circa superficies curvas*, which was published in 1827 at Göttingen, and the appearance of which may be said to mark an epoch in the history of infinitesimal geometry.

From this moment the infinitesimal method received such an impulse in France as it had not yet experienced. Frenet, Bertrand, Molins, J. A. Serret, Bouquet, Puiseux, Ossian Bonnet, and Paul Serret developed the theory of gauche curves. Liouville, Chasles and Minding combined with them in methodically following up Gauss's memoir. Jacobi's integration of the differential equation of the geodesic lines of the ellipsoid inspired many investigations. At the same time the problems studied in Monge's *Applications de l'Analyse* were largely developed. The determination of all surfaces having their lines of curvature plane or spherical completed in the most happy manner some of the partial results already obtained by Monge. It was now that one whom Jacobi considered the most penetrating of geometers, Gabriel Lamé (who like Charles Sturm had begun with pure geometry, and had already made the most interesting contributions to that science in a little work published in 1817, and in Memoirs which appeared in the *Annales de Gergonne*), utilised the results obtained by Dupin and Binet on the system of confocal surfaces of the second degree. He, conceiving the idea of curvilinear co-ordinates of space, created an entirely new theory which was destined to receive applications of the most varied kind in mathematical physics.

XI

Here, again, in this infinitesimal branch of Geometry we find the two tendencies which I pointed out in connection with the geometry of finite quantities. Some, among whom we must place J. Bertrand and O. Bonnet, endeavour to construct an autonomous method directly based on the use of infinitely small quantities. Bertrand's great *Traité de Calcul différentiel* contains several chapters on the theory of curves and surfaces which, in a measure, illustrate this conception. Others follow the usual analytical paths, and content themselves with clearly recognising and throwing into relief the elements which should figure in the foreground. This is what Lamé did when he introduced his theory of *differential parameters*. Beltrami followed this lead when he extended with great ingenuity the use of these differential invariants to the case of two independent variables, *i.e.*, to the study of surfaces.

Just now there seems to be a return to the mixed method, the origin of which is to be found in the works of Ribaucour under the name of *perimorphie*. The rectangular axes of Analytical Geometry are retained, but they are moveable, and connected with the system under discussion in whatever appears to be the most convenient manner. In this way most of the objections that can be levelled at the method of co-ordinates disappear. Thus are united the advantages of what is sometimes called

Intrinsic Geometry to those which result from the use of regular analysis. But, this analysis is by no means abandoned. The complicated calculations which it almost always involves in its application to the discussion of surfaces and of rectilinear co-ordinates disappear in most cases if we utilise the conceptions of the invariants and co-variants of the quadratic differential functions which we owe to the researches of Lipschitz and Christoffel, inspired by the work of Riemann in non-Euclidean Geometry.

XII

The results of such investigations are almost apparent at once. The notion that Gauss already possessed of geodesic curvature, but which he had not published, was given by Bonnet and Liouville; the theory of surfaces of which the radii of curvature are functions the one of the other, inaugurated in Germany by two propositions, which are worthy of a place in the pages of Gauss's memoir, has been enriched by Ribaucour, Halphen, Sophus Lie, and by others with many theories. Some relate to these surfaces regarded from the general point of view. Others are applied to the particular cases in which the relation between the radii of curvature takes a peculiarly simple form – to minimal surfaces, for instance, and also to surfaces of constant positive or negative curvature.

Minimal surfaces have been the object of investigations which form the most at-tractive chapter in infinitesimal geometry. The integration of their partial differential equations constitutes one of the finest discoveries of Monge; but, in consequence of the imperfect condition of the theory of imaginaries, the great geometer was unable to deduce from his formulæ any method of the generation of these surfaces, nor even of any particular surface. I shall not here return to the detailed historical sketch which I have already given in my *Leçons sur la Théorie des Surfaces;* but it is convenient to recall Bonnet's fundamental investigations, which have given us, in particular, the idea of *surfaces associated with a given surface;* Weierstrass's formulæ, which established the closest relationship between minimal surfaces and the functions of a complex variable; Lie's researches, in which he showed that Monge's own formulæ may now serve as a foundation for a fruitful survey of min-imal surfaces. By seeking to determine the minimal surfaces of very small classes or degrees we are led to the conception of the double minimal surfaces which we obtain in the *Analysis Situs.*

In this theory three problems of unequal importance have been considered. The first, relating to the determination of minimal surfaces inscribed in a given contour in a developable which is also given, has been solved by well-known formulæ, which have led to a large number of theorems. For example, every line traced on such a surface is an axis of symmetry.

The second problem, due to Sophus Lie, relates to the determination of all the algebraical minimal surfaces inscribed in an algebraical developable, when the curve of contact is not given. It has also been completely solved.

The third and most difficult is that which physicists solve by experiment, plunging a closed contour into a solution of glycerine. They have then to determine the minimal surface passing through a given contour.

The solution of this problem is clearly beyond the resources of geometry. Thanks to the resources of the most advanced analysis it has been solved for particular contours in Riemann's celebrated Memoir, and in the profound researches which followed or accompanied it. As for the most general contour, its discussion has been brilliantly commenced, and will be continued by our successors.

Next to minimal surfaces, surfaces of constant curvature were certain to attract the attention of geometers. An ingenious remark of Bonnet's connects both the surfaces, in which one or the other of the two curvatures, the mean or total curvature, is constant. Bour had asserted that the partial differential equation of surfaces of constant curvature could be completely integrated, but this does not appear to be the case. It even appears more than doubtful, if we refer to a discussion in which Sophus Lie tried in vain to apply a general method of integration of partial differential equations to the particular equation of surfaces of constant curvature. But, if it is impossible to determine in finite terms all these surfaces, at least some can be obtained characterised by special properties, such as that of having plane or spherical lines of curvature; and, by the use of a method which has succeeded in many other problems, it has been shown that from every surface of constant curvature can be derived an infinite number of other surfaces of the same nature, by clearly defined operations which only require the calculation of areas.

The theory of the deformation of surfaces in Gauss's sense has also been enriched. To Minding and Bour we owe the detailed study of the special deformation of ruled surfaces which leaves the generators rectilinear. If we could not, as I have just said, determine surfaces applicable to the sphere, at any rate we have attacked with some success other surfaces of the second degree, and, in particular, the paraboloid of revolution. The systematic study of the deformation of general surfaces of the second degree is already commenced. It is among the researches which in the near future will give very important results.

The theory of infinitely small deformation is now one of the most complete chapters in Geometry. It is the preliminary and slightly extended application of a general method which seems to have a great future before it.

Given a system of differential equations, or of partial differential equations suitable for the determination of a certain number of unknowns, it is convenient to associate with it a system of equations which I have called an *auxiliary system*, and which determines the systems of solutions infinitely near to any given system whatever of solutions. The auxiliary system being necessarily linear, its use in all researches throws valuable light on the properties of the system proposed, and on the possibility of obtaining its integration.

The theory of lines of curvature and asymptotic lines has been notably extended. Not only have these two series of lines been determined for particular surfaces, such as Lamé's tetrahedral surfaces, but also by developing Moutard's results relative to a particular class of linear partial differential equations of the second order, we have been able to generalise all that had been obtained for surfaces with plane or spherical lines of curvature, by completely determining all the classes of surfaces for which the problem of *spherical representation* can be solved. In the same way, the correlative problem relating to asymptotic lines has been solved by recognising all the surfaces of which the infinitely small deformation can be determined in finite terms. Here there is a wide field of research in which exploration has hardly begun.

The infinitesimal study of rectilinear congruences, already begun long since by Dupin, Bertrand, Hamilton, and Kummer has intervened in all these researches. Ribaucour, who took a preponderating share in these investigations, studied particular classes of rectilinear congruences, and, in particular, those which are called *isotropic*, which intervene in the happiest manner in the study of minimal surfaces.

The triply orthogonal systems employed by Lamé in Mathematical Physics have become the object of systematic research. Cayley was the first to form the partial differential equation of the third order on which the general solution of this problem had been made to depend. The system of confocal surfaces of the second degree has been generalised, and has given birth to the theory of general *cyclides*, in which we may employ the resources of Metrical Geometry, Projective Geometry, and of Infinitesimal Geometry. Many other orthogonal systems have also been discovered. Among them it is worth while to refer to the *cyclical* system of Ribaucour, in which one of the three families has circles as its orthogonal trajectories, and the more general systems for which these orthogonal trajectories are simply plane curves. The systematic use of imaginaries, which we must be extremely careful not to exclude from Geometry, has enabled us to connect all these determinations with the study of the finite deformation of a particular surface.

Among the methods which have enabled us to establish all these results, we should note the systematic use of linear partial differential equations of the second order, and the systems formed of such equations. The most recent researches show that their use will revolutionise most of the theories.

Infinitesimal Geometry was certain to affect the study of the two fundamental problems of the calculus of variations.

The problem of the shortest path on a surface was treated in the masterly papers of Jacobi and Ossian Bonnet. The study of geodesic lines has been followed up, and we have learned how to determine them for new surfaces. The theory of aggregates has intervened and has enabled us to follow these lines in their course on a given surface. The solution of a problem relating to the representation of two surfaces, one on the other, has greatly increased the interest of the discoveries of Jacobi and Liouville relative to a particular class of surfaces of which we can determine the geodesic lines. The results in this particular case led to the examination of a new question – the discovery of all the problems in the calculus of variations, the solution of which is given by curves satisfying a given differential equation.

Finally, the methods of Jacobi have been extended to space of three dimensions, and have been applied to the solution of a question which presented the greatest difficulties – the study of the minimum properties of a minimal surface passing through a given contour.

XIII

Among those who have contributed to the development of Infinitesimal Geometry, Sophus Lie is distinguished by several important discoveries which place him in the first rank. He was not one of those who showed from childhood a very marked aptitude, and when he was leaving the University of Christiania in 1865, he was still hesitating between Philology and Mathematics. It was Plücker's works which made him fully conscious for the first time of his vocation. He published, in 1869, his first paper on the interpretation of imaginaries in Geometry, and by 1870 he was in possession of the ideas which guided the whole of his career.

At that time I often had the pleasure of meeting him, and of talking to him in Paris, which he was visiting with his friend, F. Klein. A course of lectures given by M. Sylow revealed to him the full importance of the theory of substitutions. The two friends studied this theory in C. Jordan's great treatise. They were fully conscious of the important rôle that it was called upon to play in many branches of mathematical science in which it had not been as yet applied. They were both fortunate enough to succeed in giving to mathematical research the direction which appeared to them to be the most fruitful.

In 1870 Sophus Lie presented to the *Académie des Sciences* of Paris an extremely interesting discovery. Nothing is less like a sphere than a straight line, and yet Lie had imagined a singular transformation which made a sphere correspond to a straight line, and therefore enabled him to connect every theorem relative to straight lines with a theorem relative to spheres, and *vice versâ*. In this very remarkable method of transformation each property relative to the lines of curvature of a surface gives a theorem relative to the asymptotic lines of the transformed surface. The name of Lie will remain connected with these relations, which connect one to the other the straight line and the sphere, the two essential and fundamental elements of geometrical research. He developed them in a Memoir full of new ideas which appeared in 1872.

The investigations which followed this brilliant *début* fully confirmed the hopes which he had aroused. Plücker's conception of the generation of space by straight lines, by curves, or by surfaces arbitrarily chosen, opens to the theory of algebraical forms a field which has not been explored, and to which Clebsch has barely begun to assign the limits. But by the study of Infinitesimal Geometry the value of this conception was fully shown by Sophus Lie. The great Norwegian geometer first found in it the idea of congruences and complexes of curves, and then that of *contact transformations*, the first germ of which he had found (for the case of the plane) in Plücker. The study of these transformations led him, simultaneously with Mayer, to perfect the methods of integration which Jacobi had invented for partial differential equations of the first order; in particular, it throws a dazzling light on the most difficult and obscure parts of the theories relative to partial differential equations of a higher order. It enabled Lie, in particular, to integrate all the cases in which Monge's method of characteristics is fully applicable to equations of the second order with two independent variables.

By continuing the study of these special transformations, Lie was led to construct progressively his masterly theory of continuous groups of transformations, and to bring to light the important rôle played by the idea of groups in geometry. Among the essential elements of these researches I may point out the infinitesimal transformations, the idea of which was his exclusively.

Three great treatises published under his direction by skilful and devoted collaborateurs contain the essential part of his work, and its applications to the theory of integration, to that of complex unities, and to non-Euclidean Geometry.

XIV

Thus by an indirect path I have reached the non-Euclidean Geometry which is daily assuming greater importance in the researches of geometers. If I were the only speaker who is to address you on Geometry, I should be delighted to recall to your minds everything that has been done in this subject from the days of Euclid or, at any rate, from the time of Legendre up to the present day. Discussed in turn by the greatest geometers of last century, the question has progressively enlarged its borders. It began with the celebrated postulate of parallels; it ends with the geometrical axioms as a whole.

Euclid's *Elements*, which have resisted the wear and tear of so many centuries, at least will have the honour before they disappear of inspiring a long series of admirably connected treatises, which will contribute in the most efficacious manner to the progress of Mathematics. And at the same time they will provide philosophers with a sound and well-defined starting point for the study of the origin and formation of our knowledge. I am sure that my distinguished collaborator will not forget among the problems of the present day that which perhaps is the most important, and to which he has devoted himself with so much success. To him I leave the task of developing it with all the fulness it assuredly merits.

I have just spoken of the elements of Geometry. We must not forget the expansion they have undergone in the last five hundred years. The theory of polyhedra has been enriched by the beautiful discoveries of Poinsot on starred polyhedra, and by those of Möbius on single-faced polyhedra. Methods of transformation have widened the exposition of those elements. It may be said that the First Book now contains the ideas of translation and symmetry, that the second contains the theory of rotation and displacement, and that the third is based on inversion and the theory of homothetic figures.

But it must be recognised that it is owing to Analysis that the *Elements* have been enriched by their finest propositions. To the highest Analysis we owe the inscription of regular polygons of 17 sides, and analogous polygons. To it we owe the long sought-for proofs of the impossibility of the squaring of the circle, and the impossibility of certain constructions with the ruler and the compass. Finally, it is to Analysis that we owe the first rigorous proofs of the maximum and minimum properties of the sphere. Geometry will now intervene where it has been preceded by Analysis.

What will the Elements of Geometry be like at the end of the century which has just begun? Will there be one single elementary Geometry? Perhaps America, with its schools freed from the fetters of courses and traditions, will give us the best solution of this important and difficult question. Von Staudt has sometimes been called *the Euclid of the Nineteenth Century*; I should prefer to call him *the Euclid of Projective Geometry*; but will this Geometry, however interesting it may be, furnish us with the sole basis for the Elements of the future?

XV

It is now time to bring to a close a summary that has already lasted too long, and yet there is a large number of interesting investigations which I have been compelled to pass over in silence. I should have liked to have told you of Geometries of any number of dimensions, the conception of which goes back to the early days of Algebra, but the systematic study of which was only begun some 60 years ago by Cayley and Cauchy. This subject of research has found favour in your country, and I need not remind you that your illustrious President,* after having shewn himself a worthy successor of Laplace and Le Verrier, in a space which he with us considers as enjoying three dimensions, has not disdained to publish in the *American Journal of Mathematics* considerations of the greatest interest on the geometries of *n* dimensions. There is one objection, and one only, which may be levelled at researches of this kind, and it has already been raised by Poisson: the absence of any real basis, of any *substratum* which will enable us to present visibly, and as it were palpably, the real results obtained. The extension of the methods of Descriptive Geometry, and especially the use of the conceptions of Plücker on the generation of space will contribute to deprive this objection of much of its validity.

I should also like to have spoken to you of the method of equipollences, of which we find the germ in the posthumous works of Gauss, of Hamilton's Quaternions, of the methods of Grassmann, and in general of systems of complex unities, of *Analysis Situs*, so intimately connected with the theory of functions, of the so-called Kinematic Geometry, of the theory of Abaci, of Geometrography, of the applications of Geometry to Natural Philosophy and to the Arts. But I am afraid lest some analyst, as happened once before, may accuse Geometry of wishing to annex every field of research.

My admiration for Analysis, now so fruitful and so powerful, prevents my even thinking of such a step. But if any such reproach can be formulated in these days, it should not be addressed to Geometry, but, I think, to Analysis. The circle in which at the beginning of the nineteenth century the study of Mathematics appeared to be enclosed has been broken on all sides. The old problems are presented to us in a new form; new problems arise, and are investigated by legions of workers. The number of those who study Pure Geometry has become prodigiously restricted. Here is a danger against which we must guard. Let us not forget that if Analysis has acquired means of investigation which sometimes play it false, it largely owes them to the conceptions introduced by Geometers. Geometry must not remain buried, as it were, under its triumph. In its school we have learned, and our successors will have to learn, that we must never blindly trust to methods that are too general; that we must look at questions in themselves, and find in the particular conditions of each problem, either a direct road to a simple solution, or the means of applying in an appropriate manner the general processes which it is the duty of science as a whole to accumulate. As Chasles writes in the preface to his *Aperçu Historique:* "The doctrines of Pure Geometry often offer, and in a host of questions, that simple and natural method which, penetrating to the origin of truths, lays bare the mysterious chain which unites them one to the other, and makes them individually known in the most complete and luminous manner."

---

* Professor Simon Newcomb.

Let us then cultivate Geometry, which has advantages all its own, and has no desire to outstrip its rival at every point. Besides, even if we tried to neglect Geometry, it would not be long before it found, in the applications of Mathematics, as it found once before, an opportunity of regeneration and development. It is like the giant Antaeus, who grew stronger every time he touched his Mother Earth.

Desert Island Theorems

Group B: Elementary Euclidean Geometry

B1

Varignon's Theorem

EDITOR'S CHOICE

Varignon's Theorem: If W, X, Y and Z are the midpoints of the sides of an arbitrary quadrilateral $ABCD$, then $WXYZ$ is a parallelogram.

This astonishing result is attributed to Pierre Varignon (1654–1722), a French mathematician who attempted to reconcile continental infinitesimal methods with Euclidean geometry. I never came across it at school or at university but when I

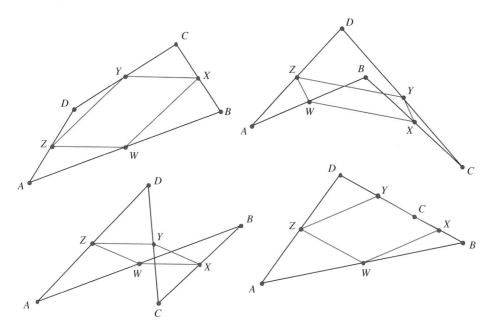

started teaching a colleague showed it to me, together with an elementary vector proof. Even with the theorem demonstrated, I could not resist sketching freehand

Chris Pritchard, Mathematics Teacher, McLaren High School. Member of the Mathematical Association Teaching Committee.

quadrilaterals at random, locating the midpoints by eye and completing each parallelogram. And even today, in an idle moment, I find myself doodling in this curious way, perhaps still seeking a quadrilateral for which Varignon's Theorem does not hold.

So what is so special about the theorem, over and above its sheer unexpectedness? It has a certain robustness in that the four vertices can be replaced by any four points in space and the midpoints still form the vertices of a plane quadrilateral. It is also a rich starting point for other inquiries: Does the theorem hold for re-entrant or crossed quadrilaterals? Is there a connection between the area of the original quadrilateral and that of the inscribed parallelogram? How is the theorem refined if we start with a kite? What can we learn by considering triangles as degenerate quadrilaterals? Above all, Varignon's Theorem epitomises simplicity, economy and elegance.

Notes

(a) If we begin with four points in space, this theorem is most easily proved by vector methods. In plane geometry, we begin by constructing AC. Then by the midpoint theorem (in $\triangle ADC$), ZY is parallel to AC and half its length. Similarly, in $\triangle ABC$, WX is parallel to AC and half its length. Hence, ZY and WX are parallel and equal. So $WXYZ$ is a parallelogram. This proof is equally valid for the re-entrant (concave) and crossed quadrilaterals.

(b) The area of parallelogram $WXYZ$ is half the area of quadrilateral $ABCD$. This result is proved in a number of texts, including that of Coxeter and Greitzer. They consider Varignon's Theorem to be 'so simple that one is surprised to find its date of publication to be as late as 1731'.

(c) If we imagine one side, CD say, of quadrilateral $ABCD$ to be reduced to zero length, its midpoint is located at one vertex of triangle ABC. This allows us to view the midpoint theorem as a special case of Varignon's Theorem.

(d) If the original shape is a kite, then the inscribed quadrilateral is a rectangle (a parallelogram with an axis of symmetry).

(e) The website: http://members.aol.com/Windmill96/varig/varig.html affords an opportunity to explore the Varignon parallelogram interactively.

Further Reading

1. H. S. M. Coxeter & S. L. Greitzer, *Geometry Revisited* (Yale, 1967).
2. Michael de Villiers, A sketchpad discovery involving areas of inscribed polygons, *Mathematics in School* **28**, 1 (March 1999), 18–21. (This splendid article promotes the use in geometry teaching of dynamic geometry software, such as the *Geometers Sketchpad* or *Cabri Géometre*. Varignon's Theorem is his starting point.)

B2

Varignon's Big Sister?

CELIA HOYLES' CHOICE

The Editor picked out Varignon's Theorem as his desert island choice. He mentioned how it was surprising and elegant. He also mentioned that it could serve as a starting point for other inquiries. My choice is a theorem that arose from exactly such an investigation. I am not sure if it has a name!

A group of us had decided to work with a dynamic geometry system to explore Varignon's Theorem in a way that might be appropriate for secondary school students. Our first attempt was simply to replicate the traditional approach – construct the quadrilateral, join the midpoints, drag and notice the parallelogram! Nice but uninspiring. We then decided to experiment. We constructed a general quadrilateral $ABCD$ and then a set of line segments, PQ, QR, RS and ST which originated at a general point, P, but were then constructed so that the vertices of the original quadrilateral were the midpoints of each line segment, as illustrated in Figures B2.1 and B2.2. As we moved P around we noticed an invariant: the vector $\overrightarrow{PT}$ was fixed (in the figures, the length happens to be 3.293).

Proving this conjecture using vectors is not hard. So why did I choose this theorem? It was because when I noticed the invariant vector and could explain it, my perspective on the original theorem completely changed. I now see Varignon as a particular case of a more general theorem, which can be stated informally as follows: every quadrilateral has associated with it a fixed vector, $\overrightarrow{PT}$, which just happens to be zero when the original quadrilateral is a parallelogram.

Editor's Note

You can explore Varignon's 'big sister' at www15.addr.com/~dscher/quad.html

Celia Hoyles, Mathematics Educationalist, Institute of Education, University of London. Chair of Joint Mathematical Council of the U.K.

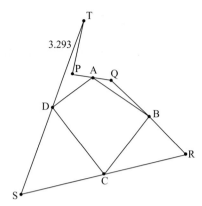

Fig. B2.1. The original construction.

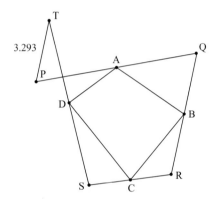

Fig. B2.2. *P* has moved, but $\overrightarrow{PT}$ is the same length, and seems to be in the same direction.

B3

The Mid-Edges Theorem

TONI BEARDON'S CHOICE

Here we discuss two very different geometrical situations in which results are proved using precisely the same algebra. For the first of these, consider any set of n points in the plane, no three of which are collinear. Is it possible to construct a polygon such that the given points are the midpoints of the edges of the polygon? If so, is the polygon unique? The answer is 'yes' on both counts when n is odd but in general 'no' when n is even. We give the geometrical constructions and proofs using vectors which can be generalized, in different ways, to higher dimensions.

It is always possible to construct a triangle given only the midpoints of its edges. Given the points a, b and c then the vertex p (see Figure B3.1) of the required triangle is found by constructing the line segment from a to p parallel, and equal in length, to the line segment from c to b. Thus the vector p is given by $p = a + b - c$. Once one vertex has been located, the other vertices of the polygon are determined. This construction depends on the fact that line segments joining the midpoints of the edges of a triangle are parallel to, and half the lengths of, the edges of the triangle.

If a quadrilateral $pqrs$ exists with a, b, c and d as the midpoints of the edges then the line segment from s to q is parallel to, and twice the length of, both the line segments a to b and d to c so that if $abcd$ is not a parallelogram (see Figure B3.2a) then no such quadrilateral can exist. If $abcd$ is a parallelogram then there are infinitely many such quadrilaterals and two super-imposed examples are shown in Figure B3.2b.

To construct the pentagon $pqrst$ given the five points a, b, c, d and e we draw the line segment from a to x parallel and equal to the line segment from b to c and then the line segment from x to p parallel and equal to the line segment from d to e (see Figure B3.3). This locates the point p and the rest of the pentagon follows. The construction depends on the fact that if x is the midpoint of the line joining s to p then the line segment from x to p is parallel and equal to the line segment from

Toni Beardon, Faculty of Education, University of Cambridge. NRICH Project Leader.

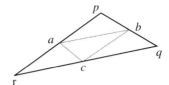

Fig. B3.1.

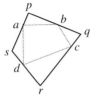

Fig. B3.2a.

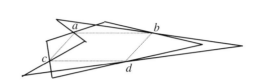

Fig. B3.2b.

d to e and similarly the line segment from a to x is parallel and equal to the line segment from b to c as both these line segments are parallel to and half the length of the line segment from q to s.

This construction depends on the relationships

$$p + q = 2a$$
$$q + r = 2b$$
$$r + s = 2c$$
$$s + t = 2d$$
$$t + p = 2e$$

Hence $p = a - b + c - d + e$. The method generalizes for any odd number of points and Figure B3.4 shows the construction for a nonagon.

Another geometrical approach is to take an arbitrary point x, and successive reflections: $2a - x$, the reflection in a, then $2b - 2a + x$, $2c - 2b + 2a - x$, $2d - 2c + 2b - 2a + x$, $2e - 2d + 2c - 2b + 2a - x$, the reflections in b, c, d and e respectively. Equating the final reflected point with the original point x gives the solution which is the midpoint of x and the last of these reflections $2e - 2d + 2c - 2b + 2a - x$. It is clear that this method will generalize for an odd number of points but that with an even number of points $a_1, a_2, \ldots, a_{2n}$ a solution only exists where $a_1 - a_2 + a_3 - \cdots - a_{2n} = 0$.

Theorem Given n points in space (in any dimension) if n is odd then there exists a unique polygon with the given points as the midpoints of its edges. If n is even then either there is no polygon with this property or there are infinitely many polygons with this property.

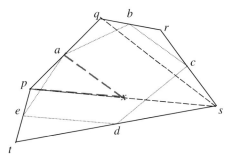

Fig. B3.3.

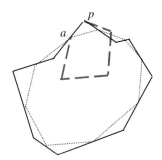

Fig. B3.4.

Proof Suppose the given points are $a_1, a_2, a_3, \ldots, a_n$. If a polygon $p_1 p_2 p_3 \ldots p_n$ with the required property exists then

$$p_1 + p_2 = 2a_1$$
$$p_2 + p_3 = 2a_2$$
$$p_3 + p_4 = 2a_3$$
$$\cdots$$
$$\cdots$$
$$p_n + p_1 = 2a_n.$$

We can write this set of equations in matrix form:

$$\begin{pmatrix} 1 & 1 & 0 & 0 & \cdots & 0 & 0 \\ 0 & 1 & 1 & 0 & \cdots & 0 & 0 \\ 0 & 0 & 1 & 1 & \cdots & 0 & 0 \\ \vdots & \vdots & \vdots & \vdots & \ddots & \vdots & \vdots \\ 0 & 0 & 0 & 0 & \cdots & 1 & 1 \\ 1 & 0 & 0 & 0 & \cdots & 0 & 1 \end{pmatrix} \begin{pmatrix} p_1 \\ p_2 \\ p_3 \\ \vdots \\ p_{n-1} \\ p_n \end{pmatrix} = 2 \begin{pmatrix} a_1 \\ a_2 \\ a_3 \\ \vdots \\ a_{n-1} \\ a_n \end{pmatrix}.$$

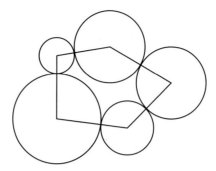

Fig. B3.5.

It is easily seen (by expanding by the first column) that the determinant of the matrix on the left has the value 2 for all odd values of n and the value zero for all even values of n. Hence there is a unique solution whenever n is odd and either no solution or infinitely many solutions whenever n is even.

Now consider a different question. Given any polygon, is it possible to draw circles with centres at the vertices so that each circle just touches its two neighbours forming a 'necklace' of circles, as in Figure B3.5 for example?

Suppose this can be done for an n-gon where the lengths of the sides are $a_1, a_2, \ldots, a_n$. We want to find the radii of the circles $r_1, r_2, \ldots, r_n$. If a solution exists then

$$r_1 + r_2 = a_1$$
$$r_2 + r_3 = a_2$$
$$r_3 + r_4 = a_3$$
$$\ldots$$
$$\ldots$$
$$r_n + r_1 = a_n.$$

This gives the same system of equations as before. If n is odd then a unique solution exists and if n is even then either there is no solution or there are infinitely many solutions. In some cases the values of r turn out to be negative where one circle lies inside another and touches it internally.

Another way to generalize this problem is to regard the midpoint of a segment as its centroid. Thus we could, for example, take four points in 3-space and seek a tetrahedron such that the centroids of its faces are at given points. There is always a unique solution in this case. We leave the reader to explore the situation for more than four points.

The unity of mathematics is shown by these two geometrical applications arising from the same algebra. Secondary school students can tackle these problems for small values of n working with coordinates in two dimensions. The author has worked with masterclasses where the students have used dynamic geometry applets to investigate both problems. See the NRICH website www.nrich.maths.org from February and March 2000:

http://nrich.maths.org/mathsf/journalf/feb00/stage4.html\#357
http://nrich.maths.org/mathsf/journalf/mar00/stage4.html\#366

Matrix algebra is not needed. This provides an interesting context for an extension of routine classwork on simultaneous equations to three, four, five and more variables. The masterclass students have been very excited to be able to find the radii of the circles by solving sets of linear equations (the simplest possible equations with unit coefficients) and coming to appreciate the Mid-Edges Theorem in its generality.

B4

Van Schooten's Theorem

DOUG FRENCH'S CHOICE

Van Schooten's Theorem is both simple and surprising and, like all good theorems, it can be proved in a variety of ways. An equilateral triangle ABC is inscribed in a circle and each vertex is joined to a point P, which can move along the minor arc BC. The surprise is that $AP = BP + CP$, a result which can be discovered readily in the classroom using dynamic geometry software and then proved. I give three proofs, the first two of which are readily accessible at school level.

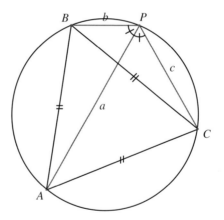

Proof 1 Let $AP = a$, $BP = b$, $CP = c$ and let the sides of the equilateral triangle ABC be d. Angle $APB =$ angle $APC = 60°$ since both angles are in the same

Doug French, Mathematics Educationalist, University of Hull. Chair of the Mathematical Association Teaching Committee.

segments of the circle respectively as angles C and B of the equilateral triangle ABC. Using the cosine rule in triangles APB and APC we then have:

$$d^2 = a^2 + b^2 - ab \quad \text{and} \quad d^2 = a^2 + c^2 - ac.$$

Subtracting and rearranging gives:

$$a(b - c) = b^2 - c^2$$
$$a = b + c, \quad \text{if} \quad b \neq c$$

In the case where $b = c$, the two triangles are right angled and it is easy to see that

$$a = 2b = 2c.$$

Proof 2

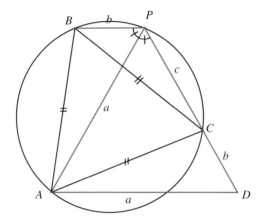

If a line is drawn from A to meet PC produced at D, with angle $PAD = 60°$, then the triangle APD is equilateral and $PD = a$. It is now easy to show that triangles APB and ADC are congruent and hence that $CD = b$.

It then follows, by considering PD, that $a = b + c$.

Proof 3 It is interesting to note that Van Schooten's Theorem follows directly from Ptolemy's Theorem, which is not well known these days. Indeed I did not encounter it at school in the 1950s. Ptolemy's Theorem states that in a cyclic quadrilateral the product of the diagonals is equal to the sum of the products of the two pairs of opposite sides.

Considering the cyclic quadrilateral $ABPC$, this gives $ad = bd + cd$ from which it immediately follows that $a = b + c$.

Perhaps there is a case for including Ptolemy's Theorem in the school geometry curriculum!

Editor's Note

Tony Crilly and Colin Fletcher have chosen Ptolemy's Theorem as their Desert Island Theorem.

B5

Ceva's Theorem

ELMER REES' CHOICE

Let A, B, C be the vertices of a triangle and λ, μ, ν be numbers. If P, Q, R are points on BC, CA, AB respectively and dividing them in the ratio μ/ν, ν/λ, λ/μ respectively, then the lines AP, BQ, CR are concurrent.

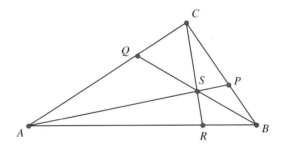

To my mind a good proof should not only verify the truth of the theorem but help to understand its statement. The proof that follows is a particularly good one from this point of view. But first, let me set the scene.

Consider a set of points and represent one of them as a pair $(\mathbf{X}, m)$ where the point is the end point of the vector $\mathbf{X}$ and m is the mass at that point. A pair of such weighted points can for many purposes be replaced by a single weight at their centre of gravity. If $(\mathbf{X}_1, m_1)$ and $(\mathbf{X}_2, m_2)$ are such points, they can be replaced by $(\mathbf{X}, m)$ where $m = m_1 + m_2$ and $m\mathbf{X} = m_1\mathbf{X}_1 + m_2\mathbf{X}_2$. This 'addition' has a number of standard algebraic properties:

1. It is commutative.
2. It has additive 'identity' elements (for each $\mathbf{X}$, $(\mathbf{X}, 0)$ is such an element).
3. Every point $(\mathbf{X}, m)$ has an inverse $(\mathbf{X}, -m)$, if negative weights are allowed.
4. It is associative.

Elmer Rees, Geometer, University of Edinburgh.

These are straightforward to verify but, as is often the case, it is a little more messy to verify associativity than the other properties. However, it is the verification of associativity which yields a proof of Ceva's Theorem.

Proof Let **S** be the vector such that

$$(\mathbf{S}, \lambda + \mu + \nu) = (\mathbf{A}, \lambda) + (\mathbf{B}, \mu) + (\mathbf{C}, \nu).$$

The right hand side equals $(\mathbf{A}, \lambda) + (\mathbf{P}, \mu + \nu)$ and so the end point S of **S** lies on the line AP. Similarly, S lies on the lines BQ and CR. So, AP, BQ and CR are concurrent at S and Ceva's Theorem is proved.

The converse (also found by Ceva) involves taking arbitrary points P, Q, R on the sides of the triangle such that the lines AP, BQ, CR are concurrent and checking that the product of the three ratios is 1. This also works well using the same addition.

Editor's Notes

1. Ceva's Theorem was discovered by Giovanni Ceva (1648–1734) and published in *De lineis rectis se invicem secantibus* in 1678. It is closely associated with Menelaus' Theorem which Ceva also rediscovered. Ceva's geometrical work was subsequently extended by Matthew Stewart (1717–1785), who succeeded Colin Maclaurin at Edinburgh University.
2. An alternative statement of Ceva's Theorem and its converse is:
   AP, BQ and CR are concurrent (at S) if, and only if, $\frac{AR}{RB} \cdot \frac{BP}{PC} \cdot \frac{CQ}{QA} = 1$.

Further Reading from the *Mathematical Gazette*

A. K. Srinivasar, On Menelaus' Theorem, Ceva's Theorem and the harmonic property of a quadrilateral, Note 2118, *Math. Gaz.* **34** (February 1950), 51–52.

John Slatterly, The nedians of a plane triangle, Note 2392, *Math. Gaz.* **38** (May 1954), 111–113.

R. M. Walker, Two theorems deduced from the theorems of Ceva and Carnot, Note 2706, *Math. Gaz.* **41** (October 1957), 206–208.

E. A. Maxwell, The theorems of Menelaus and Ceva, *Math. Gaz.* **45** (December 1961), 333–334.

D. R. Dickinson, The Theorems of Ceva and Menelaus and the principle of duality, Note 121, *Math. Gaz.* **48** (December 1964), 427–429.

E. M. Hunter, Some standard results, Note 190, *Math. Gaz.* **53** (May 1969), 159–160.

F. Chorlton, Extension of the Theorem of Menelaus, Note 3284, *Math. Gaz.* **54** (December 1970), 394–395.

W. J. Courcouf, Back to areals, *Math. Gaz.* **57** (February 1973), 46–51.

J. R. Goggins, Perimeter bisectors, Note 70.17, *Math. Gaz.* **70** (June 1986), 133–134.

Larry Hoehn, A simple generalisation of Ceva's theorem, Note 73.21, *Math. Gaz.* **73** (June 1989), 126–127.

Peter Griffin & John Mason, Walls and windows, *Math.Gaz.* **74** (October 1990), 260–269.

G. C. Shephard, The compleat Ceva, *Math. Gaz.* **83** (March 1999), 74–81.

John R. Silvester, Ceva = (Menelaus)^2, Note 84.30, *Math. Gaz.* **84** (July 2000), 268–271.

B6

The Descartes Circle Theorem

H. S. M. COXETER'S CHOICE

In the real Euclidean plane, if four circles are mutually tangent, their curvatures ε_1, ε_2, ε_3, ε_4 (the reciprocals of their radii, with suitable conventions of sign depending on external or internal contact) satisfy the equation

$$\left(\sum \varepsilon_n\right)^2 = 2 \sum \varepsilon_n^2. \tag{1}$$

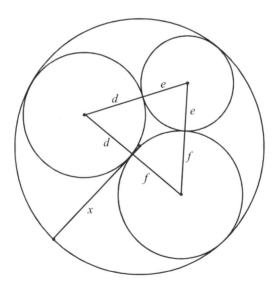

This theorem seems to have originated about 1640, when Princess Elisabeth of Bohemia considered three mutually tangent circles with radii d, e, f and sought the radius x of a fourth circle tangent to all three. In a letter of 1643, her mentor

H. S. M. Coxeter, Geometer, University of Toronto.

René Descartes [8, pp. 48–49] explained that x is given by the roots of the quadratic equation

$$ddeeff + ddeexx + ddffxx + eeffxx = 2deffxx + 2ddeffx + 2deefxx + 2ddeefx + 2ddefxx + 2deeffx$$

which Leibniz, fifty years later, would probably have expressed more concisely as

$$
\begin{vmatrix}
-d^2 & de & df & dx & 1 \\
ed & -e^2 & ef & ex & 1 \\
fd & fe & -f^2 & fx & 1 \\
xd & xe & xf & -x^2 & 1 \\
1 & 1 & 1 & 1 & 0
\end{vmatrix} = 0
\quad \text{or} \quad
\begin{vmatrix}
-1 & 1 & 1 & 1 & \varepsilon_1 \\
1 & -1 & 1 & 1 & \varepsilon_2 \\
1 & 1 & -1 & 1 & \varepsilon_3 \\
1 & 1 & 1 & -1 & \varepsilon_4 \\
\varepsilon_1 & \varepsilon_2 & \varepsilon_3 & \varepsilon_4 & 0
\end{vmatrix} = 0
$$

where ε_1, ε_2, ε_3, ε_4 are the reciprocals of d, e, f, x [4, pp. 104–105]. By ingenious reasoning, Daniel Pedoe [9, pp. 630, 634] showed how easily this can be reduced to

$$2\left(\varepsilon_1^2 + \varepsilon_2^2 + \varepsilon_3^2 + \varepsilon_4^2\right) - (\varepsilon_1 + \varepsilon_2 + \varepsilon_3 + \varepsilon_4)^2 = 0;$$

thus expressing the equation, which he named the 'Descartes Circle Theorem', in the simple form (1).

About 1840, Equation (1) was rediscovered by Philip Beecroft [1; 4, pp. vii, 14–15; 7, pp. 110–112] as a consequence of a more comprehensive investigation in which any tetrad of mutually tangent circles, with curvatures ε_1, ε_2, ε_3, ε_4, is related to another 'complementary' tetrad, with curvatures η_1, η_2, η_3, η_4, such that

$$
\begin{aligned}
2\eta_1 &= -\varepsilon_1 + \varepsilon_2 + \varepsilon_3 + \varepsilon_4, & 2\varepsilon_1 &= -\eta_1 + \eta_2 + \eta_3 + \eta_4, \\
2\eta_2 &= \varepsilon_1 - \varepsilon_2 + \varepsilon_3 + \varepsilon_4, & 2\varepsilon_2 &= \eta_1 - \eta_2 + \eta_3 + \eta_4, \\
2\eta_3 &= \varepsilon_1 + \varepsilon_2 - \varepsilon_3 + \varepsilon_4, & 2\varepsilon_3 &= \eta_1 + \eta_2 - \eta_3 + \eta_4, \\
2\eta_4 &= \varepsilon_1 + \varepsilon_2 + \varepsilon_3 - \varepsilon_4, & 2\varepsilon_4 &= \eta_1 + \eta_2 + \eta_3 - \eta_4.
\end{aligned}
$$

About 1930, the Descartes Circle Theorem was again rediscovered by Frederick Soddy (1877–1956) who had won the Nobel Prize for chemistry in 1921. After retiring from his professorship in Oxford, he returned to his birthplace on the Sussex coast and took up the geometry of circles and spheres as a hobby. He obtained Equation (1) quite independently, wrote it up as a poem, and submitted it to *Nature*, where it appears as 'The kiss precise' [10]. Here is its middle verse:

> Four circles to the kissing come,
> The smaller are the benter,
> The bend is just the inverse of
> The distance from the centre.
> Though their intrigue left Euclid dumb

> There's now no need for rule of thumb,
> Since zero bend's a dead straight line
> And concave bends have minus sign,
> *The sum of the squares of all four bends*
> *Is half the square of their sum.*

I still possess a letter of 1951 in which Soddy modestly confesses that this, and another such poem, 'The hexlet' [11], were '*tour de force*, hammered out by sheer algebra and luck. They depend on the reduction of a biquadratic equation of I think 23 terms to a quadratic by a transformation I have never really understood.'

As a graduate student in Cambridge, I was so impressed by Soddy's poem that I wrote to him to let me visit him and take him out to lunch. He accepted and accompanied me for a delightful walk along the beach. After my departure to Canada, I kept in touch with him by correspondence for the rest of his life; and his ideas influenced four of my published works [2; 3; 5; 6].

References

1. Philip Beecroft, Properties of circles in mutual contact, *The Lady's and Gentleman's Diary* (1843), 91–96.
2. H. S. M. Coxeter, Interlocked rings of spheres, *Scripta Math.* **18** (1951), 113–121.
3. H. S. M. Coxeter, Loxodromic sequences of tangent spheres, *Aequationes Math.* **1** (1968), 104–121.
4. H. S. M. Coxeter, *Introduction to Geometry*, 2nd edn. New York: Wiley, (1969). For a proof of the Descartes Circle Theorem see pp. 11–16.
5. H. S. M. Coxeter, Numerical distances among the spheres in a loxodromic sequence, *Math. Intelligencer* **19.4** (1997), 41–47.
6. H. S. M. Coxeter, Numerical distances among the circles in a loxodromic sequence, *Nieuw Archief voor Wiskunde* **16** (1998), 1–9.
7. H. S. M. Coxeter, Five spheres in mutual contact, *J. Geometry and Graphics* **2** (2000), 109–114.
8. René Descartes, *Œuvres* IV (Correspondence 1643–1647) (Paris: Leopold Serf, 1901), 45–50.
9. Daniel Pedoe, On a theorem in geometry, *Amer. Math. Monthly* **74** (1967), 627–640.
10. Frederick Soddy, The kiss precise, *Nature* **137** (1936), 1021.
11. Frederick Soddy, The hexlet, *Nature* **138** (1936), 958. For details of Soddy's life see Linda Merricks, An invisible man: On writing biography, *History Workshop Journal* **37** (1994), 194–204.

Editor's Notes

By way of an elementary example, take mutually tangent circles of radii 1, 2, and 3 units respectively. Two circles are drawn so as to touch each of the three, one externally, the other internally, as shown.

Surprisingly, for these particular circles, the centres form the vertices of a Pythagorean triangle (of sides 3, 4, 5 units).

Here $d = 1$, $e = 2$, $f = 3$ and so $\varepsilon_1 = 1$, $\varepsilon_2 = \frac{1}{2}$, $\varepsilon_3 = \frac{1}{3}$, $\varepsilon_4 = \frac{1}{x}$.

Substituting into (1) and rearranging yields $23x^2 + 132x - 36 = 0$.

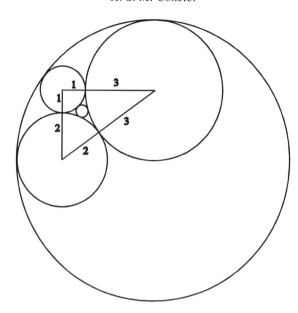

The roots of this quadratic equation, $\frac{6}{23}$ and -6, yield the radii of the kissing circles, the signs indicating whether the kiss is internal or external.

Further Reading from the *Mathematical Gazette*

D. W. Allen, Rectangling the circles, *Math. Gaz.* **53** (February 1969), 54.

Seamus Bellow, The ever-decreasing circles, *Math. Gaz.* **81** (March 1997), 104–108.

J. M. Child, Some interesting sets of circles, *Math. Gaz.* **32** (May 1848), 52–58.

Michael Fox, Chains, froths and a ten-bead necklace: Systems of circles and spheres, *Math. Gaz.* **84** (July 2000), 242–259.

A. Gardiner, $\frac{1}{1.2} + \frac{1}{2.3} + \frac{1}{3.4} + \cdots + \frac{1}{n(n+1)} = \frac{n}{n+1}$, *Math. Gaz.* **67** (March 1983), 50–52.

Mark Harvey, Ever decreasing circles and inversion, *Math. Gaz.* **82** (November 1988), 472–475.

K. E. Hirst, The kiss precise, *Math. Gaz.* **53** (October 1969), 305–308.

E. H. Neville, A curious rectangle, *Math. Gaz.* **22** (July 1938), 288–291.

P. N. Ruane, The curious rectangles of Rollett and Rees, *Math. Gaz.* **85** (July 2001), 208–225. This article offers an historical overview of kissing circles, including their treatment in the pages of the *Mathematical Gazette* over the years. It contains numerous additional references.

B7

Three Squares Theorem

BILL RICHARDSON'S CHOICE

This is a very neat and surprising result which has a long pedigree in Mathematical Association publications. The result concerns the diagram below.

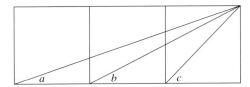

In the diagram, formed from three squares, there is a very simple relation between the angles, namely:

$$a + b = c.$$

Clearly $c = 45°$ so this means that $a + b$ is $45°$.

No doubt a 'modern student' would use a calculator and be happy for the total of a and b to be 45. But, that is not a proof. So what proofs are there? The treatments in the references fall broadly into two main camps. As is to be expected in the descendent of the Association for the Improvement of Geometrical Teaching, the majority (1, 2, 5, 8, 9, 10) are geometrical. Three (3, 6, 7) are essentially trigonometrical although vectors make an appearance in (3). I present two proofs to give a flavour but an interested reader will gain much from consulting the references.

First Proof

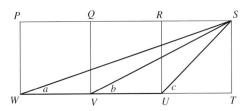

Bill Richardson, Chair of Council, The Mathematical Association. President (1996–1997) and Editor in Chief of Mathematical Association Publications (1997–2000).

Let each square have side x units.

Then, applying Pythagoras' Theorem where needed,

$$WU = 2x; \qquad US = \sqrt{2}x, \quad \text{so} \quad \frac{WU}{US} = \frac{2x}{\sqrt{2}x} = \sqrt{2}$$

$$US = \sqrt{2}x; \qquad VU = x, \quad \text{so} \quad \frac{US}{VU} = \frac{\sqrt{2}x}{x} = \sqrt{2}$$

$$WS = \sqrt{10}x; \qquad VS = \sqrt{5}x, \quad \text{so} \quad \frac{WS}{VS} = \frac{\sqrt{10}x}{\sqrt{5}x} = \sqrt{2}$$

Thus $\triangle$'s VUS, SUW are similar and so $\angle WSU = \angle SVU = b$ but, an exterior angle of a triangle is equal to the sum of the opposite interior angles i.e.

$$\angle SUT = \angle UWS + \angle WSU$$
$$c = a + b$$

Second Proof

$$a + b = \tan^{-1} \frac{1}{3} + \tan^{-1} \frac{1}{2} = \tan^{-1} \frac{\frac{1}{3} + \frac{1}{2}}{1 - \frac{1}{3} \times \frac{1}{2}} = \tan^{-1} 1 = c,$$

making use of the standard result

$$\tan^{-1} x + \tan^{-1} y = \tan^{-1} \frac{x + y}{1 - xy}.$$

References

1. Roger North, Four heads are better than one, *Math. Gaz.* **57** (December 1973) 334–336.
2. Douglas Quadling, The story of the three squares, *Math. Gaz.* **58** (October 1974) 212–215 (Editor's Note).
3. D. M. Hallowes, $\tan^{-1} \frac{1}{2} + \tan^{-1} \frac{1}{3} = 45°$, Note 62.8, *Math. Gaz.* **62** (March 1978) 53–54.
4. D. B. Eperson, Three right-angled triangles, *Mathematics in School* **12**, 1 (January 1983), 27, 35.
5. Bob Burn, The Orton-Flower tessellation, *Math. Gaz.* **74** (December 1990) 372–373.
6. A. J. G. May, Angles whose sum is $\pi/4$, *Math. Gaz.* **53** (May 1969) 157.
7. C. Pritchard, Bolstering a pillow problem, *Mathematics in School* **25**, 2 (March 1996) 21.
8. Jack Oliver, Two trisection problems, *Mathematics in School* **26**, 1 (January 1997) 9.
9. Tony Barnard, Letter to editor, *Mathematics in School* **26**, 5 (November 1997) 47.
10. Johnston Anderson, The three-square problem – a double take, *Mathematics in School* **27**, 3 (May 1998) 6.

B8

Morley's Triangle Theorem

DAVID BURGHES' CHOICE

When asked to choose a favourite geometry theorem, I had no hesitation in naming Morley's famous theorem which states that,

'the three points of intersection of the adjacent trisectors of the angles of ANY triangle form an equilateral triangle'.

My choice was not influenced by the proof but by the surprising nature of this result, i.e, from chaos, that is, from *any* triangle, you always obtain the most symmetric one! I hasten to add that I am not a Pure Mathematician, and that for those who are, this result may not be a surprise, but for the average, intelligent amateur mathematician, it undoubtedly is unexpected. When deciding on my choice, I had not considered the problem of obtaining an elegant proof. A few hours in my university library provided many varied and, at first sight, rather complicated strategies. Most of the proofs start with the equilateral triangle and 'build out', but this seems to me to be rather contrived, so here, with thanks to Alexander Bogomolny and Roger Webster, is a reasonably straightforward proof which requires no high level mathematical knowledge.

Proof Our method is to start with the angles and sides of triangle ABC, and find identical expressions for the sides of triangle RP, PQ and QR.

For simplicity, we let the angles at A, B and C be $3a$, $3b$ and $3c$. Note that this means that

$$\boxed{a + b + c = 60°} \tag{1}$$

Consider the radius of the circle circumscribing triangle ABC. Let O be its centre and r its radius and, without loss in generality, we can take $r = 1$. It is easy

David Burghes, Director, Centre for Innovation in Mathematics Teaching, University of Exeter.

195

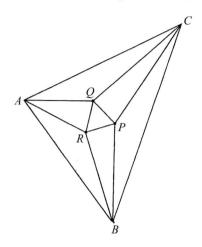

to see that

$$AB = 2r \sin 3c$$

i.e. $\boxed{AB = 2 \sin 3c}$

Similarly,

(2)

$$\boxed{BC = 2 \sin 3a, \, CA = 2 \sin 3b}$$

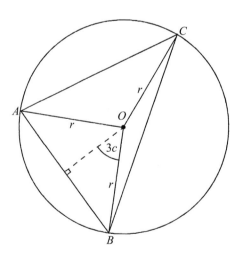

Returning to our first diagram, using the sine rule, we have

$$\frac{BP}{\sin c} = \frac{BC}{\sin(180 - b - c)}$$
$$= \frac{2 \sin 3a}{\sin(b + c)} \quad \text{using (2)}$$

$$= \frac{2\sin 3a}{\sin(60 - a)} \quad \text{using (1)}$$

So
$$\boxed{BP = \frac{2\sin 3a \sin c}{\sin(60 - a)}} \tag{3}$$

This can be rewritten by noting that

$$\sin(60 + a)\sin(60 - a) = (\sin 60\cos a + \cos 60\sin a)(\sin 60\cos a - \cos 60\sin a)$$

$$= \left(\frac{\sqrt{3}}{2}\cos a + \frac{1}{2}\sin a\right)\left(\frac{\sqrt{3}}{2}\cos a - \frac{1}{2}\sin a\right)$$

$$= \frac{3}{4}\cos^2 a - \frac{1}{4}\sin^2 a$$

$$= \frac{3}{4} - \sin^2 a$$

and

$$\sin a \sin(60 + a)\sin(60 - a) = \frac{3}{4}\sin a - \sin^3 a = \frac{1}{4}\sin 3a$$

Hence, substituting for sin 30 in (3) above,

$$\boxed{BP = 8\sin c \sin a \sin(60 + a)}$$

Similarly

$$\boxed{BR = 8\sin a \sin c \sin(60 + c)} \tag{4}$$

and we can now use the cosine law in triangle BPR to give

$$RP^2 = BP^2 + BR^2 - 2 \times BP \times BR \times \cos b$$

or

$$RP^2 = 64\sin^2 c \sin^2 a \sin^2(60 + a) + 64\sin^2 a \sin^2 c \sin^2(60 + c) -$$
$$2 \times 64 \times \sin^2 a \sin^2 c \sin(60 + a)\sin(60 + b)\cos b$$
$$= 64\sin^2 c \sin^2 a[\sin^2(60 + a) + \sin^2(60 + c) - 2\sin(60 + a) \times$$
$$\sin(60 + b)\cos b] \tag{5}$$

Now consider the family of triangles formed by using the angles $60 + a$, $60 + c$ and b (note that $(60 + a) + (60 + c) + b = 120 + a + b + c = 180$).

Without loss of generality, we can assume a circumscribed circle of radius 1, so that its sides are $\sin b$, $\sin(60 + a)$ and $\sin(60 + c)$. Now using the cosine law in the triangle gives

$$\sin^2 b = \sin^2(60 + a) + \sin^2(60 + c) - 2\sin(60 + a)\sin(60 + c)\cos b \tag{6}$$

Substituting (6) in (5) gives

$$RP^2 = 64 \sin^2 c \sin^2 a \sin^2 b$$

and, of course, we will also, in a similar way, obtain identical expressions for PQ^2 and QR^2. So

$$\boxed{RP = PQ = QR = 8 \sin a \sin b \sin c}$$

and triangle RPQ is equilateral.

So we have not only a surprising theorem, but also (for me, at least) an interesting and challenging proof!

Editor's Notes

1. For further details about Frank Morley's theorem, see Alexander Bogomolny's website: www.cut-the-knot.com/triangle/Morley/CenterCircle.html It appears that Morley probably discovered the theorem in the 1890s, alluded to it in passing in a paper of 1900 but only laid claim to it much later. It was fully elucidated in the journal of the *Mathematical Association of Japan for Secondary Mathematics* (vol. 6, December 1924).

2. David Burghes used the 'desert island' theme for his inaugural lecture at Exeter University. The lecture was published by the Institute of Mathematics and Its Applications: David Burghes, Desert Island Mathematics, *Bull. IMA* **18** (1982), 130–139.

III

Pythagoras' Theorem

3.1

Introductory Essay: Pythagoras' Theorem, A Measure of Gold

JANET JAGGER

> Geometry has two great treasures: one is the Theorem of Pythagoras; the other, the division of a line into extreme and mean ratio. The first we may compare to a measure of gold; the second we may name a precious jewel.
>
> Johannes Kepler, *Mysterium Cosmographicum*, 1596

> Neither thirty years, nor thirty centuries, affect the charm of Pythagoras' theorem, it is as dazzlingly beautiful now as it was in the day when Pythagoras first discovered it.
>
> Charles Dodgson (Lewis Carroll)

Pythagoras' Theorem is perhaps the only residual memory of school geometry for many people, with some, sadly, remembering only the name itself. This chapter explores some of the charms and mysteries of Pythagoras, the man, and the richness of his theorem. It includes an introductory biography by Rouse Ball, published in 1915, a number of demonstrations and proofs with some extensions and associated results.

As Rouse Ball reminds us, Pythagoras' mathematical achievements are far more than just this theorem. He created the first systematic exposition of geometry and his most far-reaching discovery was that there were incommensurable numbers; a proof of this, that is thought to be his, is still taught to school and university students. Pythagoras' main work in arithmetic concentrated upon integers and ratios, i.e. rational numbers. Irrational numbers such as $\sqrt{2}$ could be constructed, e.g. as the diagonal of a unit square, but not measured, so that these numbers were embedded in geometrical methods; commercial arithmetic involved the use of the abacus and was for the market place.

Walter Rouse Ball was a Fellow of Trinity College, Cambridge from 1878 to 1905. His authority as an historian was founded on his highly regarded, interpretive *Essay on Newton's 'Principia'*. He also wrote an eminently readable popular history of mathematics which went through many editions. But his most successful book, both in his own day and since, is *Mathematical Recreations and Essays*, first published in 1892 and running to fourteen editions. The last four of these editions were revised by Coxeter, a notable contributor to this centenary volume. No doubt, the then editor of the *Mathematical Gazette* was delighted when Rouse Ball agreed to pen this well- rounded biographical sketch of Pythagoras during the dark days of World War I.

Pythagoras' Theorem provides a fundamental relationship between the squares on the sides of a right-angled triangle, *viz* in any right-angled triangle, the sum of the areas of the squares on the two shorter sides is equal to the area of the square on the longest side or hypotenuse. Algebraically, this is $c^2 = a^2 + b^2$.

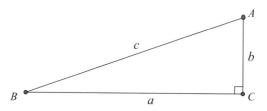

Four very different demonstrations and proofs of Pythagoras' Theorem are presented in the articles in this chapter and I shall provide a further two in this introduction. I begin with the very first dissection proof which was given by Thâbit ibn Qurra; he was a ninth-century Mesopotamian mathematician and scientist well versed in the mathematics of the Greeks. His dissection is encapsulated by the following diagram:

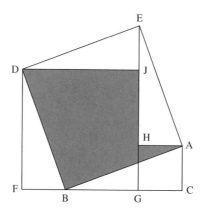

Triangle ABC is the right-angled triangle, the three squares are $ACGH$, $GFDJ$ and $ABDE$. Thâbit refers to this method as that of 'reduction and composition' or, more explicitly, as the 'method of reduction to triangles and re-arrangement by juxtaposition'. The sum of the first two squares is given by adding triangles DFB and BCA to the shaded pentagon $AHJDB$; the square on the hypotenuse is obtained by adding the triangles HAE and DJE to the pentagon, and the four triangles are identical. (Further details are to be found in an article by Sayılı [1].)

The most famous dissection proof was first published by the London stockbroker, Henry Perigal, in 1873 (see [2]) and this is generalised by A. W. Siddons in a short article of 1932. Dissection demonstrations fall into two categories, those which require the constituent shapes to be translated and rotated and those for which translation alone suffices. The Thâbit-Perigal-Siddons demonstration belongs to that latter, arguably more appealing, group but, as you will see, we are reminded of the most pleasing configuration of all by the sliding blocks of Roger Baker's article.

Some proofs of Pythagoras' Theorem involve the properties of similar triangles. This was certainly the approach of Einstein as a young lad before he set eyes on the *Elements*, as he explained in an autobiographical sketch:

I remember that an uncle told me the Pythagorean theorem before the holy geometry booklet had come into my hands. After much effort I succeeded in 'proving' this theorem on the basis of the similarity of triangles . . . for anyone who experiences [these feelings] for the first time, it is marvellous enough that man is capable at all to reach such a degree of certainty and purity in pure thinking as the Greeks showed us for the first time to be possible in geometry.

(Albert Einstein: Philosopher-Scientist, P A Schilpp (ed.), 1951)

The proof by Oliver involves the in-circle of the right-angled triangle and the areas of triangles of which the right-angled triangle is comprised. In Hoehn's extension of Pythagoras (discussed again later), he gives a beautifully simple proof of Pythagoras' Theorem itself, being derived in one line from the secant-tangent property of circles. Readers wishing to explore a wider range of proofs should refer to Loomis [3] or to another article by Hoehn written for *The Mathematics Teacher* [4], or else, check out the website at www.cut-the-knot.com/pythagoras

Until the 1960s, many schoolchildren were taught Euclid's proof of Pythagoras' Theorem. This is Proposition 47 of the first book of the *Elements*, and in previous centuries was simply referred to as Euclid I, 47. The proof involved the diagram known, variously, as the bride's chair, the windmill, the peacock's tail or the Franciscan cowl, as noted by Webster in this chapter.

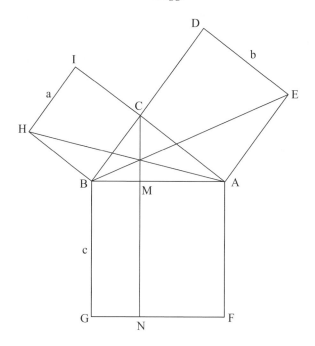

Here is a sketch of Euclid's proof:

First note that square $ACDE$ and triangle EAB share a common base EA and have the same height AC. Therefore, the area of the square $ACDE$ is twice the area of the triangle EAB. I write this in the abbreviated form

$$\text{square } ACDE = 2 \times \triangle EAB.$$

With the line CF added to the diagram, it can be seen that triangles EAB and ACF are congruent, i.e. identical in shape and size. Thus,

$$\text{square } ACDE = 2 \times \triangle ACF.$$

Triangle ACF and rectangle $AFNM$ share the same base AF and have the same height AM. Therefore,

$$\text{square } ACDE = 2 \times \triangle ACF = \text{rectangle } AFNM.$$

Similarly, with the extra line CG, it can be proved that

$$\text{square } BCIH = \text{rectangle } BGNM.$$

Thus: square $ACDE$ + square $BCIH$ = square $ABGF$, proving Pythagoras' Theorem.

The converse of Pythagoras' Theorem is also true: in any triangle, if the area of the square on one side is equal to the sum of the areas of the squares on the other two sides, then these two sides contain a right angle and the triangle is right-angled. Perfectly logically, Euclid stated and proved the converse immediately after the theorem: so this is Euclid I, 48. The converse may be proved as follows:-

We have a $\triangle ABC$ with $BA^2 = a^2 + b^2$. Construct a line CD at right angles to BC and equal in length to CA. We have to prove that angle BCA must be $90°$, though initially we draw it so that it does not look like a right angle.

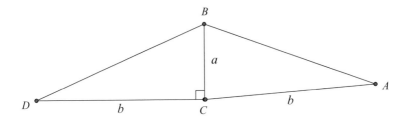

In $\triangle BCD$,

$$BD^2 = a^2 + b^2 \quad \text{(Pythagoras' Theorem)}$$

We also know that

$$BA^2 = a^2 + b^2 \quad \text{(given)}$$

Hence

$$BD = BA.$$

So $\triangle BCD$ and $\triangle BCA$ are congruent, having sides of the same length.
Therefore $\angle BCA = \angle BCD = 90°$ and $\triangle BCA$ is right-angled at C as required.

We have now explored the theorem and its converse, which incidentally are the last propositions in Book I, and now turn to the various extensions, both modern and old, and other associated results that have been published in *Mathematical Association* journals over the past century.

An obvious extension arises because if $c^2 = a^2 + b^2$ then $kc^2 = ka^2 + kb^2$ for some scale factor k. As a result, we can replace the squares in Pythagoras' Theorem by any shapes on the three sides providing that they are geometrically similar, and the theorem still holds (Euclid VI, 31). For instance, we may consider semicircles drawn on the three sides, or perhaps equilateral triangles. About 440 BC, Hippocrates of Chios – not to be confused with the physician Hippocrates of

Cos – chose similar segments of circles. He took an isosceles right-angled triangle ABC with $AC = BC$ and inscribed the triangle in a semicircle on AB as diameter.

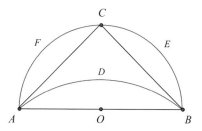

A segment is drawn on AB, geometrically similar to the segments AFC and CEB on AC and CB respectively. Pythagoras' Theorem is then used to calculate the area of the crescent enclosed by the curves ADB and ACB, i.e. the lune $AFCEBD$, as follows:

Since the three segments are similar, Pythagoras' Theorem gives

$$\text{area segment } AFC + \text{area segment } CEB = \text{area segment } ADB$$

Now,

$$\begin{aligned} \text{area lune } AFCEBD &= \text{area semicircle on } AB - \text{segment } ADB \\ &= \text{area semicircle on } AB - (\text{segments } AFC \text{ and } CEB) \\ &= \text{triangle } ABC. \end{aligned}$$

Hippocrates was able to find the area of this triangle, since it was half the square on BC (or AC) and thus knew the area of the lune. (For an alternative view of Hippocrates' approach, based on an amended figure, see William Dunham's treatment [5], pp. 1–26.)

The Greeks struggled in vain to construct squares equal in area to a given circle using only their designated instruments, compasses and a straight-edge but Hippocrates' first quadrature of a curvilinear figure gave them hope that one day they would indeed 'square the circle'. Whilst the futility of their attempts would not be revealed until the nineteenth century, a great deal of interesting mathematics was produced in the process.

The most commonly used extension of Pythagoras' Theorem is the Cosine Rule. This was given by Euclid as Propositions 12 and 13 in Book II of the *Elements*. It was formulated in a purely geometrical way, of course. Proposition 12 deals with the obtuse-angled triangle:

In obtuse-angled triangles, the square on the side subtending the obtuse angle is greater than the squares on the sides containing the obtuse angle by twice the rectangle contained by one of the sides about the obtuse angle, namely that on which the perpendicular falls, and the straight line cut off outside the perpendicular toward the obtuse angle.

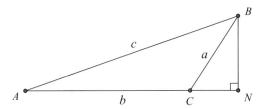

The 'rectangle' in this case is the area of the rectangle formed by the lengths AC and CN, and Euclid's statement of the Cosine Rule is equivalent to:

$$c^2 = a^2 + b^2 + 2 \times AC \times CN.$$

Proposition 13, for the acute-angled triangle is similar. Euclid proved these propositions by applying Pythagoras' Theorem to two right-angled triangles, just as we do now. In this chapter, Bibby and French explore the three cases of the Cosine Rule, for the acute-angled, right-angled and obtuse-angled triangles from a purely geometric viewpoint which gives a deeper qualitative, rather than quantitative, understanding of this theorem.

Thâbit ibn Qurra, who was mentioned above in connection with the first dissection proof of Pythagoras' Theorem, also gave an extension which is not well known but is true for any triangle, right-angled or not. Consider first an obtuse-angled triangle ABC, with points X and Y lying on BC so that $\angle AXB = \angle AYC = \angle BAC$. Then Thâbit's Theorem states that $AB^2 + AC^2 = BC(BX + YC)$. This is readily proved using similar triangles.

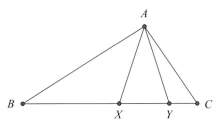

If $\angle BAC = 90°$, then the points X and Y coincide and we have the standard Pythagoras' Theorem. If it is acute, then X and Y are on BC in the order B, Y, X, C so that line segments BX and BY overlap. The theorem still holds.

Larry Hoehn's extension – his 'Pythagorean-like formula' – involves an isosceles triangle BDA with a line BC meeting AD in C.

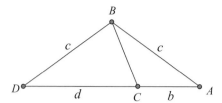

The formula is $c^2 = a^2 + bd$. With $b = d$, i.e. BC perpendicular to AC, this becomes Pythagoras' Theorem itself. Hoehn explains that he found his theorem when looking at the many proofs of Pythagoras' Theorem, and he includes his elegant proof mentioned above.

An obvious three-dimensional version of Pythagoras' Theorem is that the length, L, of the diagonal of a rectangular box of dimensions a, b and c is given by $L^2 = a^2 + b^2 + c^2$. A more interesting three-dimensional version is found in the twin articles by Hull and Perfect. It states that:

If a tetrahedron $OABC$ is rectangular at the vertex O, so that all three edges meeting at O are mutually perpendicular, then

$$\alpha^2 = \beta^2 + \gamma^2 + \delta^2,$$

where α, β, γ and δ denote the areas of the faces opposite the vertices O, A, B and C respectively.

Hull gives two different proofs and Perfect provides a beautiful vector version. I gave this theorem to my students and one of them produced a neat proof requiring only the 'half base times height' formula for the area of a triangle. Here it is, with minor re-drafting:

Using Hull's Figure 3.8.1, choose the plane perpendicular to AB which passes through OC. Then if AB cuts this plane in N, both ON and CN are perpendicular to AB. Let $CN = h$, $ON = k$ and $AB = c$.

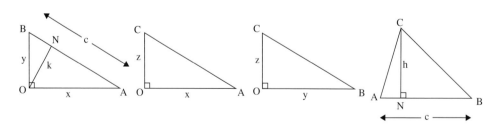

$$\triangle ABC = \frac{1}{2}ch \quad \text{and} \quad h^2 = z^2 + k^2$$

$$(\triangle ABC)^2 = \frac{1}{4}c^2h^2 = \frac{1}{4}c^2(z^2+k^2) = \frac{1}{4}c^2z^2 + \frac{1}{4}c^2k^2$$
$$= \frac{1}{4}z^2(x^2+y^2) + \frac{1}{4}c^2k^2 = \frac{1}{4}z^2x^2 + \frac{1}{4}z^2y^2 + \frac{1}{4}c^2k^2$$
$$= (\triangle OAC)^2 + (\triangle OBC)^2 + (\triangle OAB)^2 \quad \text{as required.}$$

Pythagoras' Theorem itself has suggested a number of other theorems, which are not direct extensions. For instance, in the second of his articles, Hoehn looks at the squares **inside** a right-angled triangle and discovers a neat theorem about their perimeters. Dixon extends the bride's chair by drawing two more squares on the outside. He examines the **difference** in area between two squares given by the expression $a^2 - b^2$ instead of the **sum** of two squares, $a^2 + b^2$. He also generalises this for a non-right angled triangle and obtains two further interesting results.

Webster presents a simple result about the areas of triangles associated with the bride's chair and, as he points out, there is no requirement that the original triangle be right-angled. It is interesting to note that if we begin with an equilateral triangle, then the whole figure takes the form of a hexagon which is equiangular but not regular. There are not many examples of figures that are equiangular but not regular, except of course the ubiquitous rectangle. These figures help to illustrate that regular polygons are equilateral and equiangular. Warburton extends Webster's article and develops two properties of the altitudes of the central triangle and the medians of the surrounding triangles; the second of them involves both scalar and vector products.

Pythagoras' Theorem is still very important in many different areas of science and technology, from the very ordinary setting of woodwork to advanced research in engineering. In pure mathematics, it provides the distance function for the Euclidean space in which we live. As Jacob Bronowski put it in his BBC television series, *The Ascent of Man*, [6, pp. 160–161]:

To this day, Pythagoras' theorem remains the most important single theorem in the whole of mathematics. That seems a bold and extraordinary thing to say, yet it is not extravagant; because what Pythagoras established is a fundamental characterisation of the space in which we move ... [T]he numbers that compose right-angled triangles have been proposed as messages which we might send out to planets in other star systems as a test for the existence of rational life there.

Bronowski's 'numbers that compose right-angled triangles', are *Pythagorean triples*: sets of three integers that satisfy the relationship $c^2 = a^2 + b^2$. If we relax the condition that the numbers are integers, there is an uncountable infinity of 'trios' which satisfy the equation. One such trio consists of the lengths of the sides of three regular polygons inscribed in the same circle, namely the pentagon, the hexagon and the decagon. This pleasing theorem appears as Proposition 10 in the last book of Euclid's *Elements*, Book XIII.

The reader will readily appreciate the beauty and the variety of the geometrical results in the following articles, all of them linked to Pythagoras and his theorem.

References

1. Aydın Sayılı, Thâbit ibn Qurra's generalization of the Pythagorean Theorem, *Isis* **51** (1960), 35–37.
2. Henry Perigal, On geometrical dissections and transformations, *Messenger of Mathematics* **2** (1873), 103–105.
3. Elisha Scott Loomis, *The Pythagorean Proposition* (Washington, DC: National Council of Teachers of Mathematics, 1968).
4. Larry Hoehn, The Pythagorean Theorem: An infinite number of proofs, *The Mathematics Teacher* **90**, 6 (September 1997), 438–441; with correction in the same journal, **91**, 1 (January 1998), 73.
5. William Dunham, *Journey Through Genius: The Great Theorems of Mathematics* (Penguin, 1991).
6. Jacob Bronowski, *The Ascent of Man* (BBC Publications, 1973).

3.2

Pythagoras

WALTER ROUSE BALL

The *Mathematical Gazette* has commenced a series of articles, adapted for the use of school-teachers, on the great mathematicians of former days. No series of the kind could be complete were Pythagoras omitted for it was he who raised mathematics to the rank of a science, made it a part of liberal education, and evolved a scheme on which it continued to be studied in Europe for more than 2000 years. But though his claim to inclusion is indisputable, the materials for describing his work are sadly lacking. We have a treatise by Aristotle on Pythagorean teaching, but the biography by the wife of Pythagoras is lost, and only fragments of the sketches by Aristoxenus, Dikaiarchus, and Philolaus exist, while the memoirs by Laertius Diogenes, Iamblichus and Porphyry, on which we have mainly to rely for personal details, are late and include much that is palpably untrue. The reputation of Pythagoras far exceeded that of any of his contemporaries, and so great was it that in the credulous ages miracles and magical feats were freely attributed to him; later these additions tended to make men go to the other extreme, and doubt the truth of everything written about him. Within recent years, however, the available materials have been subjected to critical examination, and the researches of T. Gomperz of Vienna, J. Burnet of St. Andrews, G. J. Allman of Dublin, and P. Tannery of Paris enable us to speak with more confidence than was formerly possible.

What we know for certain about Pythagoras's life comes to little more than saying that he was born about 570 B.C., his father being Mnesarchus, a prosperous tradesman of Samos; that he was a contemporary of Polycrates, the Tyrant of that State; that he travelled widely, and acquired great fame for his learning and moral teaching; that finally he settled at Croton in South Italy, where he founded an Order, whose members obtained for a time paramount political power; and that on the development of bitter opposition to his School he retired to Metapontum, where he died about 501 B.C. To these bald statements we may add that his opponents

First published in *Mathematical Gazette* **8** (January 1915); pp. 5–12.

at Croton finally triumphed, but that later his followers re-established themselves as a philosophical Society, and for a century or more profoundly influenced Greek thought.

There are, however, many traditions about Pythagoras and while frankly recognizing that the evidence is not good, I give a somewhat fuller outline, credible and even probable, of his career. Subject to this caution, we may say that Pythagoras came of Tyrrhenian stock: his father was a prosperous goldsmith and lapidary of Samos; his mother a clever woman, who, at his birth dedicated her boy to the service of the Gods. His father had extensive business connections in Asia Minor, and, probably through these, the lad made the acquaintance of various philosophers of the Ionian School. On growing up he went to Egypt, taking with him letters of recommendation from Polycrates to Amasis the reigning Pharaoh and by the aid of the latter monarch secured admission to the College of Priests at Memphis or Diospolis, where he mastered the secrets of Egyptian science and religion. Thence he went to Babylon, and there learnt something of Persian and Indian thought.

Pythagoras had a burning zeal for knowledge. We may be certain that he was attracted by the geometrical discoveries of the Ionian School, which included the earliest attempts to give general proofs of geometrical propositions covering all particular instances, as also by the formulae and arithmetical rules known to the Egyptians and Babylonians. But I suspect that he was even more interested in the systems of religion as expounded by Egyptian, Assyrian, Persian, Jew, and Indian priests, and in the philosophical speculations of Thales and his School on the nature and form of the world.

He returned to Samos, perhaps about 535 B.C., with a great reputation for piety and learning. It is said that about this time he visited Delphi, and secured the consent of the priests to various much-needed reforms which he suggested. If true, this is a striking testimony to his fame and influence. It may be from this incident that he was subsequently regarded as being specially under the protection of Apollo. About 529 B.C. he migrated to Sicily, whence he removed first to Tarentum and then to Croton.

It was at Croton that his life-work was done. Just before his arrival, that city had been defeated by its ancient rival, Sybaris. It was universally felt that reforms were needed. On this ground, so well prepared, the seed of his teaching bore rapid fruit. The high standard of conduct which he advocated attracted wide attention, and finally he was given an opportunity to explain his views to the Senate. With their consent he formed a Brotherhood, and its establishment was followed by a revival of public Spirit, and a cleansing of public life. Probably the Senate wished to use for their own purposes his influence with the Commonalty, and deemed it cheaply secured by their recognition of his teaching. His Order, ruled absolutely by him, was primarily a religious fraternity, but had also educational, social, scientific, and ethical sides: it was open to women as well as men. Branches were founded in neighbouring cities. Our chief interest in it today arises from its scientific achievements. The influence of a great community holding itself aloof from the general mass of citizens inevitably led to bitter opposition to which was added the personal resentment of those candidates for admission to it whom Pythagoras rejected. Somewhere about 501 B.C. Pythagoras died at Metapontum, a neighbouring city to which he had temporarily moved. Shortly afterwards there were popular risings against the Pythagoreans throughout Southern Italy, and a house at Croton in which the leading members had taken refuge was burnt, many of his followers losing their lives.

With this tragedy the political career of the Order ended. It was, however, re-established as a philosophical Society, and the chief thinkers of the time were either members of it, or like the mathematicians Hippocrates and Eudoxus (founders of the Schools of Athens and Cyzicus) and the philosophers Socrates and Plato, were greatly influenced by its doctrines. We may take the work of the four men last mentioned as opening new chapters in the history of mathematics and philosophy, and as marking the end, early in the fourth century before Christ, of the original Pythagorean School.

It will help to fix our ideas about the time at which Pythagoras lived if I add that it is said that he was an intimate friend of Hermodamus, whose grandfather had known Homer, and that in Assyria he may well have met Daniel, whose career, as told in the Old Testament, is familiar to us. We have no real evidence as to the appearance of Pythagoras, but according to tradition he was a tall, handsome, grave, bearded man, usually clad entirely in white save for a purple belt and purple headgear. Undoubtedly he was a persuasive speaker, and could sway popular audiences as well as small bodies of experts. The authority he exercised among his contemporaries in so many subjects and ways is evidence of exceptional powers: he was at once a prophet, statesman, philosopher, and man of science. The accounts of the constitution of his Order, the ranks of membership, the ceremonies of initiation and advancement, and its internal regulations, while not improbable in themselves, must be received with great caution.

Primarily Pythagoras was a religious teacher and philosopher, inculcating a stern system of morality, greatly superior to any system then known to the Greeks. For reasons which I outline below he held that the ultimate explanation of things rested on a knowledge of numbers and form, that thus the study of philosophy must necessarily be associated with the study of pure and applied mathematics, and that these latter also provided the best general mental training and discipline. I am not here concerned with his ethical or philosophical opinions, of which indeed our knowledge is vague. Even on the teaching of his School in science we cannot speak with absolute assurance, while the further fact that his instruction was entirely oral introduces the possibility that commentators have failed to distinguish his conclusions from those made later by his followers. Thus the precise extent of his discoveries is still a matter of opinion, but I think we may reasonably attribute to him the results I proceed to describe. That he created mathematics as a science there is no doubt.

We cannot say in what order Pythagoras made his discoveries, but I conjecture that he was led to the study of geometry and numbers through his researches in natural philosophy. Of these, his investigations on acoustics were the most important.

Throughout his life, Pythagoras was profoundly moved by music, and in his School he taught that it was one of the chief means for exciting and calming emotions – it purged the soul, said his followers, as medicine purges the body. By a happy chance he constructed or came across an instrument consisting of a string stretched over a vibrating board with a movable bridge by which the string could be divided into different lengths. To his delight he found that other things being equal the note given by the vibrating string depended only on its length, and that the lengths that gave a note, its fourth, its fifth, and it octave were in the ratio 6, 8, 9, and 12. Thus suddenly the whole of an intangible and artistic world seemed reduced to a question of numbers.

It was not unnatural that he should suspect that other physical phenomena were explicable by similar means. This suspicion was strengthened by his noticing that the seasons, the months, the tides, day and night, sleep and waking, the pulse, breathing, etc. are periodic. Hence he argued that whatever be their explanation, it must involve a consideration of numbers, and he is said to have wondered whether the regular rhythmic sequences of the world would not ultimately be found to be analogous to the regular breathing of animals.

He appears to have been confirmed in the idea that fortune had placed in his hands the key to the riddle of the universe by considering the manner in which geometricians think of points, lines, and planes. We arrive at these ideas by a process of abstraction, considering a plane as a boundary of a polyhedron, a line as a boundary of a polygon, and a point as a boundary of a line, but he went further and suggested, Aristotle tells us, that in fact points, lines, and planes are more real than the concrete forms or figures from which we obtain our conception of them. He identified a point with unity; and he associated a line (which is determined by two points) with the number 2; a plane (which is determined by three points) with the number 3; and a solid body with the number 4. Thus geometry seemed reducible to numbers and form.

So far Pythagoras' work rested on a sound foundation, but without evidence he extended his conclusions to other subjects, and perhaps at last came to believe that everything is connected with or typified by a particular number, and that through that number alone (if we can find it) can knowledge of the thing itself be obtained.

According to tradition, some of these guesses, connecting particular qualities and ideas with particular numbers, were that justice was associated with the number 4; that the cause of colour was to be discovered in the properties of the number 5; and that friendship and harmony were explicable by the octave. We gather from Aristotle that many of these notions were put forward by Pythagoras in the latter years of his life, and it is believed that some of the wilder ones were due to his successors. We may at any rate say that he went further in associating particular things with particular numbers than the evidence justified, and that the mistake was magnified and extended by his School.

It was inevitable that so acute a thinker should consider the explanation of the more obvious astronomical phenomena. The Ionian philosophers had thought of the earth as a disc-like body floating on water, but Pythagoras taught that the earth and moon were spheres. Whether he extended the conception to the sun and planets is uncertain, nor do we know the details of his views on the subject, but it seems probable that he sought to resolve all astronomical phenomena into the circular motion of spherical bodies. Even in this crude form the suggestion showed scientific imagination of a high order.

By way of parenthesis I should add that this conception was subsequently developed, and his successors taught that the centre of the celestial sphere was occupied by a perpetual fire round which circled nine spherical bodies, namely, the earth, the moon, the sun, the five planets then recognized, and the firmament. They explained the apparent daily motion from east to west of the sun and moon as being really due to the more rapid motion of the earth from west to east. In order to bring the number of the celestial bodies to the mystic number ten, it was supposed that another planet, known as the counter-earth circled round the central fire below the earth, always concealed from view by our unvisited hemisphere. The distances of this decade of celestial bodies were assumed to be in musical progression, and their motions were described politically as set to the music arising from their movements and known as the harmony of the spheres. This extension was subsequent to the time of Pythagoras, and the introduction of the counter-earth, for which no evidence existed, was justly condemned by Aristotle and other philosophers.

It will be seen from the above sketch that Pythagoras discussed some problems in acoustics, common periodic sequences, geometrical conceptions, and astronomical phenomena, and that his conclusions led him to seek the explanation of everything in the theories of geometry and numbers. Before his age all that was known in geometry consisted of facts obvious to every observer (such as that two intersecting circles cut in two points) and a few isolated theorems due to the Ionian School (such as that the angles at the base of an isosceles triangle are equal). Of the then extant knowledge of the theory of numbers we speak with less certainty, but most likely it included results, known only to the initiated, about certain fractions and series.

He raised both subjects to the rank of sciences. His discussion of numbers has been superseded by other and better ways of dealing with the subject, but his treatment of geometry was adopted by Euclid, and still forms the basis of modern elementary text books on it. I now proceed to describe his work on these subjects, on which today rests his chief claim to distinction.

On his geometry I have little to add to what I have said elsewhere. He probably knew and taught the substance of what is contained in the first two books of Euclid about parallels, triangles, and parallelograms, and was acquainted with a few other isolated theorems. He taught that magnitudes may be represented by lines, and hence the value of general geometrical propositions.

It is hardly necessary to say that we are unable to reproduce the whole body of Pythagoras' teaching on geometry, but we gather from the notes of Proclus on Euclid, and from occasional remarks by other writers, that it included the following propositions, and such others as are required to prove them. (i) It commenced with a number of statements about mathematical conceptions: one has been preserved in the definition of a point as unity having position. (ii) He showed that the plane space about a point can be completely filled by equilateral triangles, by squares, and by regular hexagons – results that must have been familiar wherever tiles of these shapes were in common use. (iii) He proved that the sum of the angles of a triangle is equal to two right angles (Euc. i. 32): and in the demonstration, which has been preserved, the results of the propositions, Euc. i. 13 and the first part of Euc. i. 29, are quoted. The proof is substantially the same as that given by Euclid, and it is most likely that the proofs there given of the two propositions last mentioned are also due to Pythagoras himself. (iv) He established the properties of right-angled triangles which are given in Euc. i. 47 and i. 48, and the first of these propositions has since always been definitely associated with his name. We do not know how he proved it, though we are told that his demonstration is not that given in Euclid's *Elements*, but it has been observed that the theorem follows at once from the results of Euc. ii. 2, vi. 4, and vi. 17, with all of which Pythagoras was acquainted. (v) He is credited with the discovery of the theorems, Euc. i. 44 and i. 45, on the description of a parallelogram equal to one and similar to another parallelogram: his successors were aware of the extension given in Euc. vi. 25. (vi) He gave a construction for finding the geometrical mean of two lines (Euc. ii. 14). (vii) Of the five regular solids inscribable in a sphere, he was acquainted with four, and gave constructions for them. Probably he held that the elements of which the material world is made are related to these solids. (viii) He appears to have regarded the theory of proportion as a branch of geometry, but we know nothing about his work on it, and very likely he did not discuss it generally. He may have known that similar polygons are to each other in the ratio of their homologous sides, but it seems more likely that the discovery of this and the properties of the pentagon and dodecahedron were made by his successors. We may be confident that he did not deal with the properties of the circle, though he taught that the circle was the most perfect of plane figures, and the sphere the most perfect of solids. Subsequent Greek geometry was built on the foundation thus laid by Pythagoras. No doubt the presentation of the subject was improved and made more logical and rigorous by Hippocrates and Euclid. But this does not destroy the admiration felt for the man who created the first systematic exposition of the subject.

Perhaps the most far-reaching of his discoveries was that there were incommensurable numbers – a result which profoundly affected the subsequent development of Greek mathematics. The substance of the proof attributed to him may be put as follows. If a denotes the number of units of length in the diagonal of a square and b the number of units of length in a side, we have $a^2 = 2b^2$. If each of these numbers is a fraction or integer, we can, by multiplying by the L.C.M. of their denominators, bring this relation to a form in which a and b are integers. Further, we can strike out any factor common to a and b, and thus we may take it that a and b are prime to each other. Hence if one is even the other is odd. Now a^2 is even, hence a is even, and therefore b must be odd. But if a is even, let $a = 2c$; then $4c^2 = 2b^2$ and therefore $b^2 = 2c^2$. Hence b^2 is even, and therefore b is even. Thus b must at the same time be even and odd, which is impossible. Thus a and b must be incommensurable.

It follows that it may be impossible to measure a line in terms of a given unit of measurement. This led him to distrust demonstrations which rest on the possibility of making numerical measurements, for such measurements might not be accurate. Such magnitudes can, however, be represented geometrically by lines, and hence the absorption of theoretical arithmetic in geometry, which is characteristic of Greek mathematics. This geometrical treatment is illustrated by the proofs, probably due to Pythagoras, given in the second book of Euclid of propositions like $(a + b)^2 = a^2 + 2ab + b^2$. I suspect that most of the proofs given by Euclid which involve the use of gnomons and their parts are Pythagorean in origin.

I proceed next to describe Pythagoras' researches on the science of numbers involving the investigation of properties of integers and their ratios. By way of preamble I should explain that he was not concerned with mercantile arithmetic, which necessarily formed part of the education of men of affairs, and in which the abacus was generally used. So too he was not concerned with problems in which there was a possibility of the introduction of incommensurable numbers: for these, geometrical methods were used. The particular problems with which he was concerned dealt with the properties of integers, the ratios of integers, special groups of inter-related integers, triangular integers, the factors of integers, and numbers in series.

Pythagoras commenced his discussion by pointing out the fundamental difference between geometry which deals with continuous quantities and arithmetic which

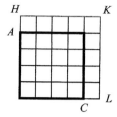

deals with discrete quantities like integers. He divided all integers into even and odd, the odd numbers being termed gnomons. Curiously enough he went on to say that the odd integers were limited or finite and the even integers were unlimited or infinite: it is useless to speculate what he meant by this. An odd number, such as $2n + 1$, was regarded as the difference of two square numbers, $(n + 1)^2$ and n^2; and the sum of the gnomons from 1 to $2n + 1$ was thus shown to be a square number, namely $(n + 1)^2$; its square root was termed a side. Products of two numbers were called plane, and if a product had no exact square root it was termed an oblong. A product of three numbers was called a solid number, and, if the three numbers were equal, a cube. All this has obvious reference to geometry, and the opinion is confirmed by Aristotle's remark that when a gnomon is put round a square the figure remains a square though it is increased in dimensions. Thus, in the figure given above, in which n is taken equal to 4, the gnomon AKC (containing 9 cells) when put round the square AC (containing 4^2 cells) makes a square HL (containing 5^2 cells).

Pythagoras is said to have discussed ratio and proportion. We must take this in a very limited sense, for it is certain that he was ignorant of the classical Greek method invented more than a century later by Eudoxus, and set out in Euclid's *Elements*. I take it that Pythagoras' work was confined to the treatment of vulgar fractions, and dealt mainly with the reduction of fractions of the form a/b to a sum of fractions each of whose numerators is unity. Such problems played a considerable part in Egyptian priestly science. For example, the priests knew that 2/97 is the sum of 1/56, 1/679, and 1/776. The fact is that the early mathematicians tried to evade the difficulty of having to consider at the same time changes in both the numerator and denominator of fractions: they liked to have all the numerators or else all the denominators equal.

He dealt with particular groups of numbers, for instance, sets like 6, 8, 9, 12, which gave the lengths of chords producing a note its fourth, its fifth; and its octave, and sets like $(2n^2 + 2n + 1), (2n^2 + 2n)$ and $(2n + 1)$, which are proportional to the sides of certain right-angled triangles. Probably he did not know the more general expressions for similar sets like

$$a, 2ab/(a + b), (a + b)/2, b \quad \text{and} \quad (m^2 + n^2), 2mn, (m^2 - n^2).$$

Pythagoras was acquainted with triangular numbers. It is not clear whether he also discussed polygonal numbers: his followers certainly did so. A triangular number represents the sum of a number of counters laid in rows on a plane: the bottom row containing n, and each succeeding row one less than the row before it; it is therefore equal to the sum of the series

$$n + (n - 1) + (n - 2) + \cdots + 2 + 1.$$

Thus the triangular number corresponding to 4 is 10: this was known as the tetrad. This is the explanation of the language of Pythagoras in the well known passage in Lucian where the merchant asks Pythagoras what he can teach him. Pythagoras replies, "I will teach you how to count." Merchant, "I know that already." Pythagoras, "How do you count?" Merchant, "One, two, three, four, ..." Pythagoras, "Stop! what you take to be four is ten, a perfect triangle, and the symbol of our oath." Pythagoras remarked that this mystic ten is the radix of notation generally adopted by mankind, and he pointed out that 10 is the lowest integer up to and including which there are an equal number of prime and composite numbers. To-day the tetrad appears in the arms of the Duchy of Cornwall and of the See of Worcester.

The Greek symbolism for numbers did not lend itself to operative uses, and was not employed for that purpose. Euclid, by choice, represented numbers by lines. The

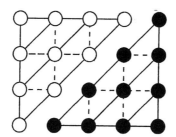

method by which Pythagoras proved properties of polygonal numbers was quite different, and rested on arranging pebbles (or beads strung on an abacus) in such a way as to give an ocular demonstration: it was still used by the Pythagoreans in the time of Eurytus, and is described by Aristotle. This method is not only interesting in the history of the subject, but offers demonstrations which are easily understood and are general in character. For instance the sum of two triangular numbers of the 4th order when expressed by pebbles arranged as shown in the annexed diagram is 4×5, i.e. 20, and thus each is equal to 10. A similar diagram shows that a triangular number of the nth order is equal to $n(n + 1)/2$. Various similar numerical results which can be at once demonstrated in this way have been given by Lucas, for instance: eight times a triangular number of the order n augmented by unity is a square of the order $2n + 1$; a pentagonal number of the order n is equal to n augmented by three times a triangular number of the order $n - 1$; an hexagonal number of the order n is equal to n augmented by four times a triangular number of the order $n-1$. As to the work of the Pythagoreans on the factors of numbers we know very little: they classified numbers by comparing them with the sum of their integral subdivisors or factors, calling a number excessive, perfect, or defective, according as it was greater than, equal to, or less than the sum of these subdivisors. Two numbers were called amicable if each was equal to the sum of the subdivisors of the other; they knew that 220 and 284 afforded an example of an amicable pair. Traces of their work on these subjects still appear in text-books. These researches may well have been expected to lead to valuable results, but in fact little came of them.

Pythagoras recognized arithmetic, geometric, and harmonic series. Perhaps he learnt about the first two of these from Babylonian sources. He may have known how to sum an arithmetic series, but probably except for this he was not acquainted with any general theorems on such series.

The subjects described above were classified by Pythagoras as dealing with numbers absolute or Pythagorean arithmetic, numbers applied or music, magnitudes at rest or geometry, and magnitudes in motion or astronomy. This quadrivium was long considered as constituting a necessary course of study for a liberal education. All his disciples were required to study these subjects, and on them was based his scheme of education and philosophy.

Pythagoras and Thales were regarded as the founders of mathematics and philosophy, but the achievements of the former were much more remarkable than those of the older man, and their great value was recognized by the Greeks. Pythagoras made geometry, said Eudemus, into a liberal science, treated its principles from an abstract point of view, and investigated its theorems in a general and intellectual manner: he further discovered incommensurables, and the construction of the regular solids. The value of his arithmetical work was equally recognized. As a brilliant mathematician, the discoverer of the fundamental principles of acoustics, an acute student of astronomical phenomena, and a great ethical and religious teacher, his career stands unique. No one will deny him the distinction of having been one of the most original and influential thinkers of the ancient world. He has, said his contemporary, Heraclitus, when inveighing against him, practised research and inquiry more than all other men. Traces of his teaching still survive in our books, and the very name of mathematics, by which we call our science, is of his invention. It is true that the whole of his teaching was not sound, and that his theories were mixed with fanciful speculations, dangerous in the case of geometry and arithmetic, which are founded on inferences unconsciously made and common to all men, but almost necessarily fatal in the applied sciences, which can rest safely only on the results of conscious observation and experiment. We must also admit that he extended his results beyond what facts warranted: later generations of Greeks recognized this, and condemned his followers for adapting facts to suit their preconceived theories. But if finally he let his imagination run away with him, we must remember that he lived in a remote age, in which science had not begun, and there were no critics to exercise a sobering influence on his wilder assumptions. Notwithstanding these defects, it is not I think too much to say that he initiated the brilliant course of Greek philosophy and science.

The fact that he played a conspicuous part in politics led to opposition to his views during his life, but after his death the loftiness of his aims was generally admitted. According to Aristoxenus, his School gloried in the fact that they sought knowledge rather than wealth. There is reason to think that the classification of men into those that trade for profit, those that compete for honours, and those who observe, comes from Pythagoras himself, and that he held up to his disciples as the highest life a disinterested search for truth.

3.3

Perigal's Dissection for the Theorem of Pythagoras

A.W. SIDDONS

The idea for the following note was suggested to me by a figure sent to me by Miss M. Charlesworth, aged sixteen.

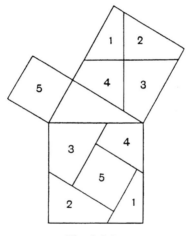

Fig. 3.3.1.

In Perigal's dissection for the theorem of Pythagoras the parallel and perpendicular to the hypotenuse are drawn through the centre of the larger of the two squares on the sides; but it is not necessary for the lines to be drawn through that point, as the figures show. Other interesting cases arise when the square on the smaller

First published in *Mathematical Gazette* **16** (February 1932); p. 44.

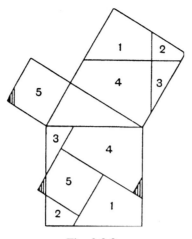

Fig. 3.3.2.

side is divided up; again there are several cases. Anyone can draw the figures for himself.

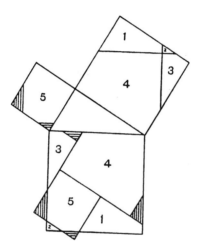

Fig. 3.3.3.

3.4

Demonstration of Pythagoras'
Theorem in Three Moves

ROGER BAKER

1. Resources

A square wooden 'picture frame' – the 'picture' being plain coloured cardboard. Four congruent wooden right-angled triangles (one with one side painted white) – the sum of the two shorter sides being equal to the side of the 'picture'.

As the triangles are slid about within the frame, the area of visible cardboard remains constant. Hence all that is required is to describe the areas of cardboard in terms of the length of the side of the white triangle.

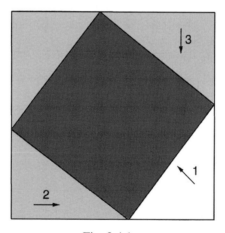

Fig. 3.4.1.

The formula for $(a + b)^2$ is, of course, also demonstrated by Figure 3.4.2. Those for $(a - b)^2$ (Figure 3.4.3) and $(a + b)(a - b)$ (Figure 3.4.4) can also be demonstrated using the fact that the area of cardboard is $a^2 + b^2$ (The triangles have been turned over so all are the same colour).

First published in *Mathematics in School* **26**, 2 (March 1997), p 27.

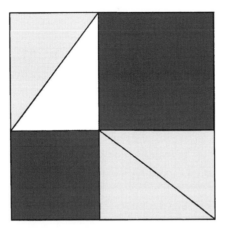

Fig. 3.4.2.

Fig. 3.4.3.

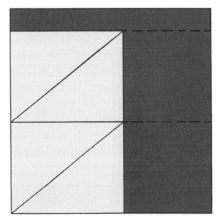

Fig. 3.4.4.

3.5

Pythagoras' Theorem

JACK OLIVER

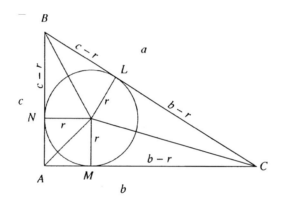

In the figure, with A the area of $\triangle ABC$

$$A = \frac{1}{2}r(a + b + c) = rs \tag{1}$$

$$b - r = MC = CL \quad \text{and} \quad c - r = NB = BL$$

and

$$BL + CL = a = (b - r) + (c - r)$$

which simplifies to

$$r = s - a \tag{2}$$

whence from (1) and (2)

$$A = s(s - a).$$

First published in *Mathematical Gazette* **81** (March 1997); pp. 117–118.

To obtain the Theorem of Pythagoras

$$A = \frac{1}{2}bc$$

so

$$s(s - a) = \frac{1}{2}bc$$
$$\frac{1}{2}(a + b + c)\frac{1}{2}(-a + b + c) = \frac{1}{2}bc$$
$$(b + c)^2 - a^2 = 2bc$$

which simplifies to

$$a^2 + b^2 = c^2.$$

3.6

A Neglected Pythagorean-Like Formula

LARRY HOEHN

Given an isosceles triangle as shown in Figure 3.6.1, we have the Pythagorean-like formula $c^2 = a^2 + bd$. If the segment a is an altitude, then $c^2 = a^2 + b^2$. This formula has surely been discovered many times, but yet it doesn't seem to appear in the mathematical literature.

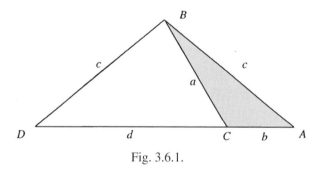

Fig. 3.6.1.

My discovery of this formula came from looking at one of the many proofs of Pythagoras' theorem. That is, given $\triangle ABC$, with a right angle at C, construct a circle with centre B and radius $BC = a$ as shown in Figure 3.6.2. By using the intersecting chords theorem, we have $(c + a)(c - a) = b^2$, so $c^2 = a^2 + b^2$.

Suppose on the other hand that $\angle C$ is not a right angle but rather an obtuse angle for $\triangle ABC$. By again constructing a circle with centre B and radius $BC = a$, the circle will intersect extended side AC at some point D as shown in Figure 3.6.3. By the intersecting chords theorem, we have $(c + a)(c - a) = ((d - b) + b)b$, so $c^2 = a^2 + bd$.

Since there are so many proofs of Pythagoras' theorem (see [1] and [2]), it is natural to suspect that this Pythagorean-like theorem can also be proved in many ways.

First published in *Mathematical Gazette* **84** (March 2000), pp. 71–73.

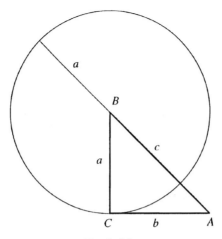

Fig. 3.6.2.

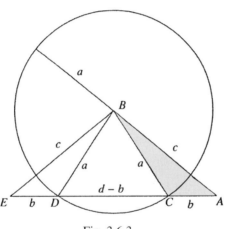

Fig. 3.6.3.

For example, by the law of cosines in Figure 3.6.4, $c^2 = a^2 + b^2 - 2ab \cos \theta$ and $c^2 = a^2 + d^2 - 2ad \cos(180 - \theta)$. Hence

$$\cos \theta = \frac{a^2 + b^2 - c^2}{2ab} \quad \text{and} \quad \cos(180 - \theta) = \frac{a^2 + d^2 - c^2}{2ad}.$$

Therefore, $(a^2 + b^2 - c^2)/2ab = -(a^2 + d^2 - c^2)/2ad$. By simplifying this result we obtain $c^2 = a^2 + bd$.

This Pythagorean-like theorem can also be derived by two applications of Pythagoras' theorem. If h is the altitude to the base of the isosceles triangle in

Larry Hoehn

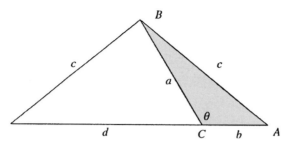

Fig. 3.6.4.

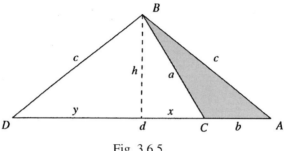

Fig. 3.6.5.

Figure 3.6.5, then $c^2 = h^2 + y^2$ and $a^2 = h^2 + x^2$. Hence

$$c^2 - a^2 = y^2 - x^2 = (y - x)(y + x) = bd, \quad \text{or} \quad c^2 = a^2 + bd.$$

Finally, we note that this formula is a special case of Stewart's theorem [3, p. 58]. 'If K, L and M are any three collinear points, and P is any other point, then both in magnitude and sign

$$PK^2 \times LM + PL^2 \times MK + PM^2 \times KL + KL \times LM \times MK = 0.'$$

Making the substitutions $P = B$, $K = D$, $M = A$, $L = C$,
we have $PK = PM = c$, $PL = a$, $KL = d$, $LM = b$, and $MK = -(d + b)$,
so $c^2 b + a^2(-(b + d)) + c^2 d + db(-(b + d)) = 0$.
This simplifies to $c^2(b + d) - a^2(b + d) - bd(b + d) = 0$, so $c^2 = a^2 + bd$.

Given the simplicity of this formula, and its close kinship to Pythagoras' theorem, it is quite curious that it is not prevalent in the mathematical literature.

References

1. Elisha Scott Loomis, *The Pythagorean proposition* (Washington DC: National Council of Teachers of Mathematics, 1968).
2. Larry Hoehn, The Pythagorean thorem: an infinite number of proofs? *Mathematics Teacher* **90** (September 1997) 438–441. (Correction to Figure 2 **91** (January 1998), 73.)

3. Howard Eves, *A Survey of Geometry* (revised Edn.) (Boston: Allyn and Bacon, Inc., 1972).

Editor's Note

In a subsequent article,'More notes on a neglected Pythagorean-like formula; *Mathematical Gazette* **85** (November 2001), 483–486. Goyala Darvasi extended Hoehn's proof and provided an alternative proof based on Heron's formula.

3.7

Pythagoras Extended: A Geometric Approach to the Cosine Rule

NEIL BIBBY

DOUG FRENCH

The motivation for this article was the desire to make the cosine rule in some way geometrically "obvious". What we have ended up with is a mixture of inductive and deductive approaches which we hope goes some way to illuminating the cosine rule, and to enhancing students' relational understanding of it. The germ of these ideas can be sown soon after students have met Pythagoras, with the question "O.K., suppose the triangle ABC is not right-angled: what can we say about a^2, b^2 and c^2 now?" The association of acuteness with the case $c^2 < a^2 + b^2$ and obtuseness with the case $c^2 > a^2 + b^2$ should soon follow:

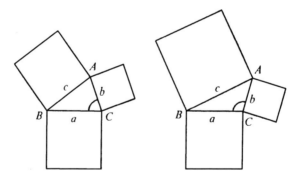

This conjures up some nice geometric imagery already; the Pythagorean case is seen to be special in that as C increases, it corresponds to a critical juncture when acuteness ends and obtuseness starts, and the relation between c^2 and $a^2 + b^2$ inverts.

A natural question to pose now is "In the obtuse case, how much bigger than $a^2 + b^2$ can c^2 become – is there some cut-off point?" Student discussion here

First published in *Mathematical Gazette* **72** (October 1988), pp. 184–188.

might involve "impossible" (a, b, c) triples, suggesting that given a and b, then c can only be chosen within certain limits. The

"acute → right-angle → obtuse"

imagery suggests we push the angle C up as far as we can:

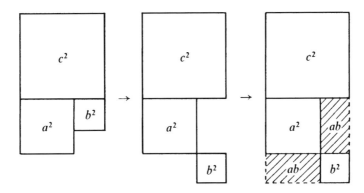

The redundant 180° triangle we end up with enables us at least to put an upper bound on the amount by which c^2 can exceed $a^2 + b^2$:

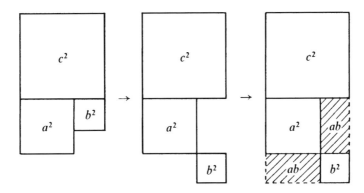

So c^2 can exceed $a^2 + b^2$ maximally by $2ab$. Considering the acute case in a similar way, the other redundant triangle ($C = 0°$) shows that $a^2 + b^2$ can exceed c^2 maximally by $2ab$, but the diagram (overleaf) for this is a bit trickier.

Both of these limiting cases are obviously more easily dealt with if students are already very familiar with the identities for $(a + b)^2$ and $(a - b)^2$. The conclusions however are that:

for $C = 0°$ $c^2 = a^2 + b^2 - 2ab$
for $C = 90°$ $c^2 = a^2 + b^2$
for $C = 180°$ $c^2 = a^2 + b^2 + 2ab.$

These three cases could now be written as

$$c^2 = a^2 + b^2 + 2ab \, f(C),$$

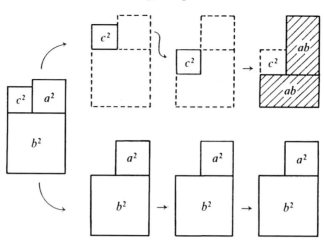

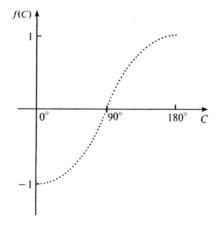

where $f(0°) = -1$, $f(90°) = 0$, $f(180°) = 1$. The students could now investigate the function f. In the first instance this could be purely empirical, where the students draw various triangles on square spotty paper for both acute and obtuse values of C, work out

$$f(C) = \frac{c^2 - (a^2 + b^2)}{2ab},$$

measure the angle C, and gradually fill out the graph:

The conjecture might now be made that $f(C) = -\cos C$. This might well be the end of the matter, but a deductive argument to support the empirically derived suggestion might now be appropriate. Only at this stage would we proffer one of the standard "proofs of the cosine rule". Better however, we think, is a geometric dissection which is a generalisation of the standard Euclidean dissection for proving Pythagoras's rule: the latter can be expressed using a shear–rotation–shear sequence of transformations on each of the rectangles formed by extending the altitude from C, thus dissecting the square of area c^2.

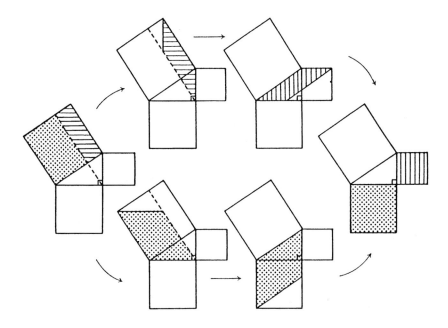

The correspondingly shaded regions are of equal area.

However, in the general case (i.e. $C \neq 90°$), *all three* altitudes can be extended to dissect their opposite squares:

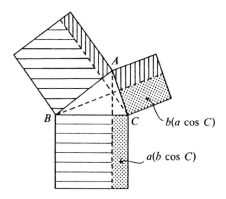

Again the correspondingly shaded regions are equal in area, and this can be demonstrated by a similar shear–rotation–shear sequence, as in the right-angle case.

So

$$c^2 = (a^2 - ab \cos C) + (b^2 - ab \cos C),$$

or

$$c^2 = a^2 + b^2 - 2ab \, \cos \, C.$$

It can be seen that in the right-angled case, two of the altitudes coincide with two of the sides of the triangle, and hence the regions of area $ab \cos C$ shrink to regions of area zero. The geometric imagery of this dissection seems remarkably direct and powerful to us – a far cry from the logico-deductive obscurity of most treatments.

3.8

Pythagoras in Higher Dimensions I: Three Approaches

LEWIS HULL

A fourth-year class (15 year olds), which I was teaching some twenty years ago, raised the question of whether there might be a three-dimensional version of Pythagoras's theorem. They were not content with the formula for the square of the diagonal of a cuboid. They regarded this as an application of the original theorem to a three-dimensional problem; what they wanted was a theorem about "a solid triangle". After some discussion we agreed that, as a right-angled triangle was a corner cut off a square, its analogue would be a corner cut off a cube.

We denoted the mutually perpendicular edges of the tetrahedron by x, y, z, as in the figure. The apparently naïve suggestion that $\triangle^2 = \triangle_1^2 + \triangle_2^2 + \triangle_3^2$, where $\triangle, \triangle_1, \triangle_2, \triangle_3$ are the areas of triangles ABC, OBC, OCA and OAB, came almost at once. We set out to prove or disprove this, with a result that certainly surprised me.

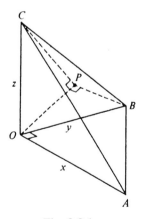

Fig. 3.8.1.

First published in *Mathematical Gazette* **62** (October 1978), pp. 206–207.

Our first attempt depended on the cosine rule:

$$\cos \angle BAC = \frac{AB^2 + AC^2 - BC^2}{2AB \cdot AC} = \frac{x^2}{AB \cdot AC}.$$

Thus

$$\triangle^2 = \frac{1}{4}AB^2 AC^2 \sin^2 \angle BAC = \frac{1}{4}AB^2 AC^2 \left(1 - \frac{x^4}{AB^2 AC^2}\right)$$

$$= \frac{1}{4}(AB^2 AC^2 - x^4) = \frac{1}{4}(y^2 z^2 + z^2 x^2 + x^2 y^2).$$

A final year student produced a vector version of this:

$$\triangle^2 = \frac{1}{4}|AB \times AC|^2 = \frac{1}{4}|(y\mathbf{j} - x\mathbf{i}) \times (z\mathbf{k} - x\mathbf{i})|^2 = \frac{1}{4}|yz\mathbf{i} + zx\mathbf{j} + xy\mathbf{k}|^2$$

$$= \frac{1}{4}(y^2 z^2 + z^2 x^2 + x^2 y^2).$$

The most interesting proof, as it seems to me, arises by analogy with the standard similar triangles proof of the two-dimensional theorem. Let P be the foot of the perpendicular from O to the plane ABC. Then triangle BPC, denoted by $\triangle_1'$, is the orthogonal projection of $\triangle_1$ on the plane ABC. Also, $\triangle_1$ is the orthogonal projection of $\triangle$ on the yz-plane. Hence, if α is the angle between the planes ABC and OBC, we have

$$\cos \alpha = \frac{\triangle_1'}{\triangle_1} = \frac{\triangle_1}{\triangle} \Rightarrow \triangle_1^2 = \triangle \triangle_1'.$$

Similarly $\triangle_2^2 = \triangle \triangle_2'$ and $\triangle_3^2 = \triangle \triangle_3'$. By addition, we get

$$\triangle_1^2 + \triangle_2^2 + \triangle_3^2 = \triangle(\triangle_1' + \triangle_2' + \triangle_3') = \triangle^2.$$

3.9

Pythagoras in Higher Dimensions, II

HAZEL PERFECT

A "right-angled tetrahedron" in three dimensions may be specified by its vertices, say $(0, 0, 0)$, $(a, 0, 0)$, $(0, b, 0)$, $(0, 0, c)$ with respect to a rectangular cartesian coordinate system. The plane through these last three points has the equation

$$\frac{1}{a}x + \frac{1}{b}y + \frac{1}{c}z = 1,$$

and its perpendicular distance d from the origin satisfies the relation

$$\frac{1}{d^2} = \frac{1}{a^2} + \frac{1}{b^2} + \frac{1}{c^2}.$$

If we denote by V the volume of the tetrahedron, by A, B, C the areas of the three faces through O, and by D the area of the face opposite O, then

$$V = \frac{1}{3}aA = \frac{1}{3}bB = \frac{1}{3}cC = \frac{1}{3}dD,$$

and therefore

$$A^2 + B^2 + C^2 = D^2 \left(\frac{d^2}{a^2} + \frac{d^2}{b^2} + \frac{d^2}{c^2} \right) = D^2.$$

The n-dimensional analogue may be proved in exactly the same way once the appropriate metric notions have been defined.

First published in *Mathematical Gazette* **62** (October 1978), p. 208.

3.10

Pythagoras Inside Out

LARRY HOEHN

There are two ways to inscribe a square in a right triangle. These are illustrated by square $DEFG$ in Figure 3.10.1 and square $CLMN$ in Figure 3.10.2. In each of these figures the squares 'cut-off' additional right triangles in which other squares can also be inscribed. In Figure 3.10.1 these squares share a side with hypotenuse AB whereas in Figure 3.10.2 these squares share a single vertex with hypotenuse AB. In each case we get smaller right triangles (shaded in the two figures) which appear to be congruent. They are, in fact, congruent as we can readily demonstrate.

We begin with the case in Figure 3.10.1 and leave the other case for the interested reader. First we observe that all of the right triangles in the figure are similar. In particular $\triangle HID \sim \triangle EJK$. Therefore $HI/EJ = ID/JK$, or $y/(s-x) = (s-y)/x$. Hence $s^2 - sx - sy + xy = xy$ or $s(s-x-y) = 0$. Since $s \neq 0$, then $s - x - y = 0$ so that $s - x = y$ and $s - y = x$. Therefore $\triangle HID \cong \triangle EJK$. Case 2 can be completed in the same manner.

We note that the three squares in each figure have a nice analogy with the Pythagorean theorem. The Pythagorean theorem can be stated as the area of the largest square constructed outward on the side of a right triangle is equal to the sum of the areas of the other two similarly constructed squares. Since $s = x + y$ or equivalently $4s = 4x + 4y$ in Figure 3.10.1 and $t = w + z$ or $4t = 4w + 4z$ in Figure 3.10.2, then we have the result that the perimeter of the largest square constructed inside a right triangle is equal to the sum of the perimeters of the other two similarly constructed squares.

First published in *Mathematical Gazette* **80** (November 1996), pp. 544–545.

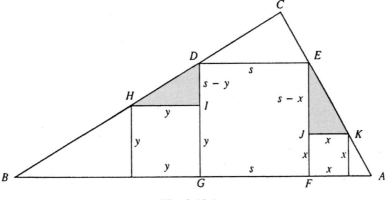

Fig. 3.10.1.

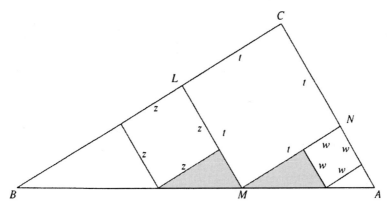

Fig. 3.10.2.

3.11

Geometry and the Cosine Rule

COLIN DIXON

I would think that most pupils would be able to give their version of Pythagoras' Theorem. It saddens me however to think that the vast majority of those pupils would not be able convince me why this amazing result is true despite several simple proofs being within their compass. Of course, there are some who get confused just stating the result, with $a^2 - b^2$ often making an unexpected appearance.

1. But What About $a^2 - b^2$?

To investigate this we refer to Figure 3.11.1, and make liberal use of the cosine rule.

Squares are drawn on the sides BC, AC, AB, XY and VW.

Then

$$
\begin{aligned}
XY^2 - VW^2 &= (a^2 + c^2 - 2ac \cos X\hat{B}Y) - (b^2 + c^2 - 2bc \cos V\hat{A}W) \\
&= a^2 - b^2 + 2ac \cos A\hat{B}C - 2bc \cos B\hat{A}C \\
&= a^2 - b^2 + 2ac\frac{a}{c} - 2bc\frac{b}{c} \\
&= 3(a^2 - b^2)
\end{aligned}
$$

so

$$
a^2 - b^2 = \frac{1}{3}[(\text{area of square on } XY) - (\text{area of square on } VW)].
$$

2. Is There a Corresponding Result for a Non Right-Angled Triangle?

Referring to Figure 3.11.2.

$$
XY^2 - VW^2 = (a^2 + c^2 - 2ac \cos X\hat{B}Y) - (b^2 + c^2 - 2bc \cos V\hat{A}W)
$$

First published in *Mathematical Gazette* **81** (Novermber 1997), pp. 439–444; extract.

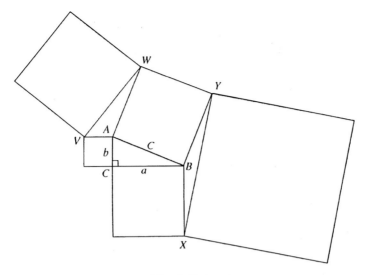

Fig. 3.11.1.

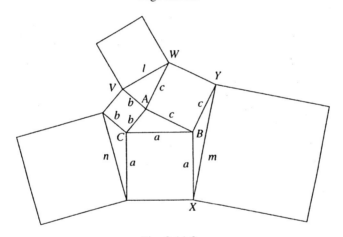

Fig. 3.11.2.

$$= a^2 - b^2 + 2ac\frac{(a^2 + c^2 - b^2)}{2ac} - 2bc\frac{(b^2 + c^2 - a^2)}{2bc}$$
$$= 3(a^2 - b^2), \quad \text{as before.}$$

3. Triangle, Squares, Triangles, Squares

Referring again to Figure 3.11.2.

$$l^2 = b^2 + c^2 - 2bc \cos V\hat{A}W$$
$$= b^2 + c^2 + 2bc \cos B\hat{A}C$$

$$= b^2 + c^2 + 2bc \frac{(b^2 + c^2 - a^2)}{2bc}$$
$$= 2b^2 + 2c^2 - a^2.$$

Similarly

$$m^2 = 2a^2 + 2c^2 - b^2 \text{ and } n^2 = 2a^2 + 2b^2 - c^2$$

so

$$l^2 + m^2 + n^2 = 3(a^2 + b^2 + c^2)$$

i.e. the sum of areas of outer squares $= 3\times$ (the sum of areas of inner squares).

4. What Can We Make of an Inner Quadrilateral?

$$\text{Area triangle } EDL = \frac{1}{2}ED \cdot DL \sin E\hat{D}L$$
$$= \frac{1}{2}DC \cdot DA \sin A\hat{D}C$$
$$= \text{area triangle } ADC.$$

Similarly

$$\text{area triangle } HBI = \text{area triangle } ABC.$$

It follows therefore that

$$\text{area triangles I, II, III, and IV} = 2 \times \text{(area quadrilateral } ABCD).$$

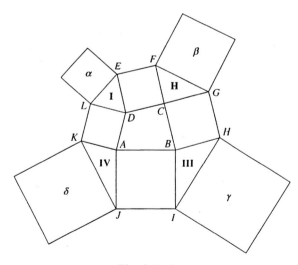

Fig. 3.11.3.

Let the outer squares be labelled α, β, γ, δ as shown.

$$\begin{aligned}
\text{Area } \alpha &= EL^2 = ED^2 + DL^2 - 2ED \cdot DL \cos E\hat{D}L \\
&= ED^2 + DL^2 + 2ED \cdot DL \cos A\hat{D}C \\
&= ED^2 + DL^2 + 2ED \cdot DL \frac{(AD^2 + DC^2 - AC^2)}{2AD \cdot DC} \\
&= ED^2 + DL^2 + (AD^2 + DC^2 - AC^2) \\
&= 2ED^2 + 2DL^2 - AC^2.
\end{aligned}$$

Similar results for areas β, γ, δ lead to

$$\begin{aligned}
\text{area } \alpha + \text{area } \beta + \text{area } \gamma + \text{area } \delta = {} &4(ED^2 + CG^2 + BI^2 + AK^2) - \\
&2(AC^2 + BD^2).
\end{aligned}$$

i.e. the sum of squares of diagonals of quadrilateral $ABCD = 2 \times$ (the sum of squares on sides of quadrilateral) $- \frac{1}{2}$(sum of squares on outer triangles).

3.12

Bride's Chair Revisited

ROGER WEBSTER

Perhaps the most famous diagram in the whole of mathematics is that of a right-angled triangle surrounded by three squares drawn on its sides. Even to those schooled only in elementary geometry, it will immediately suggest just one result, Pythagoras' Theorem. This diagram, with certain construction lines inserted (Figure 3.12.1), is that used by Euclid in his much admired proof given in *Elements* I: Proposition 47.

Over the years the diagram itself has received much attention, having been variously described as the bride's chair, the Franciscan's cowl, the peacock's tail and the windmill. It has been used on stamps, as a cover design, a framework for cartoons, and as the logo for a mathematical supply company in the United States, appearing on its mugs, carrier bags and T-shirts. Even so, the bride's chair has succeeded in keeping at least one of its secrets well hidden from public view.

Consider an *arbitrary* triangle ABC surrounded by the three squares drawn on its sides, as in Figure 3.12.2. Some vertices of the squares are joined to form the three shaded triangles shown in the figure. The area of the shaded triangle having A as one of it vertices is

$$\frac{1}{2}bc \sin(180° - A) = \frac{1}{2}bc \sin A,$$

which is the same as the area of triangle ABC itself. Clearly, the other two shaded triangles also have this area. We have thus established the following delightfully simple result.

First published in *Mathematical Gazette* **78** (Novermber 1994), pp. 345–346.

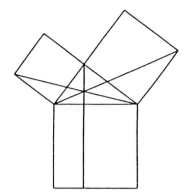

Fig. 3.12.1.

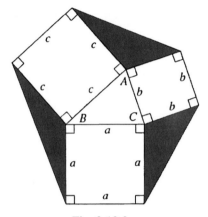

Fig. 3.12.2.

If squares are drawn on the sides of a triangle and external to it, then the areas of the triangles formed between the squares all equal the area of the triangle itself.

3.13

Bride's Chair Revisited Again!

IAN WARBURTON

The following extension to Roger Webster's article may be of interest.

The figure obtained by drawing squares externally on the sides of any triangle has the following additional properties.

Property 1 The medians AP, BQ, CR of the triangles formed externally between the squares are also the altitudes of triangles ABC (Figure 3.13.1).

Also $BC = 2AP$, $CA = 2BQ$ and $AB = 2CR$.

It is easy to demonstrate this result by rotating the triangle ALM through $90°$ about A (Figure 3.13.2a).

After rotation (Figure 3.13.2b), BC is parallel to $P'A$ and $BC = 2P'A$. Thus in the original figure BC is perpendicular to AP. Similarly AC is perpendicular to BQ and AB is perpendicular to CR.

Property 2 In Figure 3.13.3, if the lines BZ and CY meet at X, then AX is an altitude of the triangle ABC (even when angle A is not a right angle).

A proof using vectors involves only fundamental ideas but lengthy manipulation. Consider Figure 3.13.4 in which A is taken as the origin of vectors.

If we assume that the area of triangle ABC is 0.5 units then $\mathbf{b} \times \mathbf{c} = \mathbf{n}$ where $\mathbf{n}$ is a unit vector perpendicular to ABC.

Thus $\mathbf{d} = \mathbf{b} \times \mathbf{n} = \mathbf{b} \times (\mathbf{b} \times \mathbf{c}) = (\mathbf{b} \cdot \mathbf{c})\mathbf{b} - b^2\mathbf{c}$ and $\mathbf{e} = \mathbf{n} \times \mathbf{c} = (\mathbf{b} \times \mathbf{c}) \times \mathbf{c} = (\mathbf{b} \cdot \mathbf{c})\mathbf{c} - c^2\mathbf{b}$.

Now $\overrightarrow{BZ} = \mathbf{c} + \mathbf{e} - \mathbf{b} = (1 + \mathbf{b} \cdot \mathbf{c})\mathbf{c} - (1 + c^2)\mathbf{b}$ and $\overrightarrow{CY} = \mathbf{b} + \mathbf{d} - \mathbf{c} = (1 + \mathbf{b} \cdot \mathbf{c})\mathbf{b} - (1 + b^2)\mathbf{c}$.

At the point of intersection X of BZ and CY

$$\mathbf{x} = \overrightarrow{AX} = \mathbf{b} + s(1 + \mathbf{b} \cdot \mathbf{c})\mathbf{c} - s(1 + c^2)\mathbf{b}$$

First published in *Mathematical Gazette* **80** (Novermber 1996), pp. 557–558.

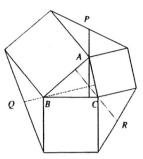

Fig. 3.13.1.

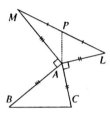

Fig. 3.13.2a.

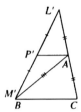

Fig. 3.13.2b.

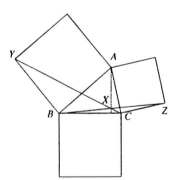

Fig. 3.13.3.

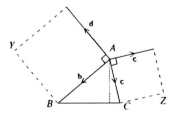

Fig. 3.13.4.

and

$$\mathbf{x} = \mathbf{c} + t(1 + \mathbf{b} \cdot \mathbf{c})\mathbf{b} - t(1 + b^2)\mathbf{c},$$

which implies that

$$(1 + \mathbf{b} \cdot \mathbf{c})s = 1 - (1 + b^2)t$$

and

$$(1 + \mathbf{b} \cdot \mathbf{c})t = 1 - (1 + c^2)s.$$

Solving these simultaneous equations leads to

$$t[(1 + \mathbf{b} \cdot \mathbf{c})^2 - (1 + b^2)(1 + c^2)] = \mathbf{b} \cdot \mathbf{c} - c^2$$

or

$$kt = \mathbf{b} \cdot \mathbf{c} - c^2$$

where

$$k = (1 + \mathbf{b} \cdot \mathbf{c})^2 - (1 + b^2)(1 + c^2).$$

Hence

$$k\mathbf{x} = k\mathbf{c} + kt(1 + \mathbf{b} \cdot \mathbf{c})\mathbf{b} - kt(1 + b^2)\mathbf{c}$$
$$\Rightarrow \quad k\mathbf{x} = k\mathbf{c} + (\mathbf{b} \cdot \mathbf{c} - c^2)(1 + \mathbf{b} \cdot \mathbf{c})\mathbf{b} - (\mathbf{b} \cdot \mathbf{c} - c^2)(1 + b^2)\mathbf{c}$$

which simplifies to

$$k\mathbf{x} = (1 + \mathbf{b} \cdot \mathbf{c})[(\mathbf{b} \cdot \mathbf{c} - c^2)\mathbf{b} + (\mathbf{b} \cdot \mathbf{c} - b^2)\mathbf{c}].$$

It follows that $k\mathbf{x} \cdot (\mathbf{c} - \mathbf{b}) = 0$, proving that AX is perpendicular to BC.

Desert Island Theorems

Group C: Advanced Euclidean Geometry

C1

Desargues' Theorem

DOUGLAS QUADLING'S CHOICE

For simplicity and elegance it is hard to beat straight-line geometry. For example, what we call Pappus' Theorem (Figure C1.1), that the points U, V, W are collinear, can be understood by an eight-year old and verified simply by using the edge of a book to draw straight lines.

My special favourite is Desargues' Theorem (Figure C1.2), that if two triangles, BCD and EFG, are in perspective from A (AEB, AFC, AGD are straight lines) then the intersections, H, I, J of the corresponding sides are collinear.

Unlike Pappus' Theorem, Desargues' Theorem still holds in three dimensions. Indeed, if you visualise Figure C1.2 as a representation on paper of a solid triangular-based pyramid $ABCD$ standing on the floor, and EFG as a section cut through it, then clearly H, I, J all lie on the line in which the plane of that section intersects the floor. It is no surprise to learn that Gérard Desargues (1593–1662) was an engineer-cum-architect, accustomed to making spatial interpretations of diagrams drawn in the flat, and that his interest in geometry was motivated by a desire to help his artist contemporaries to understand the principles of perspective.

If the theorem is true in three dimensions, then it is obviously true in two – or is it? It's certainly far tougher to prove if you restrict yourself to two-dimensional methods.

Now to earn desert island status, a theorem must have outlook value. What does Desargues' Theorem have to offer?

The statement of the theorem began with a point A, and ended with a line HIJ. Call this the 'Desargues line' of A, and denote it by a. Now start instead with another point, for example B. Without introducing any more lines or points, there are two triangles ACD and EJI, which are in perspective from B, and these produce a Desargues line HGF, denoted by b. Now start with C, ... and so on. You will find

Douglas Quadling, Teacher of Mathematics to Students and Teachers. Editor of *The Mathematical Gazette*, 1972–80. President of The Mathematical Association, 1980–81.

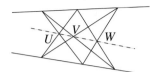

Fig. C1.1.

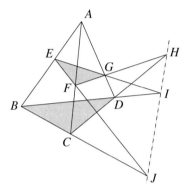

Fig. C1.2.

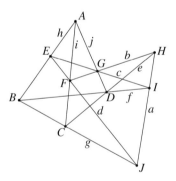

Fig. C1.3.

that you end up with Figure C1.3, in which there are 10 points and 10 Desargues lines, with three lines through each point and three points on each line. Wow!

The symmetry of these numbers suggests a dual configuration, swapping points and lines. Instead of starting with three lines h, i, j through a point A, take the three points H, I, J lying on the line a. There are two other lines through each of these points, which form two trilaterals bcd and efg; and the lines joining corresponding vertices of these trilaterals are concurrent at the point A. So the dual of Desargues' Theorem is also its converse.

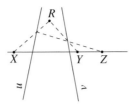

Fig. C1.4.

You could if you wished use Figure C1.3 to design a three-handed whist tournament for a group of 10 people. Each of the players $(A, B, C, \ldots, J)$ sits out in turn, and the two associated perspective triangles and the Desargues line define the three triplets of players in that round. It wouldn't be a very satisfactory was of arranging the draw – why not? But is there a better way of doing it?

The theorem also has an interesting history, both backwards and forwards from Desargues. B. L. van der Waerden, in his book *Science Awakening*, has pointed out that Pappus (about 300 A.D.) got within a whisker of Desargues' result when he showed that, in Figure C1.4, if X, Y, Z, are fixed collinear points and u, v, are fixed lines, and if variable dotted lines are drawn as shown, then the locus of the point R is another fixed line – which in fact is concurrent with u and v. If you draw the dotted lines in two different positions, you will find that this gives the Desargues figure.

Moreover, Pappus in his *Synagoge* (or 'Collection') was not claiming original results but summarising and reorganising the work of earlier mathematicians – in this instance results given by Euclid in his book *Porisms*, of which no other record now exists. So how close to Desargues' Theorem did Euclid get, nearly 2000 years earlier?

The recent history is equally intriguing. It is described by Morris Kline in his chapter on projective geometry in James R. Newman's *The World of Mathematics*:

Every printed copy of Desargues' book, originally published in 1639, was lost. Abraham Bosse, a pupil and friend of Desargues, published a book in 1648, *The Universal Method of Desargues for the Practice of Perspective*, and in an appendix to this book he reproduced Desargues' theorem and other of his results. Even this appendix was lost and was not rediscovered until 1804 . . . Fortunately a pupil of Desargues, Phillippe de la Hire, made a manuscript copy of Desargues' book. In the nineteenth century this copy was picked up by accident in a bookshop by the geometer Michel Chasles, and thereby the world learned the full extent of Desargues' major work.

It seems that we are very lucky to enjoy this accident-prone theorem.

Further Reading

Morris Kline, Projective geometry, in James R. Newman (Ed.), *The World of Mathematics*, Vol. 1 (London: Allen & Unwin, 1956) (part iv), 622–646. The above quotation is taken from p. 631.

John Fauvel & Jeremy Gray (Eds.), *The History of Mathematics: A Reader*, (Macmillan Education/Open University, 1987), 366–374.

J. V. Field & J. J. Gray, *The Geometrical Work of Girard Desargues* (Springer, 1987).

C2

Pascal's Hexagram Theorem

MARTYN CUNDY'S CHOICE

(1) The Alexandrian mathematician Menelaus (c. 110 A.D.) first proved the theorem which bears his name:

If a straight line cuts the sides of a triangle PQR at points X, Y, Z, then

$$\frac{QX \cdot RY \cdot PZ}{XR \cdot YP \cdot ZQ} = -1.$$

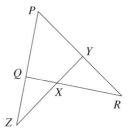

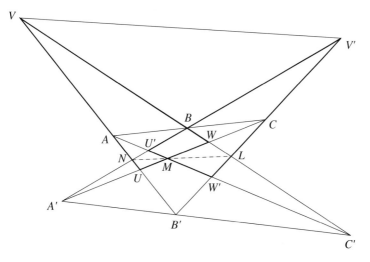

H. Martyn Cundy, Author (with A. P. Rollett) of *Mathematical Models*.

His method was presumably the same still used, by dropping perpendiculars from the vertices P, Q, R onto the line. This theorem is the basis of all that follows.

(2) Pappus, also an Alexandrian, about 200 years later (c. 340) stated and proved a theorem which has proved of enormous importance in modern discussion of the foundations and structure of geometries. Confined to Euclidean methods, he presumably used Menelaus' Theorem, and possibly the figure above.

If ABC, $A'B'C'$ are any sets of three points on any two lines, the 'cross-meets' $BC' \cap B'C$, $CA' \cap C'A$, $AB' \cap A'B$ are collinear.

Here is a proof using Menelaus' Theorem, and the figure above. The intersections shown in the figure are labelled U, V, W, U', V', W', L, M, N.

Consider the sides of $\triangle U'V'W'$ as transversals of $\triangle UVW$. The intersections are given below along with the related results:

	VW	WU	UV		
$V'W'$	L	C	B'	$\Rightarrow$	$\dfrac{VL \cdot WC \cdot UB'}{LW \cdot CU \cdot B'V} = -1$
$W'U'$	C'	M	A	$\Rightarrow$	$\dfrac{VC' \cdot WM \cdot UA}{C'W \cdot MU \cdot AV} = -1$
$U'V'$	B	A'	N	$\Rightarrow$	$\dfrac{VB \cdot WA' \cdot UN}{BW \cdot A'U \cdot NV} = -1$

Multiply these three equations together and reorganise to obtain

$$\left\{ \frac{UA \cdot VB \cdot WC}{AV \cdot BW \cdot CU} \right\} \cdot \left\{ \frac{UB' \cdot VC' \cdot WA'}{B'V \cdot C'W \cdot A'U} \right\} \cdot \left\{ \frac{UN \cdot VL \cdot WM}{NV \cdot LW \cdot MU} \right\} = -1.$$

Since ABC, $A'B'C'$ are collinear, the first two brackets are each -1; therefore the third is also -1 and LMN are collinear.

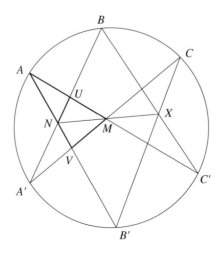

(3) In 1640, at the age of 16, Pascal replaced the outer sets of 3 points in this theorem by six points on a circle. Connecting them with six lines to form a hexagram, he arrived at the figure above, which he named his *mystic hexagram*, and proved that once again, LMN were collinear.

Here again, the proof probably rests on Menelaus' Theorem. The circle theorem which is most useful is the product formula for chords, such as $BN \cdot NA' = AN \cdot NB'$. LMN is not a transversal of any triangle in the figure, so we let MN meet BC' at X and prove CXB' collinear.

$$MVA' \text{ is a transversal of } \triangle AUN \Leftrightarrow \frac{AM \cdot UA' \cdot NV}{MU \cdot A'N \cdot VA} = -1.$$

$$UNA' \text{ is a transversal of } \triangle AVM \Leftrightarrow \frac{AN \cdot VA' \cdot MU}{NV \cdot A'M \cdot UA} = -1.$$

$$\text{Multiplying, } \frac{AM \cdot AN \cdot VA' \cdot UA'}{A'M \cdot A'N \cdot VA \cdot UA} = +1.$$

Now, from the circle theorem, $AM/A'M = MC/MC'$, $AN/A'N = NB/NB'$, $VA'/VA = VB'/VC$, $UA'/UA = UC'/UB$. Therefore $\frac{MC \cdot NB \cdot VB' \cdot UC'}{MC' \cdot NB' \cdot VC \cdot UB} = +1$.

Finally, BXC' is a transversal of $\triangle NMU \Leftrightarrow \frac{UB \cdot NX \cdot MC'}{BN \cdot XM \cdot C'U} = -1$, and again multiplying, we find $\frac{NX \cdot MC \cdot VB'}{XM \cdot CV \cdot B'N} = -1 \Leftrightarrow X, C, B'$ are collinear, and L is on MN.

(4) A modern approach. Once the projective properties of the conic had been discovered, the whole position of these theorems changed. Pappus' Theorem is all about lines and incidence and matters of distance and angle are really irrelevant. Geometries can be constructed in which the theorem is false, and in fact it is often taken as an axiom for the projective plane. As we have seen, it is true in the Euclidean plane, and indeed in any geometry which can be embedded in a fully projective plane. But then it immediately appears as just a special case of the theorem for a general conic, of which Pappus and Pascal have dealt with special cases. The proof of the general result then becomes extremely simple, as we now show.

If A, B, C, D, E, F are six points on any conic, their cross-meets, as defined above, are collinear.

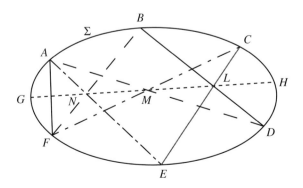

We need to know: (a) a conic is a curve of degree 2, and represents a pair of straight lines if it factorizes. (b) A conic is determined by 5 points, and the set of conics through 4 points is a linear pencil. (c) $\sum = k \sum'$ is a conic through the 4 intersection points of $\sum$ and $\sum'$ for all k.

Write $(AD) = 0$ for the line AD, and so on.

Proof Since A, B, D, F are on $\sum$, $\sum = (AF)(BD) + \lambda(AD)(BF) = 0$ for some λ. Since A, C, E, F are on $\sum$, $\sum = (AF)(CE) + \mu(AE)(CF) = 0$ for some μ.

Therefore $(AF)(BD) + \lambda(AD)(BF) = \nu\{(AF)(CE) + \mu(AE)(CF)\}$ for some ν. That is to say

$$(AF)\{(BD) - \nu(CE)\} = \mu\nu(AE)(CF) - \lambda(AD)(BF).$$

The left hand side is a line pair, consisting of AF and a line through $BD \cap CE$, which is L. The right hand side tells us that this passes through the four intersections $AE \cap AD$, $AE \cap BF$, $CF \cap AD$, $CF \cap BF$, i.e., through A, N, M, F; so it is $(AF)(MN)$ and **L, M, N are collinear**.

This proof is due to George Salmon (1879).

(5) The final generalization: the **Cross-Axis Theorem**. This is concerned with *projectivity*. There is a projectivity between two sets of points on a conic when the cross ratio of the pencil subtended on the conic by any four points of one set is equal to the cross-ratio of the pencil subtended by the corresponding four points of the other. That just means that there is a linear bijection between their corresponding parameters. Such a projectivity has two self-corresponding points (real, coincident or a complex conjugate pair), and is defined by a set of three pairs. Take the projectivity defined by the pairs $A \to A'$, $B \to B'$, $C \to C'$, let G, H be its self-corresponding points and (P, P'), (Q, Q') be any two pairs. Then $P\{P'Q'GH\} = P'\{PQGH\}$ and these pencils have PP' in common, so that the meets of corresponding rays are collinear. That is, $PQ' \cap P'Q$ is collinear with GH. The usual jargon is that the *cross-joins* of any two pairs of points meet on GH, the cross-*axis*. Taking $A' = D$, $B' = E$, $C' = F$, we have Pascal's Theorem at once.

C3

The Nine-Point Circle and Friends
(A Gourmet Four-Course Meal, with Coffee)

ADAM MCBRIDE'S CHOICE

1. Menu

The geometry of the triangle is truly remarkable. There seems to be an almost endless flow of results that are both surprising and elegant. Have we any right to expect that the three medians of a triangle are concurrent? In the first instance, why should the third median pass through the point of intersection of the other two? Why should the same be true for the altitudes, the perpendicular bisectors of the sides and the (internal) bisectors of the angles? That's only the start of a veritable feast. Here we make a particular choice from a mouth-watering menu, with the Nine-Point Circle as the main course.

2. Starter (Alphabet Soup)

Given $\triangle ABC$ (thought of as scalene to get the full effect), let

A', B', C' be the mid-points of BC, CA, AB respectively;
D, E, F be the feet of the altitudes from A, B, C respectively;
G, H, O be the centroid, orthocentre and circumcentre respectively;
U, V, W be the mid-points of AH, BH, CH respectively.

3. Entrée (Euler Line)

Here is a result which is certainly surprising and elegant.

Theorem O, G and H are collinear and G divides OH (internally) in the ratio 1:2.

Proof Let T be the point such that $\overrightarrow{OT} = 3\overrightarrow{OG}$.

Adam McBride, Mathematician, University of Strathclyde. Academic Director of International Mathematical Olympiad 2002, Glasgow. Leader of the United Kingdom IMO team in 1993, 1996–98.

We shall show that T lies on each of the altitudes of $\triangle ABC$ and hence that $T = H$. Consider Figure C3.1 below.

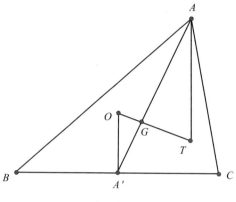

Fig. C3.1.

It is well known that G is the point of trisection of the median AA' with $A'G = \frac{1}{2}GA$.

By construction $OG = \frac{1}{2}GT$. Hence $\frac{A'G}{GA} = \frac{OG}{GT} = \frac{1}{2}$
$\therefore \triangle A'GO$ is similar to $\triangle AGT$.
$\therefore \angle OA'G = \angle TAG$.
$\therefore OA'$ is parallel to AT.
$\therefore$ The line through A and T is perpendicular to BC (as $OA' \perp BC$).
$\therefore T$ lies on the altitude of $\triangle ABC$ through A.
Similarly T lies on the altitudes through B and C. Hence $T = H$, as required.

4. Main Course (Nine-Point Circle)

Consider Figure C3.2 below.

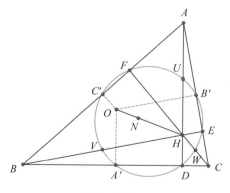

Fig. C3.2.

Start by drawing the circle through the points A', B' and C'. It looks as though this circle also passes through D, E and F. Further inspection may suggest that the circle intersects each of AH, BH and CH at its mid-point so that U, V, and W lie on the circle too. This is so. Where could the centre of the circle be? Could it be at the mid-point of OH? What could the radius be?

Theorem (The Nine-Point Circle) The points A', B', C', D, E, F, U, V, W lie on the circle with centre at N, the mid-point of OH, and radius $\frac{1}{2}R$, where R is the radius of the circumcircle of $\triangle ABC$.

Proof Use vectors with the circumcentre O as origin. (That's why we called it O!) Let $\overrightarrow{OA} = \mathbf{a}$, $\overrightarrow{OB} = \mathbf{b}$, $\overrightarrow{OC} = \mathbf{c}$. Then, by definition of O as the circumcentre, $|\mathbf{a}| = |\mathbf{b}| = |\mathbf{c}| = R$. Now,

$$\overrightarrow{OG} = \frac{1}{3}(\mathbf{a} + \mathbf{b} + \mathbf{c}) \quad \text{(by a standard result).}$$
$$\therefore \overrightarrow{OH} = \mathbf{a} + \mathbf{b} + \mathbf{c} \quad \text{(property of the Euler Line).}$$
$$\therefore \overrightarrow{ON} = \frac{1}{2}(\mathbf{a} + \mathbf{b} + \mathbf{c}) \quad \text{(as N is the mid-point of OH).}$$

Also $\overrightarrow{OA'} = \frac{1}{2}(\mathbf{b} + \mathbf{c})$ so that

$$\overrightarrow{NA'} = \overrightarrow{OA'} - \overrightarrow{ON} = \frac{1}{2}(\mathbf{b} + \mathbf{c}) - \frac{1}{2}(\mathbf{a} + \mathbf{b} + \mathbf{c}) = -\frac{1}{2}\mathbf{a}.$$

$\therefore |NA| = \frac{1}{2}R$ so that A' lies on the stated circle.
Similarly, B' and C' lie on the circle.
Next,

$$\overrightarrow{OU} = \frac{1}{2}(\overrightarrow{OA} + \overrightarrow{OH}) = \frac{1}{2}(\mathbf{a} + (\mathbf{a} + \mathbf{b} + \mathbf{c})) = \frac{1}{2}(2\mathbf{a} + \mathbf{b} + \mathbf{c}).$$
$$\therefore \overrightarrow{NU} = \overrightarrow{OU} - \overrightarrow{ON} = \frac{1}{2}\mathbf{a} \quad \text{so that} \quad |\overrightarrow{NU}| = \frac{1}{2}R.$$

Hence U lies on the stated circle, as do V and W similarly. Finally, since $\overrightarrow{A'N} = \overrightarrow{NU} = \frac{1}{2}\mathbf{a}$, $A'U$ is a diameter of the circle. Since $\angle A'DU = 90°$, D lies on the circle, as do E and F similarly. This completes the proof.

5. Dessert (Feuerbach's Theorem)

The nine-point circle is only one of many circles associated with $\triangle ABC$. Given two circles, they are most likely to intersect in two distinct points or not to intersect at all. What is the chance of their touching?

Theorem (Feuerbach, 1822) The nine-point circle of $\triangle ABC$ touches the incircle and each of the three excircles (escribed circles) of $\triangle ABC$.

The proof of this result is worth an article on its own but we shall omit the details here.

6. Coffee (The 13-Point Circle)

To the 9 points found previously, we can now add 4 more interesting points on the circle, namely the points of tangency derived from Feuerbach's Theorem.

7. The Bill

Since the service has been excellent, we should leave a tip by trying to increase the number of interesting points on our favourite circle beyond 13. That should give the reader food for thought.

C4

Napoleon's Theorem & Doug-all's Theorem
(Prelude and Centrifugue in G)

DOUGLAS HOFSTADTER'S CHOICE

1. Prelude (Napoleon's Theorem)

I. Given any triangle ABC, erect equilateral triangles pointing outwards on its three sides. The centers of these three triangles themselves form an equilateral triangle, known as the 'outer Napoleon triangle'.

II. Now erect three more equilateral triangles on the sides of ABC, but this time have them all face inwards (so they are reflections of the three previous triangles across the associated faces). Once again, the centers of these three triangles form an equilateral triangle, this one known as the 'inner Napoleon triangle'. This one is smaller than the previous one.

III. The outer and inner Napoleon triangles have the same center – namely, G, the centroid of ABC.

IV. The difference in areas of ABC's outer and inner Napoleon triangles is the area of ABC itself.

2. Centrifugue (Doug-all's Theorem)

I. Consider the set of all equilateral triangles PQR escribed about triangle ABC: that is, side PQ contains A, side QR contains B, and side RP contains C. (Incidentally, by 'side', I mean the full line, not just the line segment, so that point A isn't necessarily in between points P and Q, and so forth.) There are infinitely many such escribed equilateral triangles, and among them there is a largest one. There is also a smallest one, in a sense, but it is degenerate – namely, it is infinitely small (a point). That point is the Fermat point F of triangle ABC, which has the property that lines AF, BF, and CF meet each other in 60-degree angles.

II. Consider the set of all equilateral triangles UVW inscribed inside triangle ABC: that is, side AB contains U, side BC contains V, and side CA contains W. (Once again, by

Douglas Hofstadter, Mathematician, Cognitive Scientist & Philosopher of Science, Indiana University. *Scientific American* columnist, author of *Godel, Escher, Bach* and *Metamagical Themas*.

'side', I mean the full line, not just the line segment, so that point U isn't necessarily in between points A and B, and so forth.) There are infinitely many such inscribed equilateral triangles, and among them there is a smallest one. There is also a largest one, in a sense, but it is degenerate – namely, it is infinitely large.

III. Which is the maximal member of the family of all escribed equilateral triangles? Construct line a, the unique line through point A that is perpendicular to line AF. Similarly, let line b pass through B and be perpendicular to BF, and let line c pass through C and be perpendicular to CF. Then lines a, b, c are the sides of the maximal escribed equilateral triangle for triangle ABC.

IV. Which is the minimal member of the family of all inscribed equilateral triangles? It is the unique inscribed equilateral triangle whose sides are parallel to the sides of the maximal escribed equilateral triangle.

V. The geometric mean of the areas of ABC's maximal escribed and minimal inscribed equilateral triangles is the area of ABC itself.

VI. The centers of all the inscribed equilateral triangles form a straight line.

VII. The centers of all the escribed equilateral triangles form a circle. This circle is centered on ABC's centroid G, and indeed it is the circumcircle of ABC's inner Napoleon triangle.

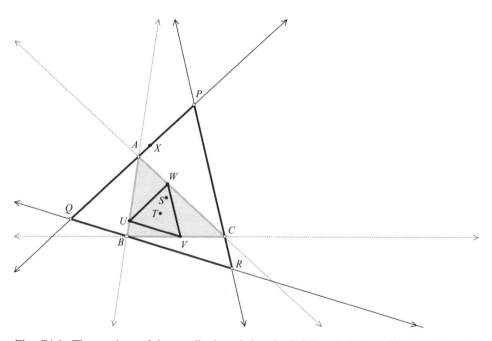

Fig. C4.1. The vertices of the escribed equitriangle PQR and the inscribed equitriangle UVW run counterclockwise and are referred to as 'direct-escribed' and 'direct-inscribed'. Note that the geometric mean of the two equitriangles' areas is always the area of ABC, independent of whether they are maximal and minimal. The centers of these two triangles are the points labeled S and T.

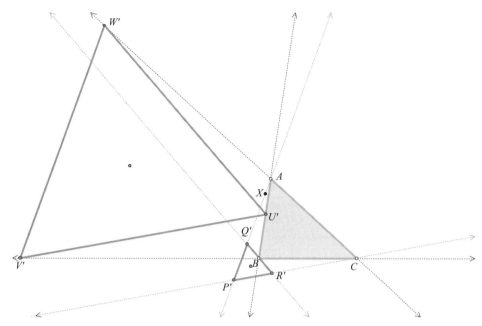

Fig. C4.2. $P'Q'R'$ is escribed about ABC but in the opposite sense to the way PQR is escribed about ABC in Figure C4.1 and is referred to as 'reverse-escribed'. Likewise, $U'V'W'$ is inscribed inside ABC but in the opposite sense to the way UVW is inscribed inside ABC and is referred to as 'reverse-inscribed'.

This cluster of results is my 'Desert Island Theorem'. I like it so much because it suggests so many more directions for exploration. I must point out that there was a time when I was avidly seeking a circle – any circle – of interest that was centered on G, and finding none. During that period, I was also investigating these inscribed and escribed equilateral triangles. When my friend Robert Boeninger asked me about the locus of centers of the escribed triangles, I discovered, to my astonishment, that it was a G-centered circle!

I must also point out that strictly speaking, there are actually two families of escribed triangles PQR. The difference is this: if side PQ contains A, side QR contains B, and side RP contains C, then vertices P, Q, and R may run around triangle PQR either in a counterclockwise manner or in a clockwise manner. The results above apply to the counterclockwise version. In the clockwise case, the circle of centers is still G-centered, but now it is the circumcircle of the outer Napoleon triangle.

Naturally, one wonders about the locus of centers of all the inscribed equilateral triangles (and here, too, there is a clockwise and a counterclockwise family). The answer is that they lie on a straight line. The clockwise and counterclockwise straight lines are parallel, meaning that if they are thought of as circles of infinite radius, they are concentric (as in the case of the escribed equilateral

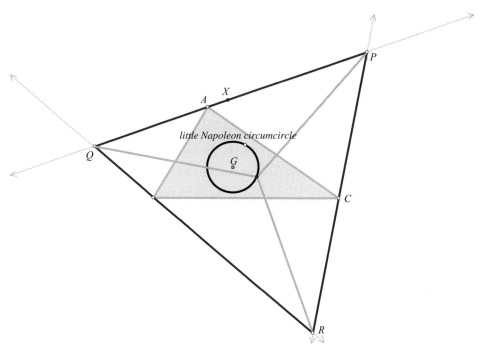

Fig. C4.3. Locus of centers of direct-escribed equitriangle PQR: little Napoleon circumcircle.

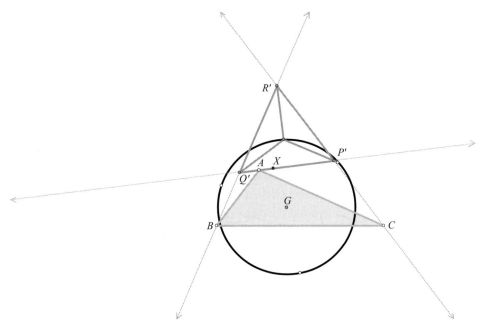

Fig. C4.4. Locus of centers of reverse-escribed equitriangle $P'Q'R'$: big Napoleon circumcircle.

triangles). Robert Boeninger discovered recently that these lines are the polars (also called the 'reciprocals') of H – the orthocenter of triangle ABC – in the two Napoleon circumcircles. (The polar of a point P with respect to a given circle is the line that passes through P', the inverse of P, and is perpendicular to line PP'.) In some sense, this beautiful pair of complementary facts forms a marvelously satisyfing closing couplet to the 'poem' that is constituted by this constellation of discoveries.

A closing 'metamathematical' comment: I have taken great care to write up this set of results in as elegant a form as possible, using a sharp division into short sections some of which have parallel grammar, because to me there is a marvelous drama to these results (and indeed, to high-quality results in mathematics in general). Indeed, I would say that, metaphorically speaking, I have written these results up as a 'poem', where what plays the role of a single line of poetry is a specific small fact, and what plays the role of rhyme is ANALOGY.

Thus, in this particular case, the first two 'lines' of Napoleon's theorem form a 'rhyming couplet' – there is a very tight analogy between them. Similarly, lines I and II of Doug-all's theorem form a tight rhyming couplet, and I took great pains to word them essentially identically in order to bring this 'rhyme'

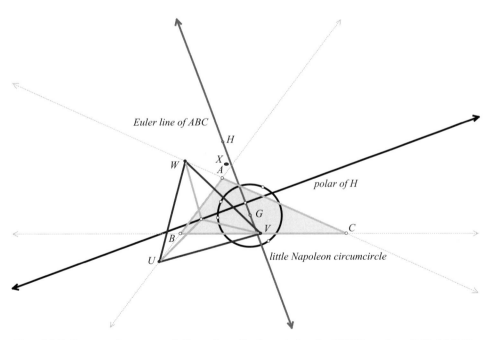

Fig. C4.5. Locus of centers of direct-inscribed equitriangle UVW: polar of H (ABC's orthocenter) with respect to ABC's little Napoleon circumcircle (the locus of centers of direct-escribed equitriangles).

out. And again, lines III and IV of Doug-all's theorem form a slightly looser but still rhyming couplet. Then comes a genuine surprise: line V of Doug-all's theorem forms what might be called an 'oblique rhyme' with line IV of Napoleon's theorem. Such a distant rhyme is very unexpected and very pleasing. Then Lines VI and VII of Doug-all's theorem once again form a rhyming couplet (with 'straight line' rhyming with 'circle', so to speak). Finally, winding things up most beautifully, there is an extremely unexpected allusion back from the second part of line VII of Doug-all's theorem to line II of Napoleon's theorem.

There is a bipartite overall structure to this 'poem' of results (i.e., it is divided into Napoleon's theorem and Doug-all's theorem). Then each part is itself subdivided into various couplets, and there are long-distance 'rhymes' between various lines in the two parts, as well. In my overall title, I alluded to Bach's preludes and fugues, because I felt that in this structure, Napoleon's theorem is prelude-like in that, when one has read the whole thing, one realizes that it is just a 'warm-up', and I felt that Doug-all's theorem is fugue-like in that it has several strands that are tightly interwoven with motifs that keep on returning in different 'keys'. (And keep in mind that in writing this 'poem', I didn't even mention the clockwise and counterclockwise families, or the second Fermat point, which for the clockwise family plays the role that the first Fermat point plays for the counterclockwise

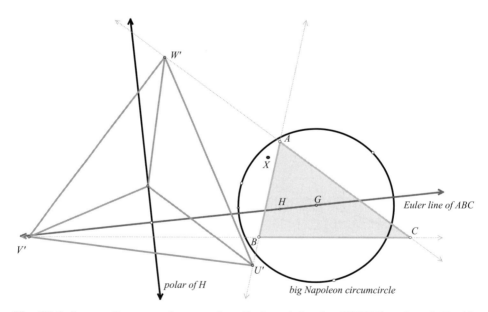

Fig. C4.6. Locus of centers of reverse-inscribed equitriangles $U'V'W'$: polar of H with respect to ABC's big Napoleon circumcircle (the locus of centers of reverse-escribed equitriangles).

family. Had I mentioned these things, the 'poem' might have been 14 lines long – a 'sonnet' – but perhaps a little too heavy.

2.1. Coda

There has been some doubt expressed here and there in the mathematical community as to whether Napoleon's theorem was really discovered by the great French general and emperor (was he sufficiently mathematically talented?), but I personally see no reason to suspect that the appellation misrepresents the truth. Similarly, there has occasionally been some doubt expressed about whether Doug-all's theorem was really discovered by the great French general and prime minister (was he sufficiently mathematically talented?), but why should one have any doubt about such a thing?

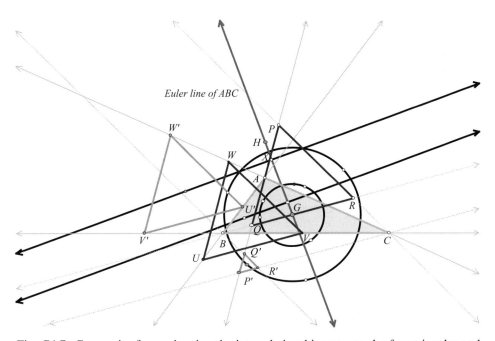

Fig. C4.7. Composite figure showing the interrelationships among the four triangles and the four loci describing their centers.

C5

Miquel's Six Circle Theorem

AAD GODDIJN'S CHOICE

Look at this configuration as a closed chain of four circles.

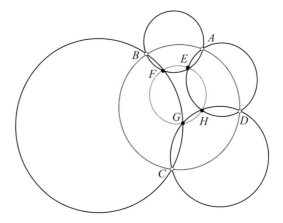

There are eight intersection points; in the figure you find a white and a black intersection point for each pair of intersecting circles. Miquel's Six Circle Theorem asserts that, if the four white points are on a circle, the four black points are also on a circle. The theorem can be more symmetrically reformulated into a statement about six circles, with already seven intersection points of three circles each arranged in the way the figure shows. The theorem guarantees the eighth intersection point of three circles. It's a beautiful circles-only theorem, which can be proven in several ways.

Sixteen-year old Dutch student, Janneke Elbers, used cyclic quadrilaterals in her proof but showed also the basic joy of finding a proof on her own. Initially, she wrote down separate angle relations, using only the four quadrilaterals on the four circles of the closed chain. It brought her nowhere. Suddenly she shouted: "But these are also on a circle!" The cross and line mark that moment. From that point on the proof ran like water and things were done in a few minutes.

Aad Goddijn, Mathematics Educationalist, Freudenthal Institute, University of Utrecht, The Netherlands.

bewys: $\angle EHG + \angle EFG = 180°$

$\angle EHG = 360 - (180 - \angle A_1) - (180 - \angle C_1)$

$\angle EFG = 360 - (180 - \angle A_2) - (180 - \angle C_2)$

$360 - (180 - \angle A_1) - (180 - \angle C_1) + 360 - (180 - \angle A_2) - (180 - \angle C_2)$

$\angle A_1 + \angle A_2 + \angle C_1 + \angle C_2 = 180°$

$360 - (180 - \angle A_1 - \angle A_2 - \angle A_1 - \angle C_2) - 180 + 360 - 180 - 180$

$360 - 0 - 180 + 360 - 180 - 180 = 180$

$\angle EHG + \angle EFG = 180$

dus ☐

Frustration, sudden insight, careful verification: a school example of the basic process of mathematical research!

Editor's Notes

1. Aad Goddijn was a member of the Freudenthal Institute's team that developed aspects of the geometry curriculum adopted by Dutch schools in recent years. The innovative feature of this curriculum is its interweaving of modern applications (Voronoi diagrams) with classical subjects (theory of cyclic quadrangles, conics). In keeping with the philosophy of Hans Freudenthal, attention is paid to the process of obtaining a proof and to the attendant heuristics.

2. The Six Circle Theorem was published by Auguste Miquel probably about 1840. Rather more is known about a related theorem, which focuses on the 'Miquel Point'. This theorem appeared in 'Liouville's Journal' under Miquel's name in 1838 (*J. math. pures appl.* [1] **3** (1838), 486), though William Wallace and Jakob Steiner had stated it earlier: Four straight lines in the plane intersect at six points and produce four triangles. The circumcircles of the four triangles are concurrent (at the Miquel Point). Incidentally, the Miquel Point is the focus of the only parabola to touch all four of the original lines. See the web page: www.geocities.com/Paris/Rue/1861/miquel.html

3. There is a Dutch website at which still another related theorem can be explored interactively: www.pandd.demon.nl/miquel.htm

C6

Eyeball Theorems

ANTONIO GUTIERREZ' CHOICE

Given two circles with centres **A** and **B**, draw the tangents **AC**, **AD**, **BF**, and **BE** from the centres of the circles as shown in the diagram. Then the chords **MN** and **PQ** are equal in length, where **M**, **N**, **P** and **Q** are as shown in the diagram.

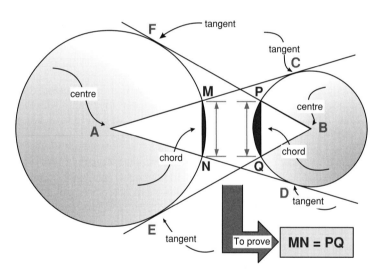

I found that MN = PQ about 35 years ago whilst exploring new problems concerning circles, tangents and chords. I am convinced that this result has been discovered in almost all cultures. However, a recent Internet search surprisingly turned up only a few relevant references to the 'Eyeball Theorem'.

This would be the type of theorem to take on a desert island to exercise the mind because it is elementary, beautiful, and surprising. It is designed to improve visual abilities, intuition and proof, thinking and reasoning skills, and above all the geometric inspiration without which no discovery can be made. Like Archimedes,

Antonio Gutierrez, Geometer and Web Designer, Lima, Peru.

274

we can draw our sketches in the sand, but without the assistance of an ancient Roman soldier, please!

In what follows, I present three proofs of the Eyeball Theorem and a new Eyeball-to-Eyeball theorem with corollary. All are described with graphics, as a reflection of how a humble geometer thinks. What is the process by which geometers scan, focus, drill down, and zoom?

Proof In sympathy with the noble qualities of 'simplicity, economy and elegance', this proof reveals the content and the context of the Eyeball Theorem and, by elucidating new relations between geometric objects like the cyclic quadrilateral FMPC, leads to further discoveries such as the new Eyeball-to-Eyeball Theorem.

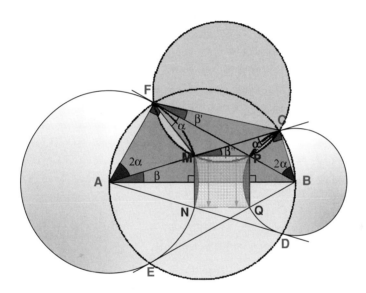

1. $\angle$AFB $= \angle$ACB $= 90°$.
   ∴ AFCB is a cyclic quadrilateral.
   ∴ $\angle$BAC $= \angle$BFC (i.e. $\beta = \beta'$).
2. Let $\angle$MCP $= \alpha$, so that
   $\angle$FBC $= 2\alpha$ (alternate segment theorem, then, angle at centre is twice angle at circumference).
   ∴ $\angle$FAC $= 2\alpha$ (angles in the same segment).
   ∴ $\angle$MFP $= \alpha$
   ∴ FMPC is a cyclic quadrilateral
   and $\angle$PFC $= \angle$PMC (i.e. $\beta' = \beta''$).
3. So $\beta = \beta''$, and hence MP $\parallel$ AB.
   Similarly, NQ $\parallel$ AB and ▭ MPQN is a rectangle
4. ∴ MN $=$ PQ.

Alternative Proofs I have recently found this simple proof which uses similarity of triangles. Like the first proof it leads to further discoveries, such as a formula for **MN**.

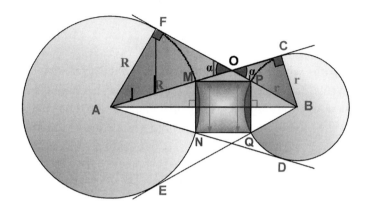

1. $\triangle$AFO $\sim$ $\triangle$BCO (corresponding angles equal).
2. $\therefore \dfrac{OA}{R} = \dfrac{OB}{r}$ (corresponding sides in proportion).
3. $\therefore \triangle$MOP $\sim$ $\triangle$AOB.
   $\therefore \angle$OMP $= \angle$OAB.
   $\therefore$ MP $\parallel$ AB and, similarly, QN $\parallel$ AB.
4. $\therefore$ ▨▨▨ MPQN is a rectangle.
5. $\therefore$ MN $=$ PQ.

Alternatively, use can be made of trigonometry, in this way: If R and r are radii and d is the distance between centres A and B,

then, MN $= 2$R $\sin \angle$MAB $= 2$R $\sin \angle$CAB $= \dfrac{2R \cdot r}{d}$.

Similarly, PQ $= 2$r $\sin \angle$PBA $= 2$r $\sin \angle$FBA $= \dfrac{2r \cdot R}{d}$.

Therefore MN $=$ PQ.

Eyeball-to-Eyeball Theorem Eureka! Exploring the Eyeball Theorem, I saw the forest through the trees and found an elementary theorem that I had never seen before. Euclid, my friend, please include this theorem in your *Elements*.

Given a quadrilateral ABCD inscribed in a circle of centre O, circles of centres A and B intersect in E and M, circles of centres B and C intersect in F and N, circles of centres C and D intersect in G and P, circles of centres D and A intersect in H and Q. If E, F, G, and H lie on the circle of centre O then MNPQ is a rectangle.

Incidentally, the Eyeball Theorem for eyeballs with centres A and C is a special case of the Eyeball-to-Eyeball Theorem when AC is a diameter of circle O.

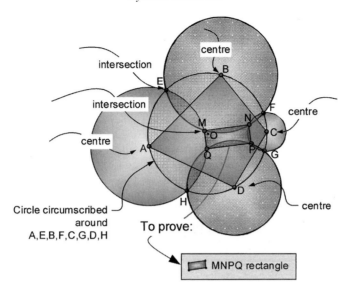

intersection

centre

intersection

centre

centre

Circle circumscribed
around
A,E,B,F,C,G,D,H

centre

To prove:

MNPQ rectangle

I proved the Eyeball-to-Eyeball Theorem by standing on the shoulders of Auguste Miquel, of whose Six Circle Theorem it is a special case. From the outset, the chances that the proof would be 'simple, economic and elegant' were not great, but with a little inspiration, I at least managed to find one which is elementary and which perhaps would have gained the approval of Euclid.

Proof

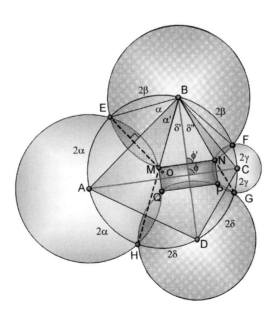

1. The angle formed by the interesting chords AC and BD is measured by half the sum of the intercepted arcs.

   That is, $\phi = \frac{(2\beta + 2\gamma) + (2\alpha + 2\delta)}{2} = 90°$.

2. Let $\angle EBA = \alpha$; then $\angle EOA = 2\alpha$ (angle at centre is twice angle at circumference), arc EA = arc AH and hence $\angle ABH = \alpha$.

3. By symmetry, from the circles with centres A and B, $\angle EBA = \angle MBA = \alpha$.

   $\therefore$ B, M and H are collinear.

   Similarly, B, N and G are collinear.

4. By a similar argument, $\angle MBD = \angle NBD = \delta$ (i.e. $\delta' = \delta'' = \delta$).

   Using the symmetry of isosceles $\triangle MBN$, $\phi' = 90°$.

   $\therefore$ MN $\parallel$ AC.

   Similarly, PQ $\parallel$ AC.

   By a similar argument, MQ $\parallel$ NP $\parallel$ BD.

5. Since AC $\perp$ BD, 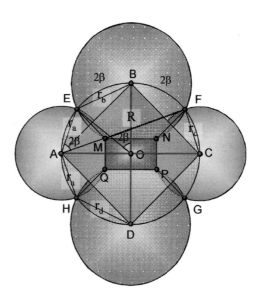 MNPQ is a rectangle.

Corollary Consider the particular case in which AC and BD are diameters 2R of the circle centred at O. If the radii of the circles centred at A, B, C and D are r_a, r_b, r_c and r_d respectively, then:

   Area ▭ MNPQ = EM^2.

Proof

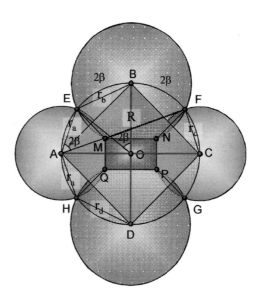

1. By symmetry $r_a = r_c$.

   Similarly, $r_b = r_d$.

2. Also by symmetry

EM = FN = GP = HQ.

3. Making use of the trigonometric relationships in the second alternative proof of the Eyeball Theorem:

$$\text{Area} \quad \square \quad \text{MNPQ} = \text{MQ} \cdot \text{MN} = \frac{2r_a \cdot r_c}{AC} \cdot \frac{2r_b \cdot r_d}{BD} = \frac{4r_a \cdot r_a \cdot r_b \cdot r_b}{2R \cdot 2R} = \left(\frac{r_a \cdot r_b}{R}\right)^2.$$

4. $\triangle EAM \sim \triangle EOB$ (corresponding angles equal).

$$\therefore \frac{r_a}{R} = \frac{EM}{r_b} \Rightarrow EM = \frac{r_a \cdot r_b}{R}.$$

In step 3,

$$\text{Area} \quad \square \quad \text{MNPQ} = \text{EM}^2.$$

Editor's Note

A visit to Antonio Gutierrez' website cannot be recommended too highly. It is devoted in part to the elementary synthetic geometry of Euclid and in part to the geometry of the indigenous peoples of his native Peru – the Inca and the Nazca. Discover this profusion of colour, sound and movement at: http://agutie.homestead.com/

IV

The Golden Ratio

4.1

Introduction

RON KNOTT

One of the simplest definitions of the *golden ratio number* ϕ (phi) is *a number whose square is* 1 *more than the number itself.* This gives rise to the quadratic equation

$$\phi^2 = \phi + 1$$

which is easily solved to give two solutions:

$$\phi = (\sqrt{5} + 1)/2 = 1.61803\ldots \quad \text{and} \quad \phi = (\sqrt{5} - 1)/2 = -0.61803\ldots$$

Dividing our quadratic through by ϕ also tells us that phi is a *number which is* 1 *more than its reciprocal.* From these follow many more fascinating phi facts but our interest here is to look at some of the geometric properties of the golden section although this will illustrate and reveal some more interesting numeric attributes of phi too.

1. Origins and Names

The origin of "golden" in such terms as *golden ratio, golden number* and *golden section* is still a mystery, according to Roger Herz-Fischler in *A Mathematical History of the Golden Number* (Dover, 1998). The term "golden section" was in use in Germany by 1835 as *goldene Schnitt.* Pacioli's book *Divina Proportione* (1509) finds connections between properties of God and properties of the golden section and so calls it the *divine proportion.* The golden section as a process of sectioning a line was well-known to the ancient Greeks and its properties are discussed in Book XIII of Euclid's *Elements.* The two largest Platonic solids (convex solids with identical flat faces and straight sides) are the icosahedron with 20 equilateral triangle faces and the dodecahedron with 12 pentagonal faces. Both of these have

Ron Knott, Former Mathematician and Computing Science Lecturer, University of Surrey. Web Page Design Consultant and creator of the popular *Fibonacci and Golden Section* website at www.mcs.surrey.ac.uk/Personal/ R.Knott/Fibonacci coordinates that are expressed most simply in terms of the golden section. Indeed, it can be argued that one of the aims of Euclid's great work (or at least of Book XIII, the last book) is to prove that there are only five such solids and to find some of their properties.

2. The Line and Its Golden Ratio Points

Euclid describes a method for finding a point G on a line AB where the ratio of the two line segments, shorter to longer, $AG : GB$, is the same as the ratio of the larger segment to the whole line, $GB : AB$, and from this we can again derive the quadratic equation above.

In Book XIII of his *Elements*, Euclid calls this *dividing a line in extreme and mean ratio* although today we might refer to it as *finding the golden section point on a line*. The fact that the quadratic equation derived from this definition gives *two ratios* shows that we can find both an internal point and an external point with the ratio property.

If AB is of unit length, then denoting either AG as x or by denoting GB as x or even by denoting $G'A$ as x, we derive related quadratics with unit coefficients and various signs for the x terms and constant. All of them involve the four quantities, $\pm 1.6180 \ldots$ and $\pm 0.61803 \ldots$, in which the decimal parts are identical, namely $(\sqrt{5} - 1)/2$. All four of these numbers have been called the golden section number although usually it is only the positive values that are referred to. Other authors will use other symbols for these quantities. Here I will refer to $0.61803 \ldots$ as the *golden section*.

3. Plane Objects

There are two isosceles triangles with sides in golden section proportions. The taller one has the two longer sides of unit length and a (shorter) golden section base with interior angles of $72°, 72°, 36°$; the other, flatter, one has a longer base (unit length) with golden section sides and internal angles of $36°, 36°, 108°$. We can therefore call these (with a musical analogy) sharp and flat.

"Sharp" triangle

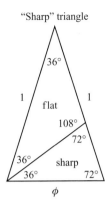

"Flat" triangle

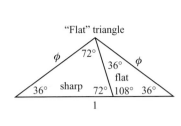

These two triangles are "recursive" in that each contains a copy of itself and the other triangle, and this structure is the basis of an interesting set of tessellations called Penrose Tilings.

A rectangle possessing sides in the golden ratio is called a golden rectangle. It is reputed to occur in many aspects of the Parthenon building on the Acropolis in Athens, in the design structure of many famous paintings and sculptures and is also the shape of a credit card.

The pentagon is the simplest plane object to exhibit the golden section and various examples and properties are illustrated in this chapter. One of the easiest ways to construct a golden section is to use a long strip of paper (such as is used for printing till receipts) and tying an ordinary knot in it. When carefully flattened, and a final fold made, a pentagon is formed. Holding the paper up to the light reveals the pentagon's diagonals. Each of them is divided by the others at its golden section points. The five-pointed star made from the diagonals is called a *pentagram*. The following illustration which, with pun intended, I call a 'Knotty pentagram', is taken from my Fibonacci and Golden Section web site.

   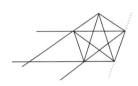

The above is an introduction to the golden section and some of the many places where it is found. The following articles explore further the properties of the golden section and its geometry.

4.2

Regular Pentagons and the Fibonacci Sequence

DOUG FRENCH

If you have ever tried to draw a regular pentagon of given edge length, you will have found that it is difficult to get an accurate diagram if you use a method which involves measuring the interior angles of 108° with a protractor. A better method would involve measuring lengths only, using compasses to construct the diagram. To do this we need to know the length of the diagonals of a regular pentagon. The procedure for drawing the pentagon is then as shown in Figure 4.2.1, where the edge length is 1 and the diagonal length is denoted by the Greek letter, ϕ ("phi").

ϕ could be calculated using sines or cosines, but here we use an alternative method based on the fact that triangles FCD and FEB in Figure 4.2.2 are similar.

Putting this in another way, triangle FEB is an enlargement of triangle FCD. The scale factor of this enlargement is ϕ, since CD = 1 and BE = ϕ. Comparing the other sides of the triangles, we can then say that:

$$\phi FD = BF$$

We note that BD = ϕ and BF = 1, and so FD = $\phi - 1$. The equation above then becomes:

$$\phi(\phi - 1) = 1$$

This is a quadratic equation which can be written as:

$$\phi^2 - \phi - 1 = 0$$

and then solved by completing the square:

$$\left(\phi - \frac{1}{2}\right)^2 - \frac{5}{4} = 0$$

First published in *Mathematics in School* **18**, 2 (March 1989), pp. 40–41.

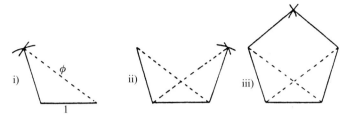

Fig. 4.2.1.

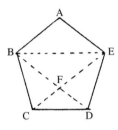

Fig. 4.2.2.

The two solutions are:

$$\phi = \frac{1}{2}(1 + \sqrt{5}) \quad \text{and} \quad \phi = \frac{1}{2}(1 - \sqrt{5})$$

or, to three decimal places:

$$\phi = 1.618 \text{ and } -0.618.$$

As far as the diagonal of the regular pentagon is concerned, the negative root is meaningless, so the diagonal length is 1.618, or, perhaps more usefully, the diagonal length is the edge length multiplied by 1.618. We now have a simple means of calculating the diagonal for a given edge and then drawing an accurate regular pentagon.

The story does not end here though, because regular polygons can grow by extending pairs of sides and joining the 5 points of intersection, as illustrated in Figure 4.2.3. A careful look at the diagram will reveal that the edge length of the larger pentagon is $\phi + 1$, with ϕ and 1 as the diagonal and edge of the smaller pentagon as before.

If we refer to the original quadratic equation $\phi^2 - \phi - 1 = 0$, we find that $\phi + 1 = \phi^2$ and, moreover, we can also say that the diagonal of the larger pentagon is ϕ^3. We then have the striking fact that the sequence of lengths:

edge, diagonal, edge, diagonal, ...

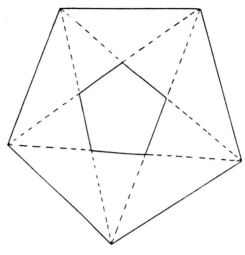

Fig. 4.2.3.

for a succession of regular pentagons, formed by extending the sides, takes the form of a simple geometric sequence:

$$1, \phi, \phi^2, \phi^3, \cdots$$

Previously we have calculated the value of ϕ as $\frac{1}{2}(1 + \sqrt{5})$. It is particularly interesting to examine the sequence above with the powers of ϕ expressed in this form.

$$\phi = \frac{1}{2}(1 + \sqrt{5})$$

$$\phi^2 = \frac{1}{2}(3 + \sqrt{5})$$

$$\phi^3 = \frac{1}{2}(4 + 2\sqrt{5})$$

$$\phi^4 = \frac{1}{2}(7 + 3\sqrt{5})$$

$$\phi^5 = \frac{1}{2}(11 + 5\sqrt{5})$$

$$\phi^6 = \frac{1}{2}(18 + 8\sqrt{5})$$

The terms in the brackets give two important sequences, known as the Lucas sequence:

$$1, 3, 4, 7, 11, 18, \ldots$$

and the Fibonacci sequence:

$$1, 1, 2, 3, 5, 8, \ldots$$

In both sequences, successive terms are found by adding the two previous terms.

1. General Formula

We next consider how we can find formulae for the general terms of these two sequences.

When ϕ was calculated there was a second root $\frac{1}{2}(1 - \sqrt{5})$, which we shall now refer to as ψ ('psi'). This generates a similar sequence to ϕ – the two sequences are placed alongside for comparison below:

$$\phi = \frac{1}{2}(1 + \sqrt{5}) \qquad \psi = \frac{1}{2}(1 - \sqrt{5})$$

$$\phi^2 = \frac{1}{2}(3 + \sqrt{5}) \qquad \psi^2 = \frac{1}{2}(3 - \sqrt{5})$$

$$\phi^3 = \frac{1}{2}(4 + 2\sqrt{5}) \qquad \psi^3 = \frac{1}{2}(4 - 2\sqrt{5})$$

$$\phi^4 = \frac{1}{2}(7 + 3\sqrt{5}) \qquad \psi^4 = \frac{1}{2}(7 - 3\sqrt{5})$$

$$\phi^5 = \frac{1}{2}(11 + 5\sqrt{5}) \qquad \psi^5 = \frac{1}{2}(11 - 5\sqrt{5})$$

$$\phi^6 = \frac{1}{2}(18 + 8\sqrt{5}) \qquad \psi^6 = \frac{1}{2}(18 - 8\sqrt{5})$$

If we add corresponding terms from the two sequences we obtain the Lucas sequence. The Fibonacci sequence is obtained by subtracting corresponding pairs of terms and then dividing by $\sqrt{5}$.

$$\phi + \psi = 1 \qquad \phi - \psi = \sqrt{5}$$
$$\phi^2 + \psi^2 = 3 \qquad \phi^2 - \psi^2 = \sqrt{5}$$
$$\phi^3 + \psi^3 = 4 \qquad \phi^3 - \psi^3 = 2\sqrt{5}$$
$$\phi^4 + \psi^4 = 7 \qquad \phi^4 - \psi^4 = 3\sqrt{5}$$
$$\phi^5 + \psi^5 = 11 \qquad \phi^5 - \psi^5 = 5\sqrt{5}$$
$$\phi^6 + \psi^6 = 18 \qquad \phi^6 - \psi^6 = 8\sqrt{5}$$

We now have formulae for the general or nth terms, L_n and F_n, of the Lucas and Fibonacci sequences.

$$L_n = \phi^n + \psi^n = \left(\frac{1}{2}(1 + \sqrt{5})\right)^n + \left(\frac{1}{2}(1 - \sqrt{5})\right)^n$$

$$F_n = \frac{1}{\sqrt{5}}(\phi^n - \psi^n) = \frac{1}{\sqrt{5}}\left[\left(\frac{1}{2}(1 + \sqrt{5})\right)^n - \left(\frac{1}{2}(1 - \sqrt{5})\right)^n\right]$$

It is instructive to look at the four sequences ϕ^n, ψ^n, L_n and F_n numerically using a calculator or a simple program on a computer. In particular, it will be noted that

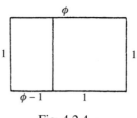

Fig. 4.2.4.

the terms of ψ^n rapidly become rather small, because ψ is numerically less than 1. The first terms in the formulae for L_n and F_n give approximations to the sequences, which improve in accuracy as n gets larger. Again, it is instructive to look at this numerically on a calculator, where the constant facility for multiplication makes it easy to generate the sequences.

The approximations are:

$$L_n \approx \left(\frac{1}{2}(1+\sqrt{5})\right)^n \quad \text{and} \quad F_n \approx \frac{1}{\sqrt{5}}\left(\frac{1}{2}(1+\sqrt{5})\right)^n$$

2. Golden Ratio

The Greeks derived ϕ from the golden rectangle, which was thought to display particularly pleasing proportions, and is, it is said, the shape of the frontage of the Parthenon in Athens. ϕ is known as the golden ratio, the ratio of the sides of a golden rectangle. A golden rectangle is such that when a square of the same width is removed the remaining rectangle is also golden. This is illustrated in Figure 4.2.4.

Since length is width multiplied by ϕ, it follows that:

$$\phi(\phi - 1) = 1$$

This gives the quadratic equation:

$$\phi^2 - \phi - 1 = 0$$

which is precisely the equation derived earlier in relation to the diagonal length of a regular pentagon.

These two sources of the golden ratio – the regular pentagon and the golden rectangle – are brought together very nicely in the icosahedron. The icosahedron is a regular polyhedron with 20 faces in the form of equilateral triangles. If 3 golden rectangles are fitted together so that they are mutually perpendicular, their 12 vertices form the vertices of a regular icosahedron. This is illustrated in Figure 4.2.5.

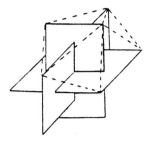

Fig. 4.2.5.

In an icosahedron, 5 equilateral triangles meet at a vertex. Such a set of triangles forms a pyramid with a regular pentagon as its base. One of these pentagons is shown by dotted lines in Figure 4.2.5. We note that the edge lengths are the same as the shorter edges of the golden rectangles and that the one diagonal of the regular pentagon that is shown is the longer edge of one of the golden rectangles. So the icosahedron provides a nice link between regular pentagons and golden rectangles.

4.3

Equilateral Triangles and the Golden Ratio

J. F. RIGBY

This article could be subtitled 'Thoughts on contemplating a model of five tetrahedra inscribed in a dodecahedron'; it is an attempt to communicate the pleasure that ensues when logical reasoning is combined with a visual delight in geometrical figures.

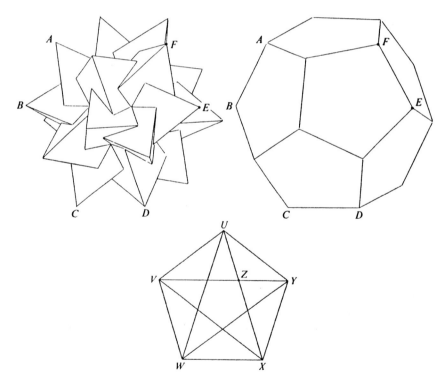

The golden ratio or golden number, ϕ, is the ratio of the lengths of a diagonal and a side of a regular pentagon. If the regular pentagon illustrated has sides of length

First published in *Mathematical Gazette* **72** (March 1988), pp. 27–30; extract.

1 and diagonals of length ϕ, then $VZ = 1$ (since calculations of the angles of triangle ZVU soon show that it is isosceles). Hence $ZY = \phi - 1$ and, as triangles WUX and YXZ are similar,

$$\phi = \frac{UW}{WX} = \frac{XY}{YZ} = \frac{1}{\phi - 1}.$$

Thus $\phi^2 - \phi - 1 = 0$ and hence $\phi = \frac{1}{2}(1 + \sqrt{5}) = 1 \cdot 618 \cdots$.

A rectangle whose sides are in the golden ratio is called a *golden rectangle*. I cannot recall any obvious example of a 'natural' occurrence of a golden rectangle in elementary plane geometry. Such a rectangle is merely used by Euclid as a step in his construction of a regular pentagon. The golden ratio belongs naturally to pentagons rather than to rectangles; it is therefore not clear why and when the golden rectangle first came to be regarded as the most pleasing shape of rectangle. (See, for example, Matila Ghyka's *The geometry of art and life*, Dover 1977, for a detailed discussion of real and imagined occurrences of the golden ratio in art, architecture and nature.)

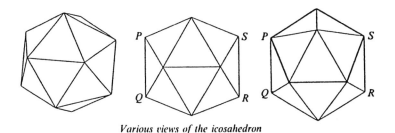

Various views of the icosahedron

However, golden rectangles do occur in three-dimensional geometry, and we shall consider one simple example. The *regular icosahedron* is one of the five Platonic regular solids, having twenty equilateral triangular faces meeting by fives at each of the twelve vertices. The rectangle $PQRS$ in the figure is a rectangle not just in the flat drawing but also in the three-dimensional solid that the figure illustrates. This can be most clearly seen if a three-dimensional model is available for inspection. The right-hand view shows more clearly that PS is a diagonal of a regular pentagon having edges of the icosahedron as its sides, and hence that $PQRS$ is a golden rectangle. (Since P and R are opposite vertices of the icosahedron, we now have a simple way of calculating the circumradius of a regular icosahedron in terms of its edge-length.)

Professor Coxeter of the University of Toronto mentioned to me in a letter the following surprising result which had been communicated to him; surprising because it involves an equilateral triangle rather than a regular pentagon.

Let ABC be an equilateral triangle inscribed in a circle, let L and M be the midpoints of AB and AC, and let LM meet the circle at X and Y as shown. Then $LM/MY = \phi$.

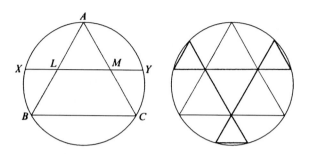

Here is a short proof. Write

$$LM = AM = MC = x \quad \text{and} \quad XL = MY = 1.$$

Then, by the intersecting chord theorem,

$$1 + x = XM.MY = AM.MC = x^2 \text{ and hence } x = \phi.$$

We can now produce the pleasing design shown on the right, in which the ratio of the sides of the larger triangles and the smaller triangles is ϕ.

Many years ago I constructed a cardboard model of five tetrahedra inscribed in a regular dodecahedron. The tetrahedra, together with the dodecahedron whose twenty vertices are the vertices of the tetrahedra, are illustrated at the beginning of this article; the twelve faces of the dodecahedron are regular pentagons. This model features in Cundy and Rollett's *Mathematical models*, Oxford 1951.

Originally I made complicated calculations to determine the exact shape of the individual pieces required in the construction. One such piece is shown by shading in the figure below, and my calculations showed that GL produced passes through

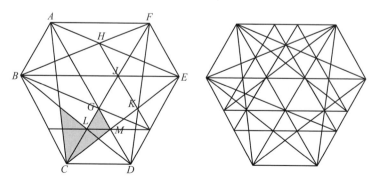

Various views of the icosahedron

C; there is presumably a simple reason for this. Once the model has been made, it is possible to examine the way in which the planes of various faces intersect each other, and to obtain (without arithmetical or trigonometrical calculations) the property just mentioned and various other geometrical properties that are helpful in the drawing of the shaded shape. I do not propose to give the details: a written description is tedious and of little help without a physical model, and those readers who make a model will I hope enjoy contemplating it and working out the details for themselves. Suffice it to say that the twenty faces of the tetrahedra form a regular icosahedron, completely invisible inside the model, whose vertices are at the 'dimples' in the model; also the points A, B, C, D, E and F in the original figure, six of the vertices of the regular dodecahedron, are coplanar and form a cyclic hexagon whose sides have lengths alternately 1 and ϕ.

The lines in the figure now provide all the information required for the construction of the shaded shape, but a further study of the various planes also shows that BG is parallel to AE, that H, J, K are collinear, and that BG, HJK and LM meet on DE. Using rotations and reflections we obtain the right-hand figure, which is fascinating because of the many unexpected points at which three or more lines meet. Golden ratios abound, and the pleasing pattern of equilateral triangles constructed earlier lies at the heart of this last figure.

4.4

Regular Pentagon Construction

DAVID PAGNI

1. Construct a circle with centre O, and radius 1.
2. Draw a diameter $\overline{AB}$.
3. Construct $\overline{DC}$ as another diameter so that $\overline{DC} \perp \overline{AB}$.
4. Bisect $\overline{OB}$, let M be the midpoint.
5. With M as centre and MC as radius, draw an arc intersecting $\overline{AB}$ at P.
6. $\overline{PC}$ is the required side of the regular pentagon circumscribed by the circle. Mark off chords of length PC around the circle.

Proof that $\overline{PC}$ is a required side of the regular pentagon.

1. Since OB $= 1$, OM $= \frac{1}{2}$ and MC $= \frac{\sqrt{5}}{2}$
2. OP $= \frac{\sqrt{5}}{2} - \frac{1}{2}$ so $(PC)^2 = (OP)^2 + 1^2$ and $\therefore$ PC $= \frac{\sqrt{10-2\sqrt{5}}}{2}$ (see Figure 4.4.1)
3. Since the central angle of a regular pentagon is $\frac{360°}{5} = 72°$, we must show that $\sin 36° = \frac{\sqrt{10-2\sqrt{5}}}{4}$. To do this we shall use a new figure with similar triangles.

In Figure 4.4.2, we have an isosceles triangle with unit sides such that its base angles are $72°$. $\overline{AD}$ bisects $\angle A$.

$$AC = CB = 1$$

$$\triangle ABD \sim \triangle CAB \quad (AAA)$$

$$\therefore \frac{AB}{BD} = \frac{AC}{AB}$$

$$\frac{AB}{1 - CD} = \frac{AC}{AB}$$

$$\frac{AB}{1 - AD} = \frac{AC}{AB}$$

First published in *Mathematics in School* **29**, 1 (January 2000), pp. 14–15.

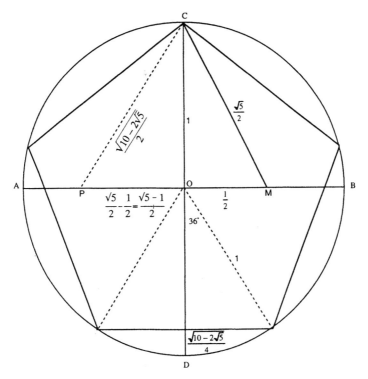

Fig. 4.4.1.

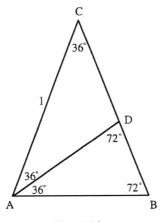

Fig. 4.4.2.

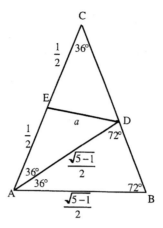

Fig. 4.4.3.

$$\frac{AB}{1 - AB} = \frac{1}{AB}$$

$$\Rightarrow AB = \frac{\sqrt{5} - 1}{2} \quad \text{(The Golden Ratio)}$$

Relabelling the triangle with this information and constructing $\overline{DE} \perp \overline{AC}$ we get Figure 4.4.3 from which we can derive the rest of the proof.

Let $ED = a$, then, using the Pythagorean Theorem:

$$a^2 + \left(\frac{1}{2}\right)^2 = \left(\frac{\sqrt{5} - 1}{2}\right)^2$$

$$a^2 = \frac{5 - 2\sqrt{5} + 1}{4} - \frac{1}{4} = \frac{5 - 2\sqrt{5}}{4}$$

$$a = \frac{\sqrt{5 - 2\sqrt{5}}}{2}$$

$$\therefore \sin 36° = \frac{a}{\frac{\sqrt{5}-1}{2}} = \frac{\frac{\sqrt{5-2\sqrt{5}}}{2}}{\frac{\sqrt{5}-1}{2}}$$

$$= \frac{\sqrt{5 - 2\sqrt{5}}}{2} \cdot \frac{2}{\sqrt{5} - 1} = \sqrt{\frac{5 - 2\sqrt{5}}{(\sqrt{5} - 1)^2}} = \sqrt{\frac{5 - 2\sqrt{5}}{6 - 2\sqrt{5}}}$$

$$= \sqrt{\frac{(5 - 2\sqrt{5})(6 + 2\sqrt{5})}{36 - 20}} = \sqrt{\frac{30 - 2\sqrt{5} - 20}{16}} = \frac{\sqrt{10 - 2\sqrt{5}}}{4}$$

Thus, since we have shown that $\sin 36° = \frac{\sqrt{10-2\sqrt{5}}}{4}$, and the side of our pentagon has length $CP \frac{\sqrt{10-2\sqrt{5}}}{2}$, CP is the correct length for the pentagon.

4.5

Discovering the Golden Section

NEVILLE REED

The golden section seems to have been first discovered by the Pythagoreans as the length of the diagonals of a regular pentagon with unit sides [1]. In the absence of hard evidence, it is fun to speculate as to how (presupposing whether!) they went on to devise a construction for the golden section. Although this is equivalent to providing a construction for the solution $\frac{1}{2} + \frac{1}{2}\sqrt{5}$ of the quadratic equation $\frac{\phi}{1} = \frac{1}{\phi - 1}$ or $\phi^2 - \phi - 1 = 0$, it seems best to argue intuitively and geometrically where possible.

Consider then an x by 1 golden rectangle (Figure 4.5.1). The defining property of the golden section shows that the small rectangle is similar to the whole rectangle. Since they are at right angles to one another, their diagonals, crossing at O and O', are perpendicular (Figure 4.5.2). Notice also that $OO' = \frac{1}{2}$.

Now suppose the smaller rectangle is slid a distance $\frac{1}{2}$ over the large one until O and O' coincide and put in AF, BC' (Figure 4.5.3). A rhombus $ABC'F$ is created

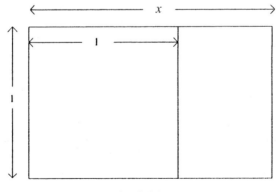

Fig. 4.5.1.

First published in *Mathematical Gazette* **81** (March 1997), pp. 115–116.

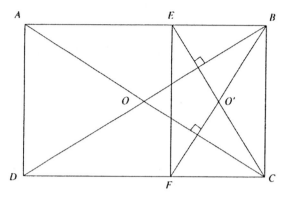

Fig. 4.5.2.

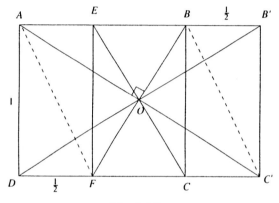

Fig. 4.5.3.

with $AB = AF = \frac{1}{2}\sqrt{5}$, by Pythagoras' Theorem in ADF, so that finally

$$x = AB + BB' = \frac{1}{2}\sqrt{5} + \frac{1}{2}.$$

The final diagram translates into a construction for the golden section which is strongly reminiscent of that given by Euclid [1].

Reference

1. Carl B. Boyer, *A History of Mathematics* (Wiley, 1968) 55–57.

4.6

Making a Golden Rectangle by Paper Folding

GEORGE MARKOWSKY

The golden ratio $((1 + \sqrt{5})2 \approx 1.618)$ is a number with many interesting mathematical properties. The purpose of this short note is to show how you can construct a golden rectangle using just a sheet of $8.5'' \times 11''$ paper[1] without any other tools. A golden rectangle is a rectangle such that the ratio of one dimension over the other is the golden ratio.

One problem in creating a golden rectangle out of an $8.5'' \times 11''$ sheet of paper is that the ratio of length/width is only about 1.29. Step 1 of the construction makes the paper narrow enough so we can create a golden rectangle from it. If for your sheet of paper the ratio of the longest dimension to the shortest dimension is greater than the golden ratio, then you can begin with Step 2.

1. Place the paper before you so its long dimension is horizontal. Make a horizontal fold in the paper in the exact middle of the sheet by folding the sheet in half. Fold the paper along the crease several times until the paper will tear readily along the crease. The final result will be a long thin rectangle $ABCD$.
2. Fold corner A so it rests on DC as shown in Figure 4.6.1. Let E be the new point created on AB where the paper is folded.
3. Using E as a guide, create a vertical crease at point E. Let F be the point where the crease touches DC. In Figure 4.6.2 the dashed line EF represents the crease, and the figure $AEFD$ is a perfect square.
4. Now fold the paper so that AD perfectly coincides with EF. This creates a vertical crease in the middle of the square. Mark the ends of this crease with G and H as shown in Figure 4.6.3.
5. Fold HC so that it runs through E and mark HC with a little fold where it intersects E as shown in Figure 4.6.4. Call this point J.
6. Make a vertical crease at J by folding the paper. Call the other end of the crease I. Figure 4.6.5 shows all the vertical creases created so far.

First published in *Mathematical Gazette* **75** (March 1991), pp. 85–87.
[1] Sheets of size $8.5'' \times 11''$ (21.6 cm $\times$ 27.9 cm) are standard in USA.

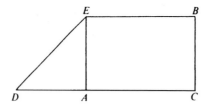

Fig. 4.6.1. The first step in making a square.

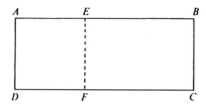

Fig. 4.6.2. Creating a perfect square.

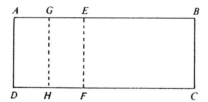

Fig. 4.6.3. Dividing the square in half.

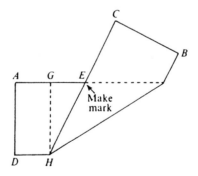

Fig. 4.6.4. Marking the edge *DC*.

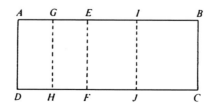

Fig. 4.6.5. All the vertical creases in *ABCD*.

Fig. 4.6.6. Finally, a Golden Rectangle.

7. Fold the paper along IJ sufficiently many times to ensure that it rips smoothly along IJ, then rip it along IJ. The final rectangle, $AIJD$ in Figure 4.6.6, is a Golden Rectangle.

To see that $AIJD$ of Figure 4.6.6 is a golden rectangle, study Figure 4.6.4 and use x to denote the distance DH. For an $8.5'' \times 11''$ piece of paper it is easy to see that $x = 2.125''$. In terms of x, GE has length x and GH has length $2x$. Since HGE is a right triangle, the Pythagorean Theorem shows that $HE = HJ = x\sqrt{5}$. Thus, the size of DJ is $x(1 + \sqrt{5})$ and since the size of AD is $2x$ the ratio DJ/AD is the golden ratio.

4.7

The Golden Section in Mountain Photography

DAVID CHAPPELL
CHRISTINE STRAKER

Recently we discovered a reference to the golden section in the guidebook *The Peak and the Pennines* by W. A. Poucher [1] who was considering the qualities possessed by a "good" photograph, particularly in the area of mountain or scenic photography. Poucher claims that the common factor in "good" photographs appears to be the positioning of the main subject on the vertical line which divides the photograph in a golden section. Poucher gives a method for constructing an approximation to this section as follows:

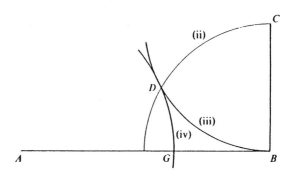

(i) AB is the given line (or width of the photograph) and BC is perpendicular to AB with $BC = \frac{1}{2}AB$.

(ii) Draw a circular arc centre B radius BC.

(iii) Draw another circular arc centre C radius BC meeting the first arc at D.

(iv) Draw a circular arc centre A radius AD meeting AB at G. Then G divides AB in approximately a golden section.

That method is not at all the kind of procedure that a photographer "in the field" could employ! Nor, of course, is it the best construction available. It should be borne in mind that Poucher is a mountaineer, not a mathematician. His construction

First published in *Mathematical Gazette* 74 (March 1990), pp. 43–45.

and the calculation of his ratio makes an interesting classroom exercise: in fact his ratio BG:GA differs from the golden section by about 0.69%: his ingenuity in deriving such a close approximation is to be admired. However his construction is more complicated than one giving the exact ratio:

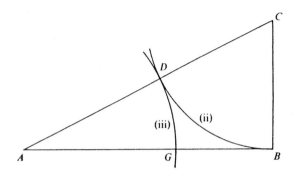

(i) AB is the given line (or width of the photograph) and BC is perpendicular to AB with $BC = \frac{1}{2}AB$.
(ii) Draw a circular arc centre C radius CB meeting AC at D.
(iii) Draw a circular arc centre A radius AD meeting AB at G. Then G divides AB in a golden section.

Of course neither construction is much practical use, nor is such accuracy likely to make a big difference to your photographs. Perhaps making the subject "a little off-centre" is enough. Alternatively, by some simple choices of mid-points we can easily divide the range in the ratio 3:5 (which differs from the golden section by about 2.9%:

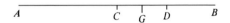

Perhaps readers will come up with their own constructions and practical applications.

Reference

1. W. A. Poucher, *The Peak and the Pennines* (London: Constable, 1978).

4.8

Another Peek at the Golden Section

PAUL GLAISTER

According to the previous note a photograph of the mountain illustrated below would look best if the vertical line through Q divided the line OP in a golden section.

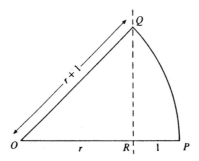

Instead, let us find r so that the volume of the mountain to the left of the vertical is equal to the volume of the mountain to the right of it. For these purposes imagine that the shape of the mountain is obtained by rotating the sector OQP through $180°$ around OP.

Left-hand piece

The half of a cone obtained by rotating OQR through $180°$ about OP:

$$\text{Volume}_L$$

$$= \frac{1}{2} \times \frac{1}{3} \times \pi (RQ)^2 \times r.$$

But

$$(RQ)^2 = (OQ)^2 - (OR)^2$$

$$= 2r + 1$$

Right-hand piece

By constructing Cartesian axes with origin O and x-axis along OP we see that the right-hand volume is given by

$$\text{Volume}_R$$

$$= \frac{1}{2} \int_r^{r+1} \pi [(r + 1)^2 - x^2] dx$$

First published in *Mathematical Gazette* **74** (March 1990), pp. 45–46.

and so the left-hand volume is

$$\frac{1}{6}\pi r(2r+1).$$

which soon reduces to

$$\frac{1}{6}\pi(3r+2)$$

For these two volumes to be equal we need

$$\frac{1}{6}\pi r(2r+1) = \frac{1}{6}\pi(3r+2)$$

which reduces to

$$r^2 - r - 1 = 0$$

whose positive solution is the much-studied golden ratio $\frac{1}{2}(\sqrt{5}+1)$. It has therefore turned out that the condition of equal volumes also means that OP should be divided in a golden section.

4.9

A Note on the Golden Ratio

A. D. RAWLINS

A curiosity that I have recently discovered may be of interest to your readers. I have looked in the standard book on this topic [1] but have not been able to find any reference to the property given below.

Consider an annulus of inner radius a and outer radius b. Inscribe an ellipse of minor axis $2a$ and major axis $2b$, so that it touches the inner and outer circle of the annulus, Figure 4.9.1.

If the area E of the ellipse is equal to the area A of the annulus, then the ratio of a to b is the golden ratio.

To see this we have

$$E = \pi ab, \quad A = \pi(b^2 - a^2)$$

so that

$$E = A \Rightarrow b^2 - ab - a^2 = 0.$$

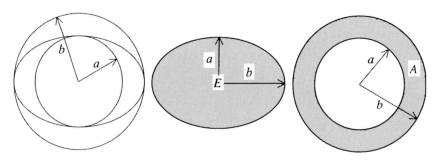

Fig. 4.9.1.

First published in *Mathematical Gazette* **79** (March 1995), p. 104.

Thus

$$b = \left(\frac{1 \pm \sqrt{5}}{2} \right) a.$$

For $a > 0, b > 0$ we have

$$\frac{b}{a} = \frac{1 + \sqrt{5}}{2} = 1.618\ldots \text{ The golden ratio!}$$

Reference

1. H. E. Huntley, *The Divine Proportion, A Study in Mathematical Beauty* (New York: Dover, 1970).

4.10

Balancing and Golden Rectangles

NICK LORD

It is always intriguing to meet old friends in new contexts. Consider the following standard-looking problem: What $x \times x$ square must be removed from the 1×1 square lamina shown in Figure 4.10.1 so that the remaining gnomon has centre of mass at the corner of the square removed?

Taking moments about the left-hand edge in the usual way, this requires that:

$$\frac{1}{2}x \times x^2 + x(1 - x^2) = \frac{1}{2} \times 1$$

$$\underset{\text{square}}{\underset{\text{removed}}{}} \qquad \underset{}{\text{gnomon}} \qquad \underset{\text{square}}{\underset{\text{original}}{}}$$

whence $x^3 - 2x + 1 = 0$ or $(x - 1)(x^2 + x - 1) = 0$.

Thus either $x = 1$ (interpret) or $x = \frac{1}{2}(\sqrt{5} - 1) = 1/\phi$ where ϕ, the golden ratio, is $\lim_{n\to\infty} F_{n+1}/F_n$ for the Fibonacci sequence $F_0 = 1; F_1 = 1; F_{n+1} = F_n + F_{n-1}$. In other words, the legs of the gnomon are golden rectangles.

In an analogous way, removing a hypercube of side length x from a unit k-dimensional hypercube results in the remainder having a centre of mass at a vertex

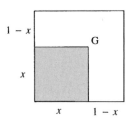

Fig. 4.10.1.

First published in *Mathematical Gazette* **79** (November 1995), pp. 573–574.

of the removed hypercube if x satisfies:

$$\frac{1}{2}x \times x^k + x(1 - x^k) = \frac{1}{2} \times 1 \quad \text{or} \quad x^{k+1} - 2x + 1 = 0.$$

Thus either $x = 1$ or $x^k + x^{k-1} + \cdots + x - 1 = 0$, an equation which has but one root x_k between 0 and 1. Again $1/x_k = \lim_{n \to \infty} G_{n+1}/G_n$ where (G_n) is the generalised Fibonacci sequence $G_0 = G_1 = \cdots = G_{k-2} = 0; G_{k-1} = 1; G_{n+1} = G_n + G_{n-1} + \cdots + G_{n+1-k}$.

In fact, for all k, $x^k + x^{k-1} + \cdots + x - 1 = 0$ is an irreducible polynomial and it is not too hard to show that the root x_k decreases with k and satisfies the bounds $\frac{1}{2} < x_k < \frac{1}{2} + \frac{1}{2k}$ so that $x_k \to \frac{1}{2}$ as $k \to \infty$.

4.11

Golden Earrings

PAUL GLAISTER

Nick Lord's interesting note on squares reminded me of a similar problem which is associated with a more aesthetic shape. Lord removed a square from the corner of a square lamina, and showed that the remaining shape has its centre of mass at the corner of the removed square when the sides of the two squares are in the golden ratio.

Figure 4.11.1 shows a circular lamina of unit radius from which is removed a circular lamina of radius x. If the centre of gravity of the remaining shape is at the edge of the removed lamina, then

$$\underset{\text{removed shape}}{x \times \pi x^2} \quad + \quad \underset{\text{remaining shape}}{2x \times (\pi 1^2 - \pi x^2)} \quad = \quad \underset{\text{original shape}}{1 \times \pi 1^2}$$

where we have taken moments about the common tangent. Simplifying this expression gives

$$x^3 - 2x + 1 = 0 \quad \text{i.e.} \quad (x - 1)(x^2 + x - 1) = 0$$

whose solutions are $x = 1, \frac{1}{2}(\sqrt{5} - 1), \frac{1}{2}(-\sqrt{5} - 1)$. Only the second solution is meaningful so that the centre of mass of the remaining shape is at the edge of the removed lamina if the radii are in the golden ratio $1 : x = 1 : \frac{1}{2}(\sqrt{5} - 1) = \frac{1}{2}(\sqrt{5} + 1) : 1$. I take the liberty of naming this the *golden earring property*! Figure 4.11.2 shows a golden earring. The generalisation to higher dimensions is straightforward, and gives either golden balls or golden hyperballs!

First published in *Mathematical Gazette* **80** (March 1996), pp. 224–225.

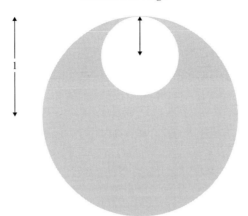

Fig. 4.11.1.

Fig. 4.11.2.

4.12

The Pyramids, the Golden Section and 2π

TONY COLLYER
ALEX PATHAN

1. Introduction

Much has been written on pyramid mysticism and in mathematics this means the relationships between the parameters of the pyramids, the golden ratio and 2π. The Greeks favoured the golden ratio in their paintings and in their classical architecture because the rectangle produced is pleasing to the eye. The Parthenon is a typical example of this. The ratio was called the 'sacred ratio' in the Papyrus of Ahmes, which gives an account of the building of the Great Pyramid at Giza about 3070 BC, but the Ancient Greeks preferred the term 'golden section'.

2. An Investigation of the Egyptian Prescription for Pyramid Building

Let us investigate the Ancient Egyptians' attitude to the Golden Section a little further. Herodotus stated that the Egyptian priests reported that the proportions of the Great Pyramid were such that the area of the square whose side is the same height as the pyramid is equal to the area of one of the sloping triangular faces (Burton, 1985; Hollingdale, 1994). In other words the area of one of the faces of the triangular sides is h^2. Suppose the height is h, the length of each side is $2b$ and a for the altitude of a face triangle, as shown in Figure 4.12.1.

Using Pythagoras' Theorem we have that:

$$a^2 = h^2 + b^2$$

giving

$$h^2 = a^2 - b^2$$

Now let us put in the condition for the areas.

The area of the base of side h = area of one of the sloping triangular faces:

$$h^2 = \frac{1}{2}(2b) \times a = ab$$

First published in *Mathematics in School* **29**, 5 (November 2000), pp. 2–5.

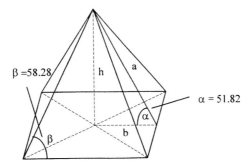

Fig. 4.12.1. Illustrating the dimensions b, a and h for a pyramid, where the angles shown are in degrees. Controlling the angle α is the key to obtaining the Golden Section throughout.

Let us now equate our two values for h^2:

$$a^2 - b^2 = ab$$

If we divide by b^2 we get:

$$\frac{a^2}{b^2} - 1 = \frac{a}{b}$$

and rearranging this gives us the quadratic equation:

$$\left(\frac{a}{b}\right)^2 - \frac{a}{b} - 1 = 0$$
$$G^2 - G - 1 = 0 \tag{1}$$

and the positive root of this equation is the Golden Section, $\frac{1+\sqrt{5}}{2} = 1.6180339$. You will notice that:

$$G = 1 + \frac{1}{G} \tag{2}$$

By checking the values of a and b with those for the Great Pyramid of Chiops at Giza, which was built about 4750 BC, we obtain a surprisingly accurate result; where $h = 481.2$ ft and $b = 377.89$ ft.

$$a = \sqrt{h^2 + b^2} = \sqrt{(481.2)^2 + (377.89)^2} = 611.85$$
$$\frac{b}{a} = 0.61762$$

and
$$G = \frac{a}{b} = 1.61762$$

Simply arranging that the area of one of the sloping triangular faces is h^2 gives rise to even more fascinating consequences when we consider three particular areas, the area of the base A_1, the area of the four sloping triangular sides A_2 and the total surface area A_3. Consider the following ratios:

$$\frac{A_3}{A_2} = \frac{A_2}{A_1} = \frac{a}{b} = G$$

Examining the first ratio gives:

$$\frac{A_3}{A_2} = \frac{(2b)^2 + 4ab}{4ab} = \frac{b}{a} + 1 = \frac{a}{b} = G$$

Examining the second ratio gives:

$$\frac{A_2}{A_1} = \frac{4ab}{(2b)^2} = \frac{a}{b} = G$$

Thus proving our initial ratio. It seems astonishing that by arranging the area of one of the faces to be the square of the height that $\frac{a}{b}$ and the ratios of these above areas gives the Golden Section.

Did the Egyptians build their pyramids to the prescription depending on the measurement of areas? It seems hardly likely to us. We suggest they must have built to a prescription that depended on easy distances to measure. We know that the Egyptians controlled the angles of their pyramids by measuring the run over the rise. In other words, they measured the cotangent of α. This is often called the *seked* or the *seqt*. Our approach is to prove that if the square of the cotangent of the angle of slope of the faces is the reciprocal of the Golden Section G, the condition expressed by Herodotus is a necessary consequence. It is simply a matter of $(\frac{run}{rise})^2$.

We have from the conditions above that $\frac{a}{b} = 0.6180339$, the reciprocal of the Golden Section or Sacred Ratio. We can find the angle of the slope, as shown in Figure 4.12.1, because $\cos \alpha = \frac{b}{a}$, from which the angle $\alpha = 51°50'$. The slope of the Great Pyramid as measured today is $51°52'$ – that is very close indeed! Additional evidence is provided from measurements on other pyramids. The first true pyramid was built at Medumi, just before the Great Pyramid, with the same slant angle, while the other two large pyramids built at Giza around 4600 BC have slant angles of $53°10'$ and $51°10'$ (Hollingdale, 1994).

How can we get this angle in terms of the run and the rise, i.e. the cotangent? The Golden Section has some very mystical properties, which we can simply investigate by finding the sin and the cot of α.

Now $\cos \alpha = 0.6180339$, from which $\cos^2 \alpha = 0.3819659$. Using $\sin^2 \alpha + \cos^2 \alpha = 1$ we can find $\sin^2 \alpha$.

$$\sin^2 \alpha = 1 - \cos^2 \alpha = 1 - 0.3819659 = 0.6180339$$

which is the reciprocal of the Golden Section again. This means that if $\cos \alpha =$ the reciprocal of the Golden Section, $\sin^2 \alpha =$ the reciprocal of the Golden Section;

and

$$\cot^2 \alpha = \frac{\cos^2 \alpha}{\sin^2 \alpha} = \frac{0.6180339^2}{0.6180339} = 0.6180339 \qquad (3)$$

From this

$$\cot^2 \alpha = \sin^2 \alpha = \cos \alpha = \frac{1}{G}$$

We cannot escape this Golden Section. From the construction of the face of the pyramid, Equations (2) and (3) are self-consistent as they represent the well-known trigonometric relationships:

$$\csc^2 \alpha = 1 + \cot^2 \alpha \qquad \sec^2 \alpha = 1 + \tan^2 \alpha$$
$$G = 1 + \frac{1}{G} \qquad \text{and} \qquad G^2 = 1 + G$$
$$G = 1 + \frac{1}{G}$$

because

$$\csc^2 \alpha = \frac{1}{\sin^2 \alpha} = G \quad \text{and} \quad \sin^2 \alpha = \frac{1}{G}$$

Therefore, returning to a practical viewpoint, if we build our pyramid such that the $\left(\frac{run}{rise}\right)^2$ is equal to the reciprocal of the Golden Section, all of these ratios come to pass. This will make $\alpha = 51°50'$ and the $\frac{run}{rise} = 0.786$, which is roughly 24 units of rise for 19 units of run, quality assurance being secured by a large number of lashes.

There is an even stranger consequence if we consider the angles in one of the triangular faces by looking at it at normal incidence. Half the triangle forms a right-angled triangle of height a and base b. This is shown in Figure 4.12.2.

From Figure 4.12.1, $\cot \beta$ is the ratio $\frac{b}{a}$, which is the reciprocal of the Golden Section again, making $\cot \beta = 0.6180339$ and $\beta = 58.28°$.

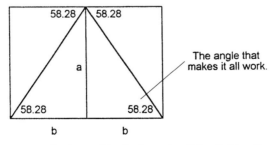

Fig. 4.12.2. The two rectangles have sides in the ratio of the Golden Section, giving the two base angles of the isosceles triangle of 58.28°. These are the base angles of the triangular faces of the Great Pyramid.

The rectangular shapes shown in Figure 4.12.2 are two rectangles such that the ratio of their sides is in the Golden Section. This gives the ratio $\frac{b}{a} = 0.6180339$ and the base angles $= \beta = 58.28°$ for each of the four triangles. Only a square rather than a triangular base permits everything to fall into place.

3. Where Does 2π Come Into It?

There is one last breathtaking consequence, which can be explained by such simple mathematics that it makes the authors believe they may be reinventing the wheel, although not having seen it before. Over the years there has been much speculation over whether the ratio of the perimeter of the base of the Great Pyramid to its height was designed to be equal to 2π as well as having these other strange properties. Several mathematics historians dismiss this as being inconsistent with the geometry known to the Egyptians at that time (Boyer and Merzbach, 1991; Wheeler, 1935). Ahmes used a value of π of $3\frac{1}{6}$ rather than our value of $3\frac{1}{7}$, and this continued in the Kahun Papyrus, and yet the ratio of the perimeter to the height of the Great Pyramid is $\frac{4 \times 440}{280} = 6.285714286$, which hardly differs from our value of $2\pi (= 6.283185307)$.

This seems very mystical indeed, but in reality it can very easily be explained. All one has to do is keep the $\frac{run}{rise} = \frac{1}{\sqrt{G}}$ which is what we described earlier.

$$\frac{run}{rise} = \frac{b}{h} = \frac{1}{\sqrt{G}}$$
$$\frac{perimeter}{height} = \frac{4 \times (2b)}{h} = \frac{8b}{h} = \frac{8}{\sqrt{G}}$$
$$= 6.289211$$

This is always true, and this value differs from the accepted value of 2π by less than 1 part in 1000.

4. Summary

With just a simple piece of trigonometry and geometry we have shown that if the ratio of $(\frac{run}{rise})^2$ is made equal to the reciprocal of the Golden Section, Herodotus' condition is immediately imposed, the eight right-angled triangles from the four isosceles triangles forming the sides of the pyramid have two sides in the ratio of the Golden Section and the ratio of the areas $\frac{A_3}{A_2} = \frac{A_2}{A_1} = \frac{a}{b} = G$ is maintained. Moreover the ratio of the perimeter of the base divided by the height is close to but never equal to 2π. All you have to do is to control the run and the rise.

References

C. B. Boyer, & U. C. Merzbach, *A History of Mathematics* (New York: John Wiley, 1991) 11.

D. M. Burton, *History of Mathematics* (London: Allyn & Bacon Inc., 1985), 56–66.

S. Hollingdale, *Makers of Mathematics* (London: Penguin Books, 1994), 48–49.

N. Peace, T. Feunteun, & M. Walker, Investigation of the great pyramid of Giza, *Mathematics in School* **26**, 1 (1994), 6–8.

N. F. Wheeler, *Pyramids and Their Purpose* (1935).

4.13

A Supergolden Rectangle

TONY CRILLY

Can we produce a ratio ψ to rival the golden ratio ϕ? In this article I shall construct a variant of the classical golden rectangle by choosing a different geometrical property. The rectangle described has more advanced mathematical properties than its golden relation, though it is not extravagant. In this sense it could be described as a *supergolden* rectangle.

The golden ratio ($\phi = 1.618034\ldots$) is at the apex of the aesthetics of rectangles – its link with classical architecture and its many distinctive mathematical properties led to its description by Pacioli as the *divine proportion*. Indeed, to support the theory that the golden rectangle has the greatest aesthetic appeal, and is in some sense perfect, Gustav Fechner, the German psychologist of the nineteenth century, made thousands of measurements of commonly appearing rectangles – playing cards, windows, books – and found that the most commonly occurring ratio was close to ϕ.

Fechner's conclusion is not so convincing today, if only because there is less uniformity in the rectangles we commonly see. Penguin books, for example, have moved away from their former uniform size (with sides in approximate golden ratio) to providing books in many different proportions. In the absence of a modern Fechner, a straw poll of rectangles around the house revealed the following data:

Paper format (e.g. A4)	1.414
Softback book (Penguin)	1.636
Hardback book	1.353
Mathematical Gazette	1.569
Playing card	1.547
Door (outlier)	3.046
Daily newspaper	1.558

First published in *Mathematical Gazette* **78** (November 1994), pp. 320–325.

Sunday newspaper   1.463
Weetabix box   1.316
Cornflakes packet   1.465

The classical golden rectangle appears too narrow for today's tastes. We need something more abundant. To find it, consider the geometry of the golden rectangle.

The property which defines the shape is that $EFCD$ shall be similar to $ABCD$, with $ABFE$ a square.

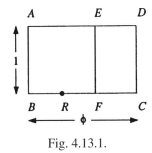

Fig. 4.13.1.

It is then easy to find ϕ exactly as a solution of the equation

$$\frac{\phi}{1} = \frac{1}{\phi - 1}$$

which rearranges to the quadratic equation

$$\phi^2 - \phi - 1 = 0,$$

having positive root

$$\phi = \frac{1 + \sqrt{5}}{2}.$$

Note that the rectangle can be constructed by ruler and compass by drawing the circle through R the midpoint of the side of the square, which then passes through E and C.

1. A Supergolden Rectangle

Now consider Figure 4.13.2. We now look for a rectangle which can be subdivided so that $ABCD$, $EHGD$ and $HFCG$ are similar.

At the outset there is the question of whether such a rectangle exists, but we shall see that it is easily determined. The ratio of the largest rectangle $ABCD$ to the smallest rectangle $HFCG$ implies

$$\frac{x}{1} = \frac{x - 1}{1 - y}$$

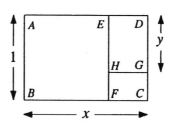

Fig. 4.13.2. The supergolden rectangle.

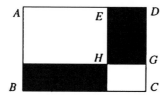

Fig. 4.13.3. Equal areas.

and reorganising this equation shows that x and y are inverse to each other: $xy = 1$. This is equivalent to saying that the shaded rectangular areas (Figure 4.13.3) are equal.

But this does not fully determine x. The value of x is fixed by comparing the rectangle $ABCD$ to the rectangle $EHGD$. This gives

$$\frac{x}{1} = \frac{y}{x-1}.$$

The fact that x and y are inverse to each other shows that x must satisfy the *cubic* equation

$$x^3 - x^2 - 1 = 0.$$

It is straightforward to check that this cubic has exactly one positive root ψ, between 1 and 2 (the other roots are complex). It is slightly easier to solve the cubic for y: $y^3 + y - 1 = 0$, since this has no squared term. The value of y is the irrational number

$$\theta = \sqrt[3]{\frac{1 + \sqrt{\frac{31}{27}}}{2}} + \sqrt[3]{\frac{1 - \sqrt{\frac{31}{27}}}{2}}$$

which has numerical value 0.68233. The value of $x = \psi$ is

$$\psi = 1.465567\ldots$$

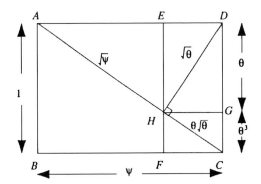

Fig. 4.13.4. Geometrical properties of the supergolden rectangle.

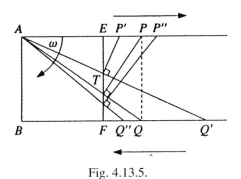

Fig. 4.13.5.

2. Dynamical Generation and a Principle of Uncertainty

The geometrical properties given above give some insight into how the supergolden rectangle might be constructed. Take any square and extend the sides AE and BF indefinitely, to form tramlines (Figure 4.13.5).

Take a line from A making an angle ω with the horizontal, mark the point where this line cuts the lower tramline. Also mark the point where the perpendicular at the intersection of this line and EF meets the upper tramline. We seek the value of ω for which the point P is directly above Q. When this happens the rectangle $ABQP$ is supergolden. [This method of construction will generate any shape of rectangle if we allow the angle at T to vary. In particular if $PTA = \tan^{-1} \phi^3$ we obtain a golden rectangle.]

There is also present here a principle of uncertainty, for since P is determined by an irreducible cubic equation, it is not possible to line up P and Q *simultaneously* by ruler and compass.

3. The Supergolden Rectangle and Fibonacci Numbers

As is well-known, the golden ratio ϕ is closely related to the Fibonacci numbers

$$1, 1, 2, 3, 5, 8, 13, 21, 34, 55, 89, 144, 233, 377, 610, 987, \ldots$$

defined by the difference equation

$$b_1 = 1, b_2 = 1, b_{n+2} = b_{n+1} + b_n.$$

The sequence then has the property that

$$\lim_{n \to \infty} \frac{b_{n+1}}{b_n} = \phi.$$

Analogously, in the case of the supergolden ratio ψ we have the cubic

$$x^3 = x^2 + 1,$$

and hence the numbers defined by the corresponding difference equation

$$a_{n+2} = a_{n+1} + a_{n-1}, \quad n = 1, 2, 3, \ldots.$$

This difference equation requires three initial values, for example 1, 1, 1, which then gives the sequence

$$1, 1, 1, 2, 3, 4, 6, 9\ 13, 19, 28, 41, 60, 88, 129, 189, 277, \ldots.$$

This is a time-delayed Fibonacci sequence, examples of which have been studied in connection with the cow problem (at the age of 3 a cow gives birth to a female and so on) [2, 3]. This is in distinction to the original Fibonacci rabbit problem, in which the births are not delayed. If we take any 5 consecutive terms of this time-delayed sequence we can use them to generate an approximate supergolden rectangle with whole number sides, an example of which is shown in Figure 4.13.6.

As with the Fibonacci numbers, the limiting behaviour of the sequence can be studied. By dividing $a_{n+2} = a_{n+1} + a_{n-1}$ by a_{n+1} we obtain

$$\frac{a_{n+2}}{a_{n+1}} = 1 + \frac{a_{n-1}}{a_n} \frac{a_n}{a_{n+1}}.$$

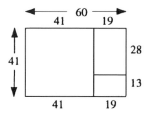

Fig. 4.13.6.

Taking limits of both sides we find that ratios of adjacent terms (larger over smaller) satisfy the equation $x = 1 + 1/x^2$. We recognise this as our cubic equation, and hence

$$\lim_{n \to \infty} \frac{a_{n+1}}{a_n} = \psi.$$

Similarly we find that

$$\lim_{n \to \infty} \frac{a_n}{a_{n+1}} = \theta.$$

More is true, and there is the rather surprising

$$\lim_{n \to \infty} \sqrt{\frac{a_{n+1}}{a_{n-1}}} = \psi.$$

Also, analogous to the basic property

$$\phi - 1 = \frac{1}{\phi}$$

satisfied by the golden ratio, we have

$$\psi - 1 = \frac{\theta}{\psi}.$$

Finally the series

$$1 + \frac{1}{\phi} + \frac{1}{\phi^2} + \cdots = \phi^2$$

is paralleled by the series

$$1 + \frac{1}{\psi} + \frac{1}{\psi^2} + \cdots = \psi^3.$$

4. In Conclusion

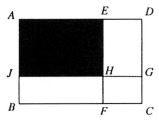

Fig. 4.13.7.

If we look at the supergolden rectangle again, we see that we have not considered the rectangles $AJHE$ and $JBFH$. Rectangle $AJHE$ is also supergolden (and the ratios of the sides of the three rectangles are ψ, ψ^2, ψ^3). Finally look at the rectangle $JBFH$. The ratio JH to JB is ψ^3, whose numerical value is $3.14\ldots$. Could it be? Well, not quite. Even with a supergolden rectangle you can't have everything!

References

1. H. E. Huntley, *The Divine Proportion* (Dover, 1970).
2. H. Langman, *Play Mathematics* (Hafner, 1962).
3. N. J. A. Sloane, *A Handbook of Integer Sequences* (seq. no. 207) (Academic Press, 1973).

Desert Island Theorems

Group D: Non-Euclidean Geometry & Topology

D1

The Four-and-a-half Colour Theorem

DEREK HOLTON'S CHOICE

This Four Colour Theorem has everything. It is easy to tell non-mathematicians. It excited Victorian England and made a splash in the *New York Times* when it was finally proved. Its first 'proof' gained fame for its author but when the fallacy of the proof was exposed no one seemed very interested. But this did lead to the proof of the Five Colour Theorem. And that contained a very nice idea. In fact, that idea remained part of the 'final' proof which was effected by the heavy use of a computer. This proof may well have been the first major result in mathematics to rely on the use of the new technology.

So the Four Colour Theorem has everything. It has public interest, a false proof, a nice idea, some good mathematics, and the use of a computer. All these surely are points to savour on the long hot days of this desert island that we're on. What then is the Four Colour Theorem?

Four Colour Theorem: The countries of any map can be coloured in four or fewer colours so that no two countries with a common boundary receive the same colour.

The first interesting thing that has to be said about the map-colouring problem is that the original conjecture, that four colours suffice, was made by a very young man. Francis Guthrie had recently graduated from University College London where he had studied under Augustus De Morgan and where his brother Frederick was still studying. Frederick brought it to the notice of De Morgan, who promptly became the conjecture's publicist.

That was 1852. No significant progress was made on the problem until 1879 when Arthur Cayley investigated the conditions under which the conjecture might be true and Alfred Bray Kempe used what became known as 'Kempe chains' to produce a 'proof'. Unfortunately, Percy John Heawood found an error in Kempe's proof but was able to use Kempe chains to prove the Five Colour Theorem – that

Derek Holton, Graph Theorist and Mathematics Educationalist, University of Otago.

five colours suffice. These provide a neat way to recolour a map when you get into certain difficulties.

And we should not overlook the work of Peter Guthrie Tait, who explored a number of ways of providing a proof, some plainly invalid. Just after Kempe published his proof, Tait was able to show that the Four Colour Theorem is equivalent to a three-edge colouring of the **boundaries** of certain maps. Every map is related to these special maps in such a way that there is no loss of generality here. The interesting thing about this is that, although we can't yet prove the Four Colour Theorem via the edge colouring route, this might be the best approach for a new proof of the Four Colour Theorem.

For almost 100 years the Four Colour problem stirred the public imagination: it was posed as a challenge to his school by a headmaster; a Bishop of London found a 'proof' during a meeting; and it was very common for amateur mathematicians to submit 'proofs' to their local mathematics department. But in mathematical circles progress was slow. Some emphasis had been put on showing that a map with n countries could be four-coloured and by the early 1970s n had reached about 40. Then in the period 1973–76, Kenneth Appel and Wolfgang Haken got the computer working and, using ideas very similar to those of Kempe, finally polished off the Four Colour Theorem.

But although the Four Colour Theorem is dead, it won't lie down. Some mathematicians feel that there ought to be a better way to prove the result. Using a computer doesn't give a 'nice' proof, with elegant ideas and a feeling for why **four** is the crucial number. The lazy days on the desert island might be just the right conditions to produce the insightful proof that some mathematicians would like to see.

Before the sun goes down, let me provide an outline proof of the Five Colour Theorem. Some subtle conditions have been omitted. A full proof can be found in the book by Holton and Sheehan cited below.

Now you can turn every map into a graph by putting a vertex in the middle of every country and joining the vertices of two countries with an edge if they have a common border. So we will be looking at colouring the vertices of this graph.

We essentially now continue using proof by induction. Using the Euler Polyhedral formula, it can then be shown that in any of the graphs above, there is a vertex of degree 1, 2, 3, 4 or 5. Let's see roughly what happens with a vertex of degree 3, that is, there are three edges coming in to v. We show the vertex v in the diagram.

Now the worst thing that could happen if we coloured the map with v removed, would be that the three vertices next to v would get three colours. In that case we still have a colour (or two) left to colour v. So the most difficult case comes when v has degree 5 (five edges come into v), We show this situation below.

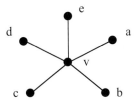

Now if not all of the five colours are used on a, b, c, d and e, then we can use the colour left over for v. So we have to suppose that all of the five colours appear on a, b, c, d and e.

We now use the idea of Kempe chains. Let's look at the colours on vertices a and c. Assume that they are coloured red and blue, respectively. Then starting at a and moving along by adjacent red and blue vertices, there is either a way of getting to vertex c or there isn't. If there isn't we recolour the vertex a in blue and swap the colours of the red and blue coloured vertices that can be reached from a. The net effect of this is that the graph without v is still five-coloured but now only four colours appear on the neighbours of v. This leaves the fifth colour for v.

So we now know that there must be a chain linking red and blue vertices going from a to c. So look at vertices b and d. Repeat the argument above. If b is coloured white and d green, there is either a white-green chain from b to d or there isn't. If there is, we recolour the white-green vertices that can be reached from b. If there isn't, we think about the white-green chain from b to d.

The question that now has to be asked is, where does this white-green chain cross the red-blue chain that started at a and ended at c? Well the two cannot cross at a vertex because vertices only have one colour. And they can't cross at an edge because of the way that the graph was constructed. So the white-green chain doesn't exist and we can five-colour the graph of the map.

Notes

1. Kempe was trying to prove a four colour theorem. So his proof was valid with vertices of degree 4. He came unstuck on vertices of degree 5 because he tried to use a double chain argument that doesn't work.

2. The proof of the Four Colour Theorem that Appel and Haken produced still used two of the ideas from the proof above. They first established that some situations were unavoidable in a map (but they had many more situations than the five vertices of degree 1, 2, 3, 4 and 5 because they could not handle a vertex of degree 5 on its own) and they then showed that they could reduce all of these

to four-colourable maps in a similar way to the way we have above. But the number of cases that they had to deal with was considerably more than our five here.

Further Reading

1. For a very good account of the early history of the Four Colour Theorem, see: N. L. Biggs, E. K. Lloyd and R. J. Wilson, *Graph Theory 1736–1936*, Clarendon Press, Oxford, 1976. The full history is given in Robin Wilson *Four Colours Suffice*, London: Penguin, 2002.
2. The full details of Appel and Haken's proof can be found in K. Appel & W. Haken Every planar map is four colorable, *Contemporary Mathematics* **98** (Providence, RI: American Mathematical Society, 1989), but you'll get a better feel for it from K. Appel and W. Haken, The solution of the four-colour map theorem, *Scientific American* **237**, 4 (1977).
3. A full proof of the Five Colour Theorem and another possible approach for the Four Colour Theorem can be found in D.A. Holton and J. Sheehan, *The Petersen Graph* (Cambridge: Cambridge University Press, 1993).

Acknowledgement

I would wish to thank Robin Wilson for his comments on an earlier draft of this article.

D2

The Euler-Descartes Theorem

TONY GARDINER'S CHOICE

Mathematics is a way of working. Hence its essence cannot be represented by theorems or problems so much as by proofs and solution methods. Thus, when making my selection, I looked for an elementary result, with a *proof* which reveals something of the subtle interplay between *imagination and ideas* and *calculation*.

The Euler-Descartes Theorem: For any convex polyhedron with F faces, E edges and V vertices, $V - E + F = 2$.

Let $\mathscr{S}$ be a sphere of unit radius. A *great circle* on $\mathscr{S}$ is the intersection of the sphere with any plane Π passing through the centre O of the sphere (as in Figure D2.1). Examples include the only horizontal great circle, the equator, and any circle of longitude (i.e. a great circle passing through the North and South poles).

Any two great circles intersect in two points A and A', which are *antipodal*. For example, any two circles of longitude intersect at the North and South poles. At a point where two great circles intersect (at A, say) we get four *angles*. The size of each angle is equal to the angle between the tangents to the two great circles at A. Hence, the four angles at A are α, $\pi - \alpha$, α, $\pi - \alpha$, and the two great circles determine four *spherical lunes* with angles α (shaded in Figure D2.2), $\pi - \alpha$, α, $\pi - \alpha$.

A spherical lune with angle α covers $(\alpha/2\pi)$ of the sphere, and so has area $(\alpha/2\pi) \times 4\pi = 2\alpha$. Thus two great circles crossing at A at an angle α form four spherical lunes of area 2α, $2(\pi - \alpha)$, 2α, $2(\pi - \alpha)$.

Given any two points A, B on the sphere, either A and B are antipodal (in which case, A, B and O are collinear), or A, B and O define a plane Π, and the intersection of the plane Π with the sphere S determines the great circle through A and B. The shorter of the two arcs joining A and B then determines the shortest distance from A to B, and the *spherical line segment* $\overset{\frown}{AB}$ joining A to B on the sphere (Figure D2.3).

Tony Gardiner, Algebraic Graph Theorist, University of Birmingham. Chair, Education Committee of the European Mathematical Society. President of The Mathematical Association, 1998–99.

333

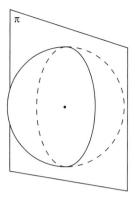

Fig. D2.1.

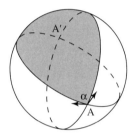

Fig. D2.2.

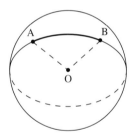

Fig. D2.3.

Any three points A, B, C on the sphere which do not lie on a single great circle determine three spherical line segments $\widehat{AB}$, $\widehat{BC}$ and $\widehat{CA}$, and hence a *spherical triangle*, $\triangle ABC$. The spherical lune which is formed by the great circle through A and B and the great circle through A and C, and which contains the segment $\widehat{BC}$, determines the angle α at A; the angles β at B and γ at C are determined similarly (see Figure D2.4).

Theorem Given a spherical triangle $\triangle ABC$ with angles α, β, γ,

$$\triangle = \text{area}(\triangle ABC) = (\alpha + \beta + \gamma) - \pi.$$

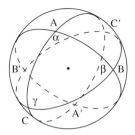

Fig. D2.4.

That is, the area $\triangle$ of any spherical triangle is equal to the excess of its actual angle sum $\alpha + \beta + \gamma$ over and above the angle sum one would expect for a triangle in the Euclidean plane.

Proof The sides $\widehat{AB}$, $\widehat{BC}$ and $\widehat{CA}$ are parts of three great circles. Let the great circles containing $\widehat{AB}$ and $\widehat{AC}$ meet at A and A'; those containing $\widehat{BA}$ and $\widehat{BC}$ meet at B and B'; and those containing $\widehat{CA}$ and $\widehat{CB}$ meet at C and C'.

Then $\overrightarrow{OA'} = -\overrightarrow{OA}$, $\overrightarrow{OB'} = -\overrightarrow{OB}$, $\overrightarrow{OC'} = -\overrightarrow{OC}$, so $\triangle A'B'C'$ is the image of $\triangle ABC$ under inversion in the origin O (that is, under the transformation $-I$). Hence, $\triangle A'B'C'$ is congruent to $\triangle ABC$.

There are two angles equal to α at A, and these determine two spherical lunes each of area 2α. Similarly, there are two angles β at B, and these determine two spherical lunes each of area 2β; and there are two angles γ at C, which determine two spherical lunes each of area 2γ. These six lunes cover the whole sphere: points outside $\triangle ABC$ and $\triangle A'B'C'$ are covered exactly once, while points of $\triangle ABC$ and $\triangle A'B'C'$ are covered exactly three times.

$$\therefore 4\pi = \text{area } (S) = 2(2\alpha + 2\beta + 2\gamma) - 4\triangle$$
$$\therefore \triangle = (\alpha + \beta + \gamma) - \pi. \qquad\qquad \text{QED}$$

Corollary 1 A spherical n-gon with angles α_i $(1 \leq i \leq n)$ has area $\sum \alpha_i - (n-2)\pi$.

In two-dimensions the size of an angle at A is given by the arc length subtended by the angle on the unit circle with centre A. Hence, the total angle at any point A is equal to 2π. In three-dimensions (Figure D2.5) the size of a *solid angle* at A is given by the area of the region it subtends on the unit sphere centred at A. Hence the total solid angle around any point is equal to 4π.

A polyhedral angle at A with n edges (and hence n faces) at A subtends a spherical n-gon on the unit sphere centred at A; the angles $\delta_i (1 \leq i \leq n)$ of this spherical n-gon are precisely the dihedral angles between adjacent faces of the polyhedron around the vertex A (Figure D2.6). Hence we have a further corollary.

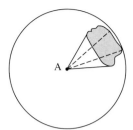

Fig. D2.5.

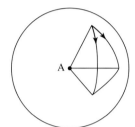

Fig. D2.6.

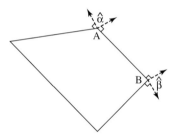

Fig. D2.7.

Corollary 2 If $\delta_i (1 \le i \le n)$ are the dihedral angles around vertex A of a polyhedron, the size of the solid angle at vertex A is $\sum_{i=1}^{n} \delta_i - (n-2)\pi$.

In two-dimensions, for any given convex n-gon we can draw at each vertex A the outward perpendicular to each edge at A. For the internal angle α at A, this determines the *external angle* $\hat{\alpha}$. Since the two perpendiculars create four angles at A, two of which are right angles, $\alpha + \hat{\alpha} = \pi$, so α and $\hat{\alpha}$ are *supplementary*. External angles $\hat{\alpha}$ and $\hat{\beta}$ at adjacent vertices A and B have neighbouring sides parallel (Figure D2.7). Hence all the external angles fit together exactly at a point: $\sum \hat{\alpha} = 2\pi$.

$$\therefore \sum \alpha = \sum (\pi - \hat{\alpha}) = n\pi - 2\pi = (n-2)\pi.$$

In three-dimensions, for a given convex polyhedral angle α with n faces around vertex A, draw the perpendicular to each face at the point A. The convex hull of these rays determines the n faces of the *external* polyhedral angle $\hat{\alpha}$ at A. The face angles φ_i around $\hat{\alpha}$ are precisely the *supplements* of the dihedral angles $\delta_i (1 \le i \le n)$ between adjacent pairs of faces around the angle α at A.

Now $\hat{\hat{\alpha}} = \alpha$ (since the rays perpendicular to the faces of the external angle $\hat{\alpha}$ are the edges of the original angle α). Hence, the dihedral angles between adjacent faces of the external angle $\hat{\alpha}$ are the supplements of the face angles θ_i around the original angle α at A. Corollary 2 then shows that

$$\hat{\alpha} = \sum (\pi - \theta_i) - (n - 2)\pi \tag{*}$$

In two-dimensions we observe that the external angles at neighbouring vertices have adjacent sides parallel, so that the external angles of any polygon have sum $\sum \hat{\alpha} = 2\pi$. In three-dimensions, consider any convex polyhedron P. If AB is any edge of the polyhedron, the two faces of P on either side of the edge AB ensure that the external angles $\hat{\alpha}$ at A and $\hat{\beta}$ at B are bounded at the ends of AB by two parallel faces. So the external angles of all the vertices of P fit together exactly at a point: hence the external solid angles of any convex polyhedron have sum $\sum \hat{\alpha} = 4\pi$.

Now sum both sides of equation (*) over the set V of all vertices of P. The sum of the LHS $= \sum \hat{\alpha} = 4\pi$. To evaluate the sum of the RHS treat the two terms on the RHS separately:

1. Summing the first term involves summing $\sum \theta_i$ over all vertices, which gives the sum of all face angles in the polyhedron. The sum of the terms each having the form $\sum (\pi - \theta_i)$ can be rearranged so as to sum the angles round each face of the polyhedron in turn. Group together all those faces having r edges, since all have the same angle sum $(r - 2)\pi$: let F_r denote the set of all faces of the polyhedron with r sides and let $F_r = |F_r|$. Summing the face angles of all faces in F_r gives

   $$\sum (\pi - \theta) = r F_r \pi - [F_r \times (r - 2)\pi] = 2\pi F_r.$$

   Summing the face angles of all faces of the polyhedron as r varies ($r \ge 3$), gives

   $$\sum (\pi - \theta) = 2\pi \sum_r F_r = 2\pi F \tag{1}$$

   where $F = \sum F_r$ is the total number of faces of the polyhedron.
2. Now sum the second term $(n - 2)\pi$ of the RHS of (*) over the set V of all vertices of P. Let V_n denote the set of vertices of P having valency n, and let $V_n = |V_n|$, $V = |V| = \sum V_n$. Then,

   $$\sum (n - 2)\pi = \pi \sum_n n V_n - 2\pi V = \pi \cdot 2E - 2\pi V. \tag{2}$$

Combining (1) and (2) yields:

$$4\pi = \sum \hat{\alpha} = 2\pi F - [2\pi E - 2\pi V]$$
$$\therefore V - E + F = 2. \qquad\qquad \text{QED}$$

Though Descartes proved a result based on the hard-to-visualise geometrical steps involving Euclidean angles in the foregoing, it would never have occurred to him to express his observations in terms of the combinatorial invariants V, E, F. Hence, while the proof is essentially Descartes', the formula remains Euler's.

D3

The Euler-Poincaré Theorem

CARLO SÉQUIN'S CHOICE

The Euler-Poincaré Theorem relates the numbers of vertices, V, edges E, faces F, cells C, etc, of graphs, polygons, polyhedra, and even higher-dimensional polytopes. It can be presented in many different ways. For a single 3-dimensional polyhedral body without any holes, Euler [1] originally stated it as:

$$V + F = E + 2. \tag{1}$$

Poincaré [2] extended the formulation to such a body in D-dimensional space:

$$N_0 - N_1 + N_2 - N_3 + \cdots N_{D-1} = 1 + (-1)^{D-1}. \tag{2}$$

Here N_i denotes an element of dimensionality i; e.g., N_0 represents the number of vertices. For $D = 3$ this formula reduces to Equation (1).

In my work in computer-aided design and solid modelling, I use this formula to check the consistency and validity of the models I want to build on rapid-prototyping machines. However, I normally deal with more complex objects that also may have holes or tunnels, and sometimes the data file contains a description of several objects. Thus I need a more general formula that can accommodate all these cases. I like the following form:

$$I - N_0 + N_1 - N_2 + N_3 - \cdots N_D = R_1 - R_2 + R_3 - \cdots R_D. \tag{3}$$

On the left-hand side is again Poincaré's list of i-dimensional building elements. However, the list has been extended by one additional term on either end: I denotes the number of individual, non-connected assemblies of such components; and the count of the building blocks now includes elements of the same dimension as the dimension in which the assembly is embedded. Note that the inclusion of these two terms allows us to get rid of the less than elegant term $1 + (-1)^{D-1}$ in Equation (2).

Carlo H. Séquin, Computer Scientist and Sculptor, University of California, Berkeley. Codeveloper of RISC concept, pioneer of Computer-Aided Design (CAD).

339

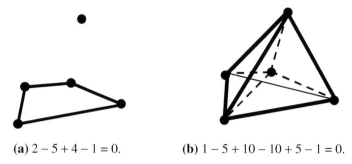

(a) $2 - 5 + 4 - 1 = 0.$ **(b)** $1 - 5 + 10 - 10 + 5 - 1 = 0.$

Fig. D3.1. An incremental construction. (a) Five partially connected points in 2D; (b) the 4-dimensional simplex featuring five 3D solid cells (N_3), and one 4D hypercell (N_4). In both cases, the relevant values inserted into Equation (3) are also shown.

On the right-hand side, we tally all the elements of various dimensions that form closed rings in and by themselves. Thus R_1 counts closed-ring edges; R_2 annular, ring-shaped faces, and R_3 solid-body handles. The exact use of these elements will become clear when we discuss an inductive construction of this formula.

Euler's theorem can be proven in many different ways. Eppstein cites 17 different proofs [3]. An instructive and inductive proof can be obtained by a simple counting process, applied while one incrementally constructs the object of interest. For instance, let's start with v isolated vertices (N_0) in a plane. These obviously form v separate, individual components (I); thus only the first two terms appear in Equation (3), and the equation trivially balances. As we start adding edges between pairs of vertices, we add terms of the third kind (N_1); but for each such connection made, the number of individual components is lowered by one, and the equation thus stays balanced.

When we close a sequence of edges into a cycle, we gain an edge without further reducing the number of connected components. However, we also gain a loop (N_2), which we must count as a new, separate entity (Figure D3.1(a)). It is counted as an element of dimensionality 2, because we can stretch a 'membrane' over the area encircled by the loop, and make this into a face. Counting these loop/face elements, formed by cycles with an empty interior, keeps the equation balanced.

A minimal cycle can be formed with a single vertex and a single edge returning to that same vertex. But what happens, if we remove the vertex and just consider a single edge forming a ring? Such a ring-edge is a special topological element indeed, and it is counted separately as R_1 on the right hand side of Equation (3).

Cycles of either kind also allow us to draw internal contours inside a face and then remove the membrane inside this inner contour, thus opening up a hole in the face. If the face surrounding the hole is subdivided by some edges so that the inner cyclic contour is edge-connected to the outer contour, then Equation (3) will readily balance. But if we remove the last connecting edge between the inner and outer contours, we need a new term to balance that loss, since the number of individual components (I) does not change, as long as the two contours are connected by a face. However, if this is the only connection between the two contours, then this is a very special face, since it forms a ring or annulus all by itself. This new topological entity is counted with the term R_2 in Equation (3). Thus for every hole in a face surrounded by an undivided annulus, we have to increment the ring-face term (R_2).

Let's assume we continue to create many new loops and turn them into faces as soon as we add the closing edge on a cycle. But now let's take our construction out of the plane and contemplate it in 3D space. We can then bend our collection of faces into a bowl and aim at eventually closing it into an orientable (two-sided) shell without any holes. Topologically, something new happens again, when we turn the last cycle into a face and thus close the last opening in the shell: at this moment, we isolate a piece of 3D space and created a separate 3D cell (N_3). Counting these cells with term N_3, balances the face (N_2) that we gained 'for free' by filling in a cycle that we had not previously counted because it was the outer surrounding contour of a graph in a plane. Equation 3 continues to work, as we create clusters of adjacent cells, as long as we count individually all cells that are separated from one another by a membrane of faces. Thus we can now handle 3D objects with vertices, edges, faces, and cells. Of course, cells could be 'filled in' and thereby be turned into solid bodies.

Something special happens again if we create a torus, i.e., a shell in the shape of a doughnut or in the shape of a handle on some other solid object. We can create such a closed solid handle from a cylindrical worm by fusing its two end-faces. In so doing, we lose these two end-faces, but what do we gain in turn? First we gain a very visible ring, which somehow divides space in such a way that we can distinguish between infinitely long threads that go through this ring and those that don't; this happens when we first merge two vertices of the two end-faces (Figure D3.2(a)). This central cycle (which could also be turned into a face) is counted in term N_2. Topologically, the situation remains the same while we completely fuse the two end-faces but retain them as an inner membrane. But when we then remove this shared inner face, we also change the nature of the space internal to the torus. We can now go around the inside of this loop without having to cross any faces. This new ring cell is accounted for in term R_3. Thus for each solid handle that we add to a solid object, we gain two loops: one is formed by the handle surface (N_2), and one is internal to the handle body (R_3). The number of such handles on an object is also called its genus: a sphere has genus zero, and a doughnut has genus 1.

When we start to look at a cluster of solid cells from 4-dimensional space, we can conceive of bending that cluster into a curved, and possibly closed, hyper-shell (N_4), in analogy to the way that we took an originally planar collection of vertices, edges, and faces, and then considered it to be a partial 3D shell of a solid object.

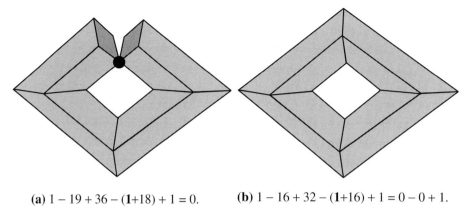

(a) $1 - 19 + 36 - (\mathbf{1}+18) + 1 = 0.$ **(b)** $1 - 16 + 32 - (\mathbf{1}+16) + 1 = 0 - 0 + 1.$

Fig. D3.2. Creating a solid handle: (a) Situation when a 'worm' is bent into a loop and creates a cycle (**1**); and (b) after the two end-faces have been merged into a single internal membrane and then removed to form a cyclic ring-shaped handle.

Similarly, when we now plug the last hole in a 4D hyper-shell composed of 3D solid cells, we gain an additional 3D solid without having to 'expend' any new vertices, edges or faces – it is just a matter how we colour that region of space – either as a hole or as a solid piece. To keep the equation balanced, we now need to count the newly generated hyper-cell (N_4) with a sign opposite to that of the 3D cells (N_3). You may suspect, that tricky issues arise as we contemplate higher-dimensional loops and handles – but this is beyond the scope of this little treatise of the Euler-Poincaré Theorem. Equation 3 should be good enough for use on a deserted island on a 3D world.

References

[1] Leonhard Euler, Elementa doctrinae solidorum – Demonstratio nonnullarum insignium proprietatum, quibus solida hedris planis inclusa sunt praedita, *Novi comment acad. sc. imp. Petropol.* **4** (1752–3), 109-140-160.
[2] Henri Poincaré, Analysis Situs, *Jour. École Polytechnique* **2**, 1 (1895).
[3] David Eppstein, Seventeen proofs of Euler's formula: $V - E + F = 2$, www1.ics.uci.edu/~eppstein/junkyard/euler/

Editor's Note

Carlo Séquin's sculpture designs and mathematical models can been viewed at the following pages of his website:

www.cs.berkeley.edu/~sequin/BIO/sculptures.html
www.cs.berkeley.edu/~sequin/SCULPTS/sequin.html

D4

Two Right Tromino Theorems

SOLOMON GOLOMB'S CHOICE

We define a "right tromino" to be the geometric figure formed by placing three equal squares in a right-angle pattern

and answer the following questions:

1. Where can a single square be removed from an 8×8 checkerboard so that the remaining 63 squares can be tiled by 21 right trominoes?
2. Where can a single square be removed from a 5×5 board so that the remaining 24 squares can be tiled by 8 right trominoes?

The questions are extremely similar, but the results, and the methods of proof, are paradigms of two very different techniques.

First Tromino Theorem *Any* single square can be removed from the 8×8 board, and the remaining 63 squares can be tiled by 21 right trominoes. In fact, we prove the following proposition $P(n)$ for all positive integers n by mathematical induction. (The 8×8 board is the case $n = 3$.)

$P(n)$: No matter where a single square is removed from a $2^n \times 2^n$ board, the rest can be tiled by right trominoes.

Proof *Case $n = 1$.* Wherever a single square is removed from the 2×2 board, clearly the rest can be tiled by a single right tromino.

Solomon Golomb, Mathematician, University of Southern California. Inventor of 'Polyominoes'.

Inductive Step. Assume $P(k)$ is true: *Wherever a single square is removed from a $2^k \times 2^k$ board, the rest can be tiled by right trominoes.* Now consider $n = k + 1$. We divide the $2^{k+1} \times 2^{k+1}$ board into four *quadrants*, each $2^k \times 2^k$. Wherever a single square is removed from the original board, that square is in one of the four quadrants, and we can complete the tiling of that quadrant by the inductive assumption.

From each of the remaining three quadrants, remove the "central square". Finish off the tiling of each of those quadrants (each $2^k \times 2^k$) by the inductive assumption. Then replace the three "central squares" that were removed by a single right tromino.

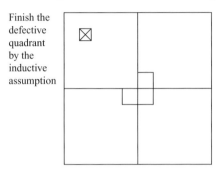

Finish the
defective
quadrant
by the
inductive
assumption

Second Tromino Theorem On the 5×5 board, if and only if the single square is removed from one of the 9 marked locations (the cells with both coordinates *odd*), the rest can be tiled with 8 right trominoes.

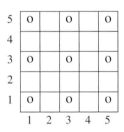

Proof First, observe that no right tromino can cover more than one of the 9 dots. If no dotted square is removed, at least 9 right trominoes will be required to cover the rest, but 9 right trominoes have a combined area of 27, whereas the entire 5×5 board has an area of only 25. If a dotted square is removed, there are only three distinct cases to consider:

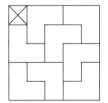
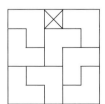
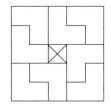

Editor's Notes

1. Polyominoes were invented by Solomon Golomb as a 22 year-old student at Harvard and were popularised by Martin Gardner, first in his *Scientific American* column and then in his books of mathematical puzzles. Golomb's eponymous book was published in 1965 (*Polyominoes*, Charles Scribner's Sons, New York; 2nd edition, revised, Princeton University Press, 1994, 1995).

2. Both of the Right Tromino Theorems were first stated and proved by Golomb. The former result, with the proof by induction, is in 'Checkerboards and Polyominoes', *American Mathematical Monthly* **61**,10 (December, 1954) and in numerous subsequent publications, including the first chapter of *Polyominoes*. As a proof by mathematical induction it is considered a model of its type and is much quoted by other writers. The latter result and proof have not previously appeared in print, though they have featured in a number of Golomb's talks.

D5

Sum of the Angles of a Spherical Triangle

CHRISTOPHER ZEEMAN'S CHOICE

The theorem about the three angles of a triangle adding up to 180 degrees can be generalised to spherical triangles, and then used to give the sum of the four solid-angles of a tetrahedron. The proof is accessible to 13-year olds. I have often taught it to them at my masterclasses.

Definition 1 A *great circle* on a sphere is the intersection of the sphere with a plane through its centre. A *spherical triangle* consists of three arcs of three great circles. Let A, B, C be the angles at the vertices, or more precisely between the tangents to the sides at each vertex (Figure D5.1). Let S and T be the surface areas of the sphere and the triangle respectively.

Theorem $A + B + C = 180\,(1 + 4T/S)$.

Example The triangle shown has 3 right-angles and so $A + B + C = 270$. Meanwhile, T occupies a quarter of the northern hemisphere and so $T/S = 1/8$ (Figure D5.2). Also note that if T is small compared with S (like a triangle on the surface of the earth) then the sum of the angles tends to 180. To prove the theorem we need a lemma.

Definition 2 Define the A-lune to be the area between the 2 great circles through A.

Lemma A-lune$/S = A/180$.

Proof of Lemma (See Figure D5.3)
Looking down on S from above A, A-lune$/S = 2A/360 = A/180$.

Christopher Zeeman, Geometer and Pioneer of Catastrophe Theory. Former Gresham Professor of Geometry.

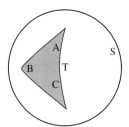

Fig. D5.1.

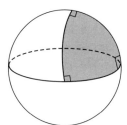

Fig. D5.2.

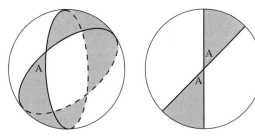

Fig. D5.3.

Proof of Theorem (See Figure D5.4) The 3 lunes cover the whole sphere, but cover the triangle 3 times, which is twice two many, and the same with the antipodal triangle. Therefore,

$$A\text{-lune} + B\text{-lune} + C\text{-lune} = S + 4T.$$
$$(A\text{-lune} + B\text{-lune} + C\text{-lune})/S = 1 + 4T/S.$$
$$(A + B + C)/180 = 1 + 4T/S, \text{ by the lemma.}$$

Multiplying by 180 gives the theorem.

Corollary In a tetrahedron,

sum of the 4 solid-angles = (sum of the 6 dihedral angles) − 1.

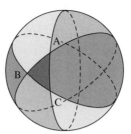

Fig. D5.4.

In particular, for a regular tetrahedron, each solid-angle has size

$$\frac{3}{2}\cos^{-1}\left(\frac{1}{3}\right) - \frac{1}{4}.$$

A proof of the corollary is given in: *Teaching and Learning Geometry 11-19: Report of a Royal Society/Joint Mathematical Council Working Group* (Chair: Adrian Oldknow), The Royal Society, July 2001.

V

Recreational Geometry

5.1

Introduction: Recreational Geometry

BRIAN BOLT

What springs to mind with a title like this? No doubt we all have different ideas: to some of us the construction of polyhedra models, to others the challenge of spatial puzzles or investigations, while many games with a spatial content can justifiably be included. Escher's patters gave a whole new meaning to tessellations and fractal tilings advance space filling to another level.

Experience of teaching mathematics at many levels has shown me the importance of spatial thinking in all branches of the subject, and the need to give students a wide range of spatial challenges to develop it. My own geometric education, after the early secondary stage, I consider was hampered by the rush into algebra to solve spatial problems, and this seemed to get worse the further I progressed. The problems in recreational geometry seemed to come like a breath of fresh air to keep my interest alive. I can clearly remember the reaction of a group of postgraduate students hoping to become secondary teachers when faced with finding the area inside an arbitrary polygon formed by an elastic band on a pinboard. They were groping for appropriate formulae relating to the coordinates of the polygon's vertices. They were humbled when told how easy the problem was for a group of primary-aged children at a Mathematics Club I had run the previous Saturday. The children were setting each other similar problems as a game and finding the area by dissecting the polygons into rectangles and right-angled triangles (which they saw as half rectangles), before I led them on to discover Pick's Theorem.

This Saturday Club ran for twelve years throughout term time with the emphasis on geometrical recreations and was formed in an attempt to counteract the very arid arithmetical syllabuses in primary schools at the time. The spatial ideas I used then have been found to stimulate serious mathematical thinking in groups of all ages and abilities. Even the various groups of retired adults I get invited to talk to

Brian Bolt, Teacher, recreational geometer and author of *A Mathematical Jamboree, The Amazing Mathematical Amusement Arcade, Mathematical Cavalcade* and *A Mathematical Pandora's Box*, all published by Cambridge University Press.

are fascinated by the spatial models, patterns and tricks I share with them. Cutting up a Möbius Band in various ways never fails, and invariably prompts that 'What happens if...?' question. I know for a fact that many of these retired people go home and experiment for themselves. Seeing one of their number trying to put two halves of a regular tetrahedron together always intrigues and amuses them. This is a model that never fails to interest children who then want to make it for themselves. Making a model of three, or six pyramids which fold to form a cube is such a good introduction to the volume of the pyramid and the time taken to make the model is well justified. Getting students to be creative rather than copying well-established nets from a book is challenging and satisfying. One task that I have found useful is to get students to design and make their own versions of *half a cube*. They learn so much from their own attempt and seeing what their peer group has made.

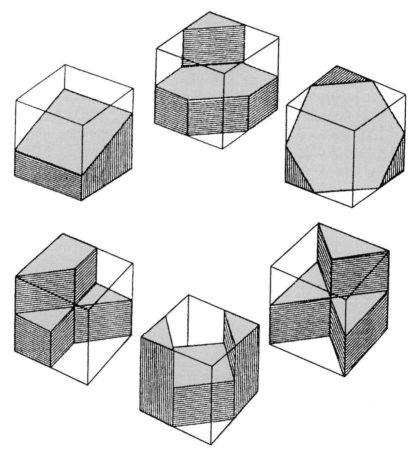

Polyhedra models can be made in a variety of ways apart from the traditional net with tabs and some of these are included in this chapter. Paper folding, plaiting or using polygons clipped together along their edges, or skeleton polyhedra formed by threading shirring elastic through drinking straws all emphasise different features. The latter is closely related to the intrinsic rigidity, or lack of it, of the steel lattice work in bridges, scaffolding, cranes and electricity pylons. A cube will only be rigid if each face has a diagonal added and it soon becomes clear that all faces must be triangles. The octahedron is rigid, and as four edges meet at each vertex its twelve edges are traversible. From this fact it is possible to make an octahedron by first threading twelve straws onto shirring elastic, tie the ends to form a loop and perform a few twists. Try it for yourself before demonstrating it in public!

The Assessment of Performance Unit (APU)[1] found that many pupils had difficulty in recognizing a square unless it was the 'right way up'. In trying to help overcome this failing I devised a simple game based on two players placing counters, in turn, on a 5 × 5 squared board. The first player to place a counter so that it formed the fourth vertex of a square with three existing counters on the board is the loser. In practice, at least at first, players often failed to spot that their opponents had lost . . . but they had the incentive to learn. Games like this are enjoyed and at the same time advance the players' spatial awareness.

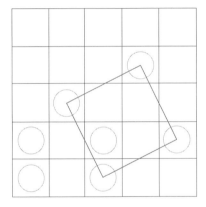

Commercial games, such as three-dimensional noughts and crosses (tic-tac-toe), or Othello, are great at developing spatial perception. The chessboard and pieces, apart from the game itself, have potential for so many interesting geometrical puzzles. Finding solutions to the largest number of pawns which can be put on an $n \times n$ board without three being in a straight line often catches out the solver because of the lines created at odd angles.

Knight's tours have been widely explored, and investigating them on shapes other than squares has many possibilities. What is the smallest rectangle, for example, on which it is possible to complete a tour? L. D. Yarborough published an interesting puzzle in which the challenge is to find the longest non-intersecting knight's tour

---

[1] APU was akin to the National Assessment of Educational Progress (NAEP) in America and the School Achievement Indicators Program (SAIP) in Canada.

on a 6 × 6 board. (By non-intersecting is meant that the path of the knight, seen as sequence of line segments, does not cross itself.) A 13-step and a 15-step solution are shown. Yarborough thought the optimum to be 16 steps but, in fact, a 17-step solution is possible. Can you find it?

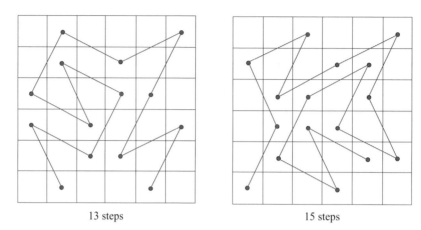

13 steps   15 steps

There is nothing like publishing a puzzle with an incomplete or incorrect solution to create interest as I have found from experience. Have you tried dissecting a square into acute-angled triangles? If so, you may understand why I initially came to the wrong conclusion!

Squares and cubes are the starting point for so many puzzles and activities in recreational geometry. One commercial puzzle that has been in my possession for many years consists of five colourful plastic shapes which you are invited to fit together to make a square. Most people find that they soon manage a square with four of the pieces, leaving out the small square, but find it impossible to make the breakthrough with all five pieces. To solve it, mathematical thinking is needed . . . or else a lot of luck!

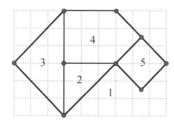

An interesting puzzle based on cubes is to consider whether it is possible to colour the faces of 27 unit cubes in such a way that they can be assembled into a 3 × 3 × 3 cube with outside faces that are all red, or all blue, or all yellow. I used this puzzle for many years before I had the insight to see how easy it could be generalized to paint n^3 unit cubes with n colours to make n different coloured super cubes. Imagine starting with an unpainted set of cubes assembled into the $n \times n \times n$ cube. Slice the cube in each of the three perpendicular directions, colouring the faces you have sliced and move them to the outside. Children's bricks with pictures on their faces are an example of this.

Steiner's solution for the minimum connector of the four points at the vertices of a square always surprises people meeting it for the first time, but little has been written about equivalent three-dimensional nets, that is until I challenged readers of the *Mathematical Gazette* to find the minimum length of web a spider need spin to connect its eight nests at the vertices of a cubical box. The New Zealander, A. Zulauf, and the American, E. J. Andrews, came up with the optimum solution and Richard Bridges extended the idea to other polyhedra [1–3].

In 1991 I came across what appeared to be a problem in elementary geometry in the newsletter of the Brunei Mathematical Society. Try as I might I could only solve it by using trigonometry. I took the problem to the mathematics department at Exeter University where it created much interest and eventually a solution based on elementary geometry was found. More recently, I used it as one of the daily problems at the Mathematical Association Annual Conference in Exeter. Several delegates sent me solutions and encouraged me to write it up for the *Mathematical Gazette*, but then someone recalled that it had already appeared there, and more than once. In fact, it first appeared in the *Mathematical Gazette* in 1923, but since then I discover it was set in the 1916 Entrance Examination for Peterhouse and Sidney Sussex Colleges, Cambridge. You clearly can't keep a good problem down. Over the years it must have baffled thousands of potential solvers. I reproduce it again here for all those readers who have yet to meet it. A very accessible article on the problem, by Diamond and Georgiou can be found in the November 2001 issue of *Mathematics in School* [4].

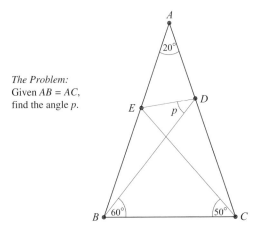

The Problem:
Given *AB* = *AC*,
find the angle *p*.

Geometrical dissections come in various guises. Tangrams are dissections of a kind where you are invited to fit a given set of shapes into various outlines with the same area. A dissection recently brought to my attention is based on a regular dodecagon. It shows beautifully that the area of a circle of unit radius is greater or equal to 3 square units. Known as Kürschak's tile, one can imagine the immense pleasure it must have given to its discoverer. Many dissection puzzles are set in the form of cutting a shape into a given number of pieces which can be rearranged to form another, often a square. Satisfying solutions to these can often be found by superimposing a tessellation of squares of the same area over a tessellation of the original shape.

One of my earliest memories of reading any mathematical book outside the school text books was a chance introduction to *Amusements in Mathematics* by Henry Dudeney and this soon had me looking for other 'recreational' titles in mathematics. Northrop's *Riddles in Mathematics* and *Mathematics and the Imagination* by Kasner and Newman followed and were influential in my decision to read mathematics at university [5, 6, 7]. Throughout my teaching career, first at a boys' school and then at the School of Education at Exeter University, I have always introduced a thread of recreational mathematics and seen it spark the interest and creative thinking of many a student. The many books I have written to stimulate mathematical thinking have had both a strong geometric and recreational bias in the belief that they would motivate others to enjoy mathematics as they still do me. The selection of ideas chosen for this chapter only scrapes the surface but I will be surprised if you do not find something here to make you think.

References

1. A. Zulauf, Shortening the home stretch I, Note 79.5, *Mathematical Gazette* **79** (March 1995), 94.
2. E. J. Andrews, Shortening the home stretch II, Note 79.6, *Mathematical Gazette* **79** (March 1995), 95.
3. Richard Bridges, Minimal Steiner trees for three dimensional networks, *Mathematical Gazette* **78** (July 1994), 157–162.
4. R. A. Diamond & G. R. Georgiou, Triangles and Quadrilaterals Revisted Part 2: The Solution, *Mathematics in School* **30**, 1 (November 2001), 11–13.
5. H. E. Dudeney, *Amusements in Mathematics* (New York: Dover, 1917).
6. Eugene Purdy Northrop, *Riddles in Mathematics: A Book of Paradoxes* (Van Nostrand, 1944; Penguin, 1960).
7. Edward Kasner & James Newman, *Mathematics and the Imagination* (New York: Simon and Schuster, 1940).

5.2

The Cube Dissected into Three Yángmǎ

JAMES BRUNTON

I was intrigued to read in Kiang's article (reproduced in this volume) about the ancient Chinese method of finding the volume of a sphere a reference to "...a familiar object known as a yángmǎ". There is also a note that "a cube can actually be cut into 3 yángmǎ" (Figure 5.2.1).

It may amuse your readers to know that the net for this solid, a square-based skew pyramid, can be constructued from a square of paper, after the manner of Origami, entirely without the aid of compasses, straight-edge or scale and that this can make a very useful visual aid for the mathematics room, since it suggests a way to formulate the volume of a pyramid.

It should be clear from Figure 5.2.2 that the net can be circumscribed by a square, $ABCD$, and is symmetrical about the diagonal AC; since F is the mid-point of AE the whole square can be folded along AC, FG, EH, $F'G'$ and $E'H'$ so long as the point E can be found. One can then construct the net. Now, E divides AD in the ratio $2:\sqrt{2}$, i.e. $\sqrt{2}:1$, and, in the $\triangle ACD$, $AC:CD = \sqrt{2}:1$.

Using Euclid VI.1 the bisector of the angle at C will meet AD in E. The procedure is illustrated in Figure 5.2.3.

1. Fold the square on the diagonal AC.
2. Fold DC on to AC to locate E on AD.
3. Fold FG, the perpendicular bisector of AE.
4. Fold EH, perpendicular to AD, to meet AC in J.
5. Fold $F'G'$, the perpendicular bisector of EJ.
6. Fold $E'H'$ perpendicular to EH through J.

For the completed development, join up as indicated.

All that remains to do, to produce the model illustrated in Figure 5.2.1, is to mark this net simultaneously onto three pieces of card, with pin-holes through the

First published in *Mathematical Gazette* **57** (February 1973), pp. 66–67.

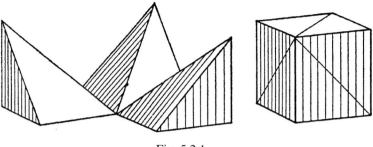

Fig. 5.2.1.

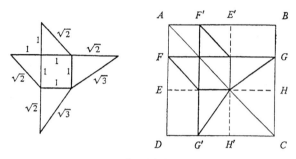

Fig. 5.2.2.

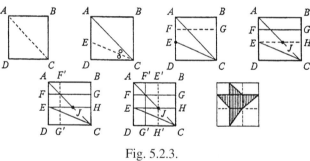

Fig. 5.2.3.

appropriate vertices, and cut out, after scoring with a blunt knife-blade, allowing flaps for sticking. If these pyramids are hinged together they make the excellent visual aid referred to above.

5.3

Folded Polyhedra

CECILY NEVILL

I have found that models which hold together by interlocking are most useful and enjoyable as sticking together a model often results in a disappointing, botched job.

Primary school children would not be allowed to "score with knife". If fold lines are ruled with a fine Ballpoint, a dry one will do, all but heavy card will fold neatly.

1. An Open Cube

Number both sides of the paper. Cut along the lines marked CUT. Fold away from you on the solid lines and towards you on the dashed lines.

Fold the middle into a cube by putting 1 inside 2 inside 3, 4 inside 5 inside 6. Fold 7 inside 1. Put 8 on the base. Fold 9 inside 4. Put 10 on the base.

2. Folding a Regular Octahedron

Use equilateral triangle grid paper of a larger size if possible.
Cut along the line marked CUT.
Fold away from you on all lines.
Hold the net with A towards you.

Put 1 under 2 to make the vertex of the octahedron at A. Put 3 under 4 and 5 under 6 to make the rest of the octahedron. Put 7 under 8. Tuck in 9.

3. Folding a Half Tetrahedron

The squares have side 4 cm, the triangles have base 4 cm and height 3.5 cm. Shave a little off edges of those shapes which will be inside, (A, B, C, D and G). Fold

First published in *Mathematics in School*, **25**, 4 (September 1996), pp. 36–44; extract.

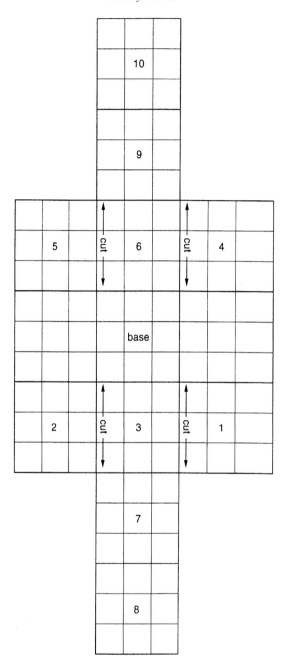

on all lines. Hold A and B and turn them until B covers A. Fold in C and D. Their ends overlap. Cover them with E. Cover B with F. Tuck G under H. Make a second solid. Fit together to make a tetrahedron.

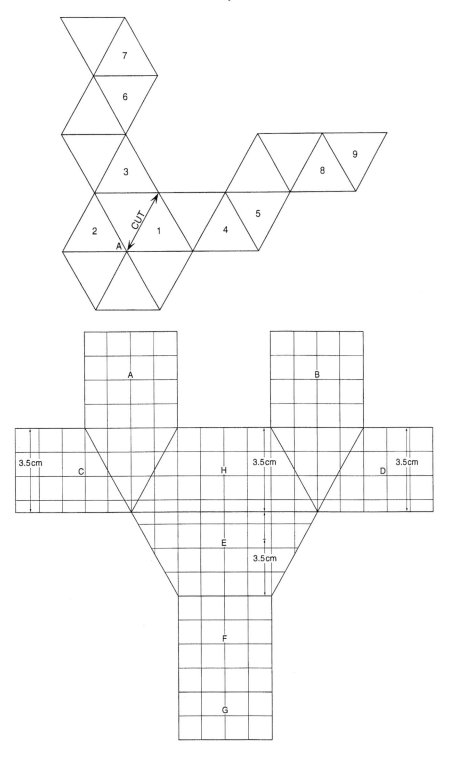

5.4

The use of the Pentagram in Constructing the Net for a Regular Dodecahedron

E.M. BISHOP

The diagonals of a regular pentagon, form a pentagram as in diagram (i) thus forming a pentagon inside the original one.

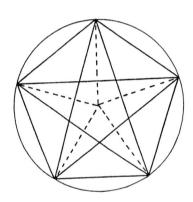 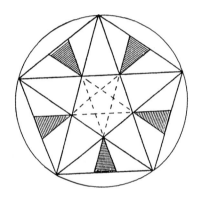

The diagonals of the inner pentagon are drawn and produced to meet the sides of the outer one as in diagram (ii). If the shaded triangles are then cut away, the resulting figure, consisting of 6 equal pentagons, forms half the net required for the dodecahedron.

The complete net is thus obtained from a small number of initial measurements. If two equal circles are drawn and the vertices of the outer pentagons found by making 5 equal angles at the centre of each circle, the sides can be immediately checked for accuracy with dividers, and the diagrams are then completed by joining points.

This method, in addition to being an easy one to carry out accurately, produces a diagram which is rich in elementary geometry and which can also be used to introduce the idea of the stellation of the solid and which through similar triangles leads to the discovery of a line divided in Golden Section.

First published in *Mathematical Gazette* **46** (December 1962), p. 307.

5.5

Paper Patterns: Solid Shapes from Metric Paper

WILLIAM GIBB

The simplest and yet most interesting solid that can be folded from metric paper is a tetrahedron;

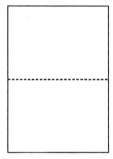

Fig. 5.5.1.

Take a sheet of metric paper and fold it in half from top to bottom. Crease it firmly and open.

Fold across top and bottom, crease and open, and pinch to mark the mid points.

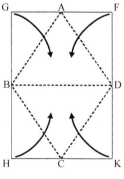

Fig. 5.5.2.

First published in *Mathematics in School* **19**, 3 (May 1990), pp. 2–4; extract.

363

Now fold up along AB, BC, CD and AD. There are not easy folds and it helps if you place a ruler along the lines before folding.

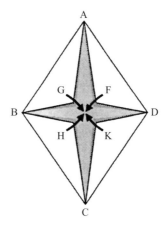

Fig. 5.5.3.

Now join the edges, AG to AF and CH to CK using tape. There should be no overlap.

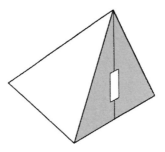

Fig. 5.5.4.

Join the remaining edges using tape to produce a tetrahedron which has exactly the same surface area as the original sheet of paper.

This tetrahedron is not regular but each face is congruent being an isosceles triangle with sides in the ratio $2:\sqrt{3}:\sqrt{3}$. There are some interesting facts that can be discovered about this tetrahedron by having several and joining them in different ways.

Firstly, the angles between the faces of the tetrahedron can be deduced. Two can be placed together on a table demonstrating that the angle between the faces in this case is a right angle.

Three can be built together using the other angle between the faces. Here the angle must be 60 degrees. A little 3-dimensional geometry, using the dimensions of the original rectangle as the starting point, will prove these results.

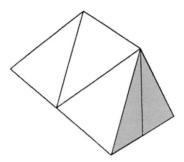

Fig. 5.5.5.

Secondly, this tetrahedron will tessellate in space and can self replicate. A larger similar tetrahedron can be built with sides twice as long by using 8 of our original tetrahedrons. 27 will build an even larger one with sides 3 times as long as the original. With a little sticky tape the tetrahedrons can be joined quite easily even if the original tetrahedrons are a little flimsy;

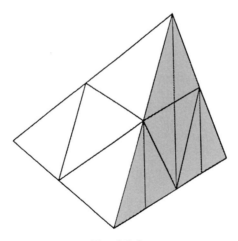

Fig. 5.5.6.

Another way to demonstrate this same relationship is to start with 8 sheets of A6 and one sheet of A4 and to fold each to give a tetrahedron. The 8 smaller tetrahedrons can now be built into a tetrahedron just like the one made from the A4 sheet;

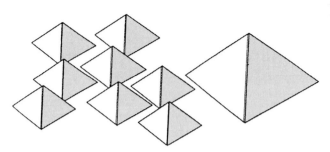

Fig. 5.5.7.

A Pyramid from Metric Paper

Two of the tetrahedrons just described can be built into a pyramid with a base in the shape of a rhombus. But it is also possible to fold a square based pyramid from A4 with no cutting or measuring. Make the folds as shown here, the dotted lines give "valley" folds and the dashed lines give "hill" folds. Then fold and glue together triangles marked with the same letters.

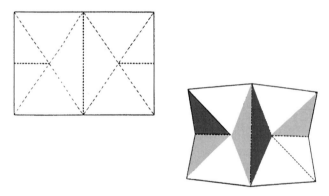

Fig. 5.5.8.

The pyramid that results has no base but has the special property that 6 can be joined to give a cube. This relationship can be used as the starting point for deducing the volume formula of a pyramid.

The cube does not have solid faces but shows the diagonal planes created if opposite edges of a cube are joined.

5.6

Replicating Figures in the Plane

SOLOMON W. GOLOMB

1. Figures Composed of Four Replicas

In [1], C. D. Langford asked for those plane figures which can be dissected into four "replicas", congruent to one another and similar to the original figure. (An equivalent formulation is that four identical figures are to be assembled into a scale model, twice as long and twice as high.) In addition to triangles and parallelograms, which always have this property (see Figure 5.6.1), he also exhibits three trapezoids (Figure 5.6.2) and three hexagons (Figure 5.6.3) with this property. Finally, Langford gives an example involving a stellated hexagon (Figure 5.6.4). He then asks if the list is complete.

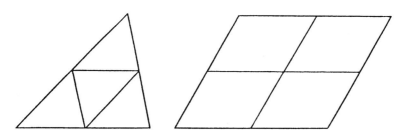

Fig. 5.6.1. An arbitrary triangle or parallelogram can be dissected into four replicas.

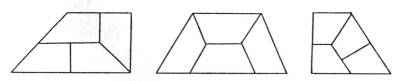

Fig. 5.6.2. Three trapezoids which can be dissected into four replicas.

First published in *Mathematical Gazette* **48** (December 1964), pp. 403–412; extract.

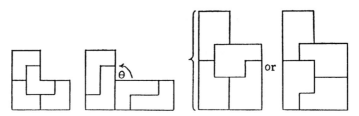

Fig. 5.6.3. Three hexagons which can be dissected into four replicas. (The second of these can obviously be done four ways.)

Fig. 5.6.4. A stellated hexagon which can be dissected into four replicas.

The first objective of this article is to extend Langford's list. In Figure 5.6.4, the two squares can be replaced by a pair of identical rectangles, and in the first hexagon of Figure 5.6.3, three quadrants of *any* rectangle could be used, rather than three-fourths of a square. The special cases of a *right* triangle, and of a parallelogram whose sides are in the ratio 2 : 1, have alternative dissections as shown in Figure 5.6.5.

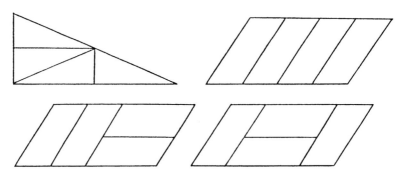

Fig. 5.6.5. Special triangles and parallelograms with alternatives to the dissections of Figure 5.6.1.

The angle θ in the second hexagon of Figure 5.6.3 need not be 90°. For $\theta \neq 90°$, the only dissection which works is:

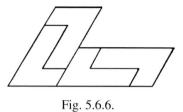

Fig. 5.6.6.

One example which Langford missed entirely is the pentagon of Figure 5.6.7, which may be regarded as composed of six equilateral triangles, or two half-hexagons, or two-thirds of an equilateral triangle.

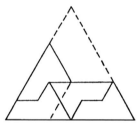

Fig. 5.6.7. A pentagon which can be dissected into four replicas.

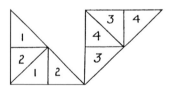

Fig. 5.6.8. Another stellated hexagon which can be dissected into four replicas.

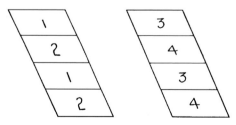

Fig. 5.6.9. A disconnected polygon which can be dissected into four replicas.

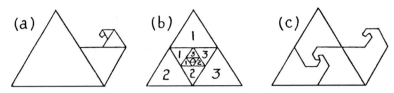

Fig. 5.6.10. Construction of a "snail" which can be dissected into four replicas.

There is also another stellated polygon, composed of two identical isosceles right triangles attached at right angles (Figure 5.6.8).

A disconnected example is obtained by taking two identical parallelograms, each having sides in the ratio 2 : 1, and placing them as if separated by a third such figure. This dissection is shown in Figure 5.6.9.

Finally, there is a non-polygonal example obtained by taking an equilateral triangle, adjoining to it a triangle one-fourth the size, adjoining to that a triangle one-fourth its size, etc. ad infinitum, as indicated in Figure 5.6.10*a*. The addendum has an area of $1/4 + 1/16 + 1/64 + \cdots = 1/3$ of the original triangle's, and the addendum is *similar* to the entire figure. It remains only to dissect the original equilateral triangle into three parts, each congruent to the addendum.

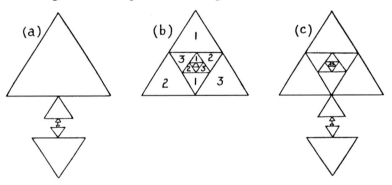

Fig. 5.6.10'. Construction of a "lamp" that can be dissected into four replicas.

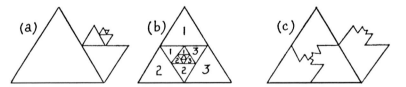

Fig. 5.6.10''. Construction of a "carpenter's plane" which can be dissected into four replicas.

This dissection is indicated in Figure 5.6.10*b*, and the entire "snail", dissected into four replicas, is depicted in Figure 5.6.10*c*.

Two other non-polygonal examples have also been found, both close relatives of the snail. These are the "lamp" in Figure 5.6.10', and the "carpenter's plane" in Figure 5.6.10''.

5.7

The Sphinx Task Centre Problem

ANDY MARTIN

1. Introduction

The Sphinx Problem is a very rich mathematical activity that can easily be missed on the first reading of the task. Where is the mathematics in simply putting together four jigsaw type pieces?

2. The Problem

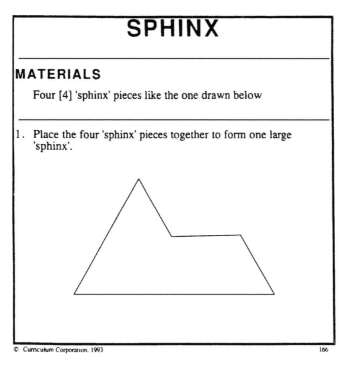

SPHINX

MATERIALS

Four [4] 'sphinx' pieces like the one drawn below

1. Place the four 'sphinx' pieces together to form one large 'sphinx'.

166

Fig. 5.7.1. Task 166 Curriculum Corporation, Australia, Task Centre Project.

First published in *Mathematics in School* **29**, 3 (May 2000), pp. 6–10.

3. Solving the Problem

A sphinx piece can be made from six equilateral triangles. So, when I first tried this task on a whole-class basis pupils constructed their own sphinx pieces using 20 mm isometric paper. Fitting the four pieces together is not easy for many pupils. Working with paper that is blank on one side and isometric grid on the other allows pupils to see easily that one of the pieces must be flipped over if the solution is to be found. The solution can also be demonstrated effectively using an overhead projector and the wooden pieces from the task.

At this point many pupils (and teachers!) believe they have finished and that there is no more mathematics in the task. However, the task is really 'the tip of a mathematical iceberg' and there are many other fruitful directions that can be developed.

4. Into the Iceberg

Let us call the larger sphinx, made with the four pieces, a size-2 sphinx. This is because its base (the longest side) is twice as long as the base of an original sphinx piece. Now we can investigate the relationship between the size of a sphinx and the number of sphinx pieces needed to construct it.

I tried this approach with a year 8 class. A colleague, Simone, from the design-technology department manufactured some sphinx pieces for me from plastic. This meant I could work at the concrete level with all the pupils. The four pieces were stuck together using tape. Now the size-2 sphinxes could be fitted together. This produces a new sphinx that pupils can readily see is size-4.

The process can be repeated to produce a size-8 and size-16 sphinx. The original size-2 sphinxes produced by the individual pupils now assemble into one large sphinx from the whole class! As we engaged in the process of construction we recorded in a table on the board what we had noticed at each stage.

Size of sphinx	Number of pieces
2	4
4	16
8	64
16	256

Now stop the construction and discuss the patterns that are emerging. My year 8 class could see that the 'size of the sphinx' keeps doubling. This created the opportunity for me to introduce the idea of the 'powers of 2' and to write these

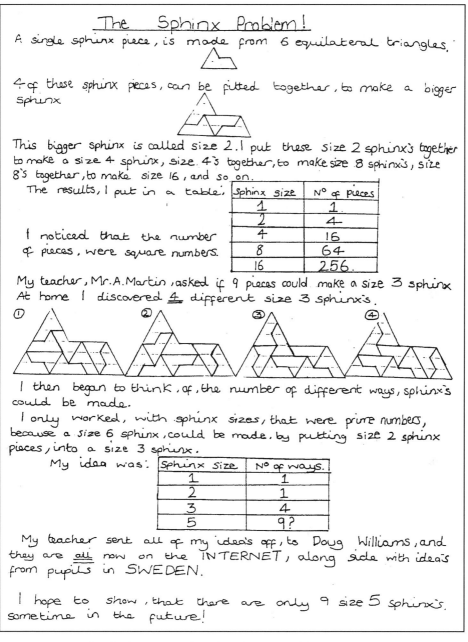

The Sphinx Problem!

A single sphinx piece, is made from 6 equilateral triangles.

4 of these sphinx pieces, can be fitted together, to make a bigger sphinx

This bigger sphinx is called size 2. I put these size 2 sphinx's together to make a size 4 sphinx, size 4's together, to make size 8 sphinx's, size 8's together, to make size 16, and so on.

The results, I put in a table:

Sphinx size	N° of pieces
1	1
2	4
4	16
8	64
16	256.

I noticed that the number of pieces, were square numbers.

My teacher, Mr. A. Martin, asked if 9 pieces could make a size 3 sphinx. At home I discovered 4 different size 3 sphinx's.

① ② ③ ④

I then began to think, of, the number of different ways, sphinx's could be made.

I only worked, with sphinx sizes, that were prime numbers, because a size 6 sphinx, could be made, by putting size 2 sphinx pieces, into a size 3 sphinx.

My idea was:

Sphinx size	N° of ways.
1	1
2	1
3	4
5	9?

My teacher sent all of my idea's off, to Doug Williams, and they are all now on the INTERNET, along side with idea's from pupils in SWEDEN.

I hope to show, that there are only 9 size 5 sphinx's. sometime in the future!

Fig. 5.7.2. Extract of work by Sarah Hutchinson (year 8).

numbers in their equivalent form using the notation of powers. As a class we predicted that our next sphinx would have a size of 'two to the power of five' or 32. Unfortunately, I did not have sufficient plastic sphinx pieces to actually construct and display the model!

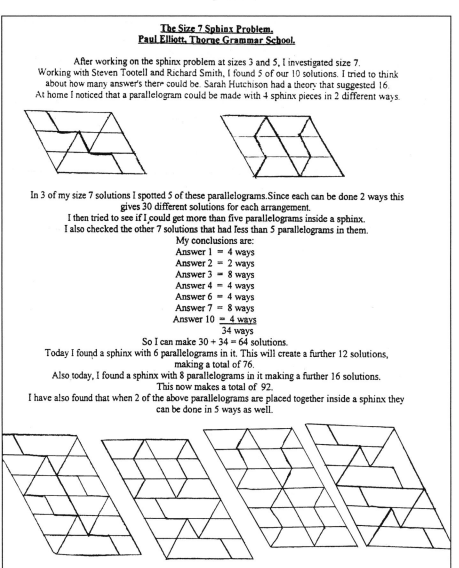

The Size 7 Sphinx Problem.
Paul Elliott, Thorne Grammar School.

After working on the sphinx problem at sizes 3 and 5, I investigated size 7.
Working with Steven Tootell and Richard Smith, I found 5 of our 10 solutions. I tried to think
about how many answer's there could be. Sarah Hutchison had a theory that suggested 16.
At home I noticed that a parallelogram could be made with 4 sphinx pieces in 2 different ways.

In 3 of my size 7 solutions I spotted 5 of these parallelograms. Since each can be done 2 ways this
gives 30 different solutions for each arrangement.
I then tried to see if I could get more than five parallelograms inside a sphinx.
I also checked the other 7 solutions that had less than 5 parallelograms in them.
My conclusions are:
Answer 1 = 4 ways
Answer 2 = 2 ways
Answer 3 = 8 ways
Answer 4 = 4 ways
Answer 6 = 4 ways
Answer 7 = 8 ways
Answer 10 = 4 ways
34 ways
So I can make 30 + 34 = 64 solutions.
Today I found a sphinx with 6 parallelograms in it. This will create a further 12 solutions,
making a total of 76.
Also today, I found a sphinx with 8 parallelograms in it making a further 16 solutions.
This now makes a total of 92.
I have also found that when 2 of the above parallelograms are placed together inside a sphinx they
can be done in 5 ways as well.

This is going to create more solutions still, so I know that Sarah's theory is wrong. I think this will
possibly create another 45 answers, but I will need to do some checking!

Paul Elliott 14/07/99

Fig. 5.7.3. Extract of work by Paul Elliott.

What about the numbers in the column headed 'number of pieces'? Do these
numbers have any patterns? The year 8 class suggested that these were going up by
a factor of four each time. I didn't exploit the possibility of linking area (number
of pieces) to length (sphinx size) because of a suggestion made by a girl called
Amanda. She said the numbers were all square numbers and this changed the

direction of our problem solving. Amanda posed the question 'If the number of pieces we have is a square number can we always build a sphinx?'.

The obvious challenge to the class was to start with nine pieces. Could they find a size-3 sphinx? This challenge proved to be a very fruitful homework. One girl, Sarah, managed to find not one, but four distinct size-3 sphinxes. I faxed Doug Williams in Australia who had been working on the same problem. He already knew of three solutions and when Sarah's other solution appeared on his web site there was a wave of enthusiasm from the class to investigate sphinxes further.

Several of the class managed to put 25 sphinx pieces together to make a size-5 sphinx. The work of Catherine, Sarah, Steven, Richard and Nicola was mentioned in further correspondence between Doug and I. Meanwhile Sarah had formulated her own hypothesis. She suggested that there would be a relationship between the size of a sphinx and the number of different ways in which it could be constructed. Her idea was:

$$\text{Size-2} \longrightarrow 1 \text{ way}$$
$$\text{Size-3} \longrightarrow 4 \text{ ways}$$
$$\text{Size-5} \longrightarrow 9 \text{ ways}$$

Sarah had decided to focus on the sphinxes whose size was a prime number. She suggested that there was a connection between these and the square numbers again. The size of the sphinx being a prime number was very important, because Sarah was able to explain that a size-6 sphinx could be built from any size-3 sphinx where all the pieces were replaced with size-2 sphinxes. In essence, any sphinx whose size was composite could be constructed from the sphinxes whose sizes were the prime factors of this number.

The challenge to this year 8 class now moved on to try and prove/disprove Sarah's claim. Could we find more, less or exactly nine different size-5 sphinxes? This search resulted in a host of different solutions, all of which were displayed in the corridor. Other pupils in year 7 and year 9 helped the search along and produced more solutions.

I revisited this task the following year with a class in year 9. Some of these pupils had started with me in year 8; others were new to the problem. A boy called Paul became captivated by the search for different solutions. He frequently stopped me around school and provided different solutions for the size-5 and size-7 sphinxes. Meanwhile news of our work had travelled via Doug Williams to Sweden. Two Swedish teachers, Per Berggen and Maria Lindroth, visited Thorne and worked with some pupils from year 9, including Paul, on this task. Success! Whilst Per and Maria were present Paul, Sarah, Nicola, Richard and Steven managed to find 10 different size-5 sphinxes. Sarah's hypothesis, now nearly a year old, had been disproved!

Paul, however, had now developed a new counting approach based on the arrangement of sphinx pieces in the shape of parallelograms of differing sizes. He conjectured that there were 137 different possibilities for a size-7 sphinx. Doug had now involved a mathematician called David Shield to write a proof for his web site. I faxed a copy of Paul's work to Australia for David to check. He spotted a flaw, which Paul then corrected and subsequently changed his conjecture to 36960 different solutions. This has resulted in mathematical correspondence between a pupil here and a mathematician in Australia. The class looks forward to the exchange of e-mail, and the ongoing developments with the task.

This is just one route into the mathematical iceberg from this task, a direction that I think this class may continue to explore. The task, however, has been extended in other directions by staff using the plastic sphinx pieces with their classes. These plastic pieces have proved to be very robust, versatile and capable of generating many other mathematical starting points. Some examples are:

5. Sphinx Perimeters

The length of the shortest side of the sphinx is defined as 1 unit. This means that a size-1 sphinx has a perimeter of 8 units (a good counting exercise for weaker pupils). The pupils are now given 2, 3, 4 ... sphinx pieces as appropriate for the class. The challenge is to build shapes with the largest/smallest perimeter. Their findings can easily be drawn on isometric paper, displayed and checked by other pupils. The activity also requires some clarification on how sphinxes can be joined together (do we allow point to point?). The final challenge is to find at least one arrangement of the pieces with each perimeter between the largest and smallest values agreed upon by the class.

6. Sphinx Symmetries

Give all pupils four sphinx pieces and challenge them to fit the pieces together to make a shape. What symmetry does the resulting shape have? This activity can make display space an active teaching area. Pupils are asked to draw as many different shapes as possible. Each one is drawn on a separate piece of paper along with the name of the pupil. The teacher can then allow another pupil, or group of pupils, to organize the display area by placing shapes together that have a common property. This can be a very fruitful way of initiating discussion on reflection symmetry only (and if so, with how many lines), rotational symmetry only (and if so, with what order) and both reflection and rotational symmetry.

7. Sphinx Tessellation

The sphinx, based on the equilateral triangle, does tessellate. The plastic sphinx pieces have been used to introduce, at the concrete stage, tessellation to lower ability pupils.

8. Professional Development for Staff

The mathematics produced from the four jigsaw pieces in this task has generated discussion and development for the staff in the school. It has helped colleagues realize how a practical activity can develop both number and algebra topics in a context that stimulates the pupil's interest. The task has produced a wealth of display material, each with a slightly different focus and many turning a display area into a teaching area! It also has provided those theoretical challenges that teachers never quite have the time to develop. Now, what would a 3-D sphinx look like?

References

This task can also be found on the web at http://www.blackdouglas.com.au/taskcentre/ where Doug Williams has used this as an introduction to a lesson on 'Patterns and Powers'. Further information about the Mathematics Task Centre Project can be found on the web at http://www.mav.vic.edu.au/PSTC/index.html

5.8

Ezt Rakd Ki: A Hungarian Tangram

JEAN MELROSE

The Hungarians seem to have a special genius for inventing intricate and challenging puzzles – Rubik's cube is merely the best known – and for devising more engaging and extending versions of standard puzzles. At ICMI 6 in Budapest in 1988 various mathematical games and puzzles were on sale. There were sets of dominoes with faces having zero to nine spots, a 3-D 'four in a row puzzle' that would serve well to introduce 3-D coordinates, various rotational puzzles whose transformations form a group, and many others too!

Amongst them was 'Ezt Rakd Ki'. It is made by POLITOYS and is a more subtle version of a tangram puzzle. The name, according to my best information, literally means 'put it out' i.e., put the pieces in the right places to make the required design. The puzzle came in a shallow plastic tray together with a booklet of 100 designs to be made and no answers given! It consists of the following rectangle of pieces:

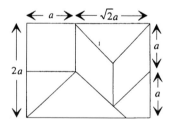

If the length of the side of the square is a then the area A of the rectangle is

$$2a(a + \sqrt{2}a) = 2a^2(1 + \sqrt{2}) = A$$

First published in *Mathematics in School* **27**, 2 (March 1998), pp. 14–15.

This is the area of a regular octagon of side a.

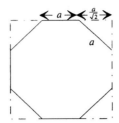

The area of the large square is $[(1 + \sqrt{2})a]^2$ and subtracting the four corner triangles, which together form a square of a^2, gives the area of the octagon, which is found to be A.

Already we have 2 potential designs. The pieces have an area of A, but do they, in fact, fit

i) into a regular octagon of side a?

ii) or into an L-shape resulting from the subtraction of the 2 squares?

PUZZLE 1

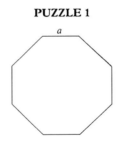

PUZZLE 2

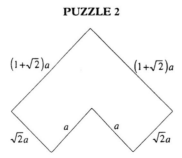

Can rectangles other than the original one be made?

A rectangle of sides a by $2a(1 + \sqrt{2})$ is impossible because fitting one rhombus leaves a space that is too narrow to be fitted.

But can the rectangle $\sqrt{2}a$ by $a(2 + \sqrt{2})$ be made?

PUZZLE 3

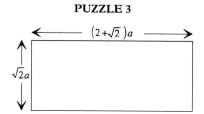

Can other rectangles be made? And how could pupils justify or prove that they had found all the possibilities and that there are no more?

It is also possible to show that a square cannot be formed by setting

$$2a^2(1 + \sqrt{2}) \equiv (p + q\sqrt{2})^2$$

and checking that p, q cannot be integers.

Neither a right-angled isosceles triangle nor the rhombus can be replicated. However, a kite can be made which is an enlargement of the kite shaped piece.

Replication of the kite gives possibilities for mathematics:-

i) The area of the original kite K can be found using the fact that it has been enlarged by linear scale factor $(2 + \sqrt{2})$ and by solving:

$$(2 + \sqrt{2})^2 K = A$$

ii) Repeated replication opens up possibilities for exploring geometric series whose common ratio is $(2 + \sqrt{2})$ by considering lengths of similar shapes and whose common ratio is $(2 + \sqrt{2})^2 = 6 + 4\sqrt{2}$ by considering area of similar shapes.

PUZZLE 4

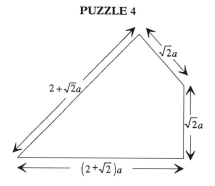

Ezt Rakd Ki is at its most confusing when the designs to be made, such as those in puzzles 5 and 6, look almost alike but in construction are quite different from each other.

PUZZLE 5

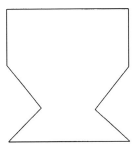

PUZZLE 6

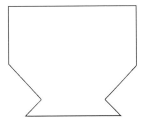

SOLUTIONS

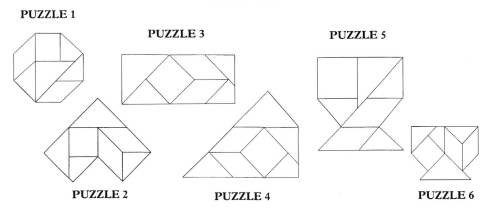

The original 7 piece tangram, whether it is a spoof designed by Sam Loyd or not, (Loyd, 1968; Read, 1965) is undoubtedly an attractive puzzle that can be used both to create interest in maths classrooms (Renshaw, 1996) and particularly to enhance pupils' understanding of area. It is good to come across a development of the original tangram that adds mathematical sophistication and that throws up opportunities for problem solving for older or more advanced pupils. In retrospect, it seems obvious to look for a dissection puzzle of a regular octagon to build on the success of the tangram which is a dissection of the square, because an octagon is the only other possibility of a shape that will give pieces whose angles are multiples of 45°.

The Hungarian devising and use of these puzzles both reflects and contributes to their good performance in mathematics in studies such as the Third International Mathematics and Sciences Study (TIMSS). The challenge for us is to enhance the quality of mathematical culture in our schools in an interesting and attractive way and our Hungarian friends point us to the importance of mathematical games and puzzles in that process.

References

S. Loyd, *The Eighth Book of Tan* (Dover, 1968).
R. C. Read, *Tangrams – 360 Puzzles* (Dover, 1965).
B. Renshaw, Tangrams, *Mathematics in School* **25**, 4 (1996).
US National Center for Educational Studies: website at www.nces.ed.gov/timss/timss95/

5.9

Dissecting a Dodecagon

DOUG FRENCH

The Mathematical Association has recently published *Pig and Other Tales* (French and Stripp, 1997), a collection of mathematical readings taken from past issues of *The Mathematical Gazette*, together with sets of comprehension questions for students. Although the readings are aimed at older students some, at least, involve topics which have interesting possibilities at a more elementary level. These ideas were inspired by one of the readings in 'Pig', an article by Alexanderson and Seydel on Kürschak's Tile, originally published in *The Mathematical Gazette* in 1978.

A regular dodecagon may be dissected into 12 equilateral triangles and 12 rhombuses as shown in Figure 5.9.1.

How are the edge lengths of the triangles, rhombuses and dodecagon related?

What are the angles of the rhombus?

What is the interior angle of a regular dodecagon?

Clearly all the edges in the diagram are of the same length. The angles at the centre are $\frac{1}{12}$ of $360° = 30°$ and so each rhombus has two angles of $30°$ and two of $150°$. Alternatively, at each of the interior points where an equilateral triangle and two rhombuses meet we can calculate the obtuse angle of the rhombus as $\frac{1}{2}(360° - 60°) = 150°$. Each interior angle of the dodecagon is $60° + 30° + 60° = 150°$. It is interesting to note for later that two consecutive edges of the dodecagon will neatly contain one of the rhombuses.

Drawing the dissection provides a variety of challenges. Using ruler and compasses it is not difficult to construct an angle of $30°$ by bisecting the $60°$ angle of an equilateral triangle. The diagram can thus be built up by constructing the angles around the centre. My diagrams have been drawn using *Cabri Géomètre* and that

First published in *Mathematics in School* **27**, 1 (January 1998), pp. 18–19.

involves using the same basic construction. Using LOGO would involve different challenges.

For students to explore some of the simple properties of the dissection it is best for them to make up a set of the 24 pieces on thin card. This gives a purpose for the drawing exercise, but, to save time, they could be provided with a photocopied version. It adds to the attractiveness of the tasks that follow if the triangles and

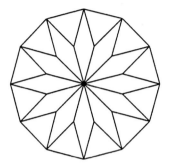

Fig. 5.9.1.

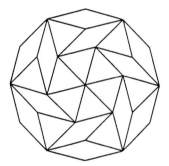

Fig. 5.9.2.

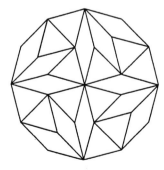

Fig. 5.9.3.

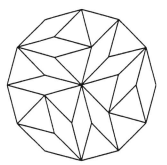

Fig. 5.9.4.

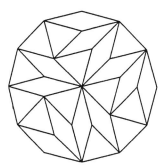

Fig. 5.9.5.

rhombuses are in different colours. For display purposes I have found it useful to use coloured gummed paper rather than card so that students can stick particular arrangements of their pieces onto card without the trouble of using glue.

One of the intriguing features of the dissection is that the 24 pieces can be rearranged to make the dodecagon in many different ways. Two examples are shown in Figures 5.9.2 and 5.9.3.

What symmetries do Figures 5.9.1, 5.9.2 and 5.9.3 have?
What other arrangements are possible and what are their symmetries?

Figure 5.9.1 has 12 lines of symmetry and rotational symmetry of order 12. Figures 5.9.2 and 5.9.3 have only rotational symmetry: order 6 in Figure 5.9.2 and order 4 in Figure 5.9.3. We might now ask, for instance, if we can create an arrangement with rotational symmetry of order 3. Figure 5.9.4 shows how this can be done and Figure 5.9.5 shows how a slight rearrangement destroys the symmetry. Figures 5.9.6 and 5.9.7 show, respectively, examples with 1 line of symmetry and with rotational symmetry of order 2.

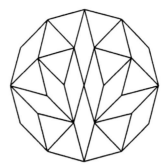

Fig. 5.9.6.

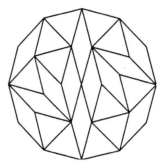

Fig. 5.9.7.

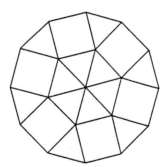

Fig. 5.9.8.

The motif consisting of two rhombuses and an equilateral triangle can be replaced by a square and an equilateral triangle, which suggests that it might be fruitful to look at the possibilities with squares included in the set of pieces. Figure 5.9.8 shows the simplest example with just squares and triangles where there are six lines of symmetry and rotational symmetry of order 6. Figure 5.9.9 includes some rhombuses to give four lines of symmetry and rotational symmetry of order 4.

The most striking property of the original configuration is the way that it gives the area of a regular dodecagon and thereby a good lower bound to the area of a

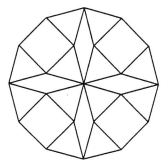

Fig. 5.9.9.

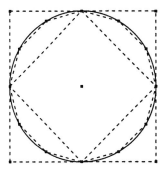

Fig. 5.9.10.

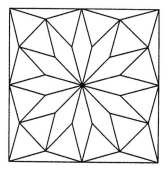

Fig. 5.9.11.

circle. Figure 5.9.10 makes it clear that a circle of radius r has an area that lies between $2r^2$ and $4r^2$, which are the areas of the two squares that are shown. It also shows that the area of a regular dodecagon inscribed in a circle is slightly less than the area of the circle.

Figure 5.9.11 shows the dodecagon with an additional triangle and two half rhombuses on each corner to make the same larger square of area $4r^2$. A simple check shows that the angles do work out correctly. The square tile is known as

Kürschak's Tile. The regular dodecagon consists of 12 rhombuses and 12 triangles, whereas the square has 16 of each. So, the area of the dodecagon is $\frac{3}{4}$ of the area of the square which is $3r^2$, a little less than the familiar πr^2 for the area of the circle.

This simple dissection does not seem to be widely known and yet, as we have seen, it has a number of interesting properties which give scope for a variety of interesting classroom discussions and activities accessible to a wide range of pupils.

References

D. W. French, & C. Stripp (Eds), *Pig and Other Tales* (The Mathematical Association, 1997).

G. L. Alexanderson, & K. Seydel, *Kürschak's Tile*, The Mathematical Gazette **62** (October 1978).

5.10

A Dissection Puzzle

JON MILLINGTON

Recently a puzzle appeared in an AT&T newspaper advertisement in which the reader was challenged to form four pieces into either a square or a triangle. An illustration revealed how the pieces were linked which considerably reduces the difficulty. Nevertheless, it is an interesting dissection puzzle which was discovered by Henry Dudeney and published in 1902. He demonstrated a mahogany model of it to the Royal Society in 1905 and included it in his first book, *The Canterbury Puzzles*, two years later.

Figure 5.10.1 shows a square made from the four pieces; (rotating them about the three numbered links gives the triangle shown in Figure 5.10.2.)

Some lengths are the same as others because:

From the triangle, $h = a$ and $d = e$ so from the square $h = a = d = e$

From the triangle, $g + b = f + c$ and from the square $c + b = f + g$, which means that $c = g$ and $b = f$

From the square, $c + b = 2h$ from which $b = 2h - c$.

So the number of seemingly different lengths that appear in Figures 5.10.1 and 5.10.2 are now reduced to those shown in Figures 5.10.3 and 5.10.4.

What, then, should be the lengths of c and m to produce a triangle for any given length of h? The answer is that c can range from 0 to $2h$ while anything from 0 to $\sqrt{5}h$ will do for m. In the extreme cases when c is zero, or when it is $2h$ and m or n is zero, then you only have three pieces but they still form a triangle.

Perhaps the most elegant dissection occurs when the triangle is equilateral, which it will be when the three angles where p meets q are all $60°$; for this the calculations below establish the values of c and m relative to h. The parallelogram with sides p

First published in *Mathematics in School* **22**, 1 (January 1993), pp. 34–35.

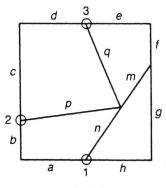

Fig. 5.10.1.

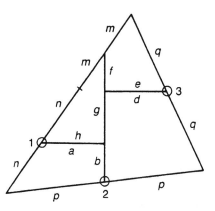

Fig. 5.10.2.

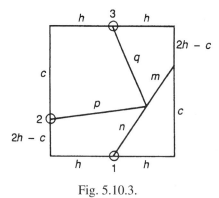

Fig. 5.10.3.

and *r* formed by the construction lines in Figure 5.10.5 appears at first glance to be rectangular, but this would be impossible since adjacent triangles outside it cannot be similar.

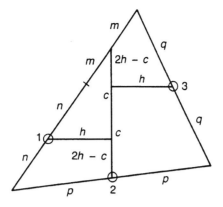

Fig. 5.10.4.

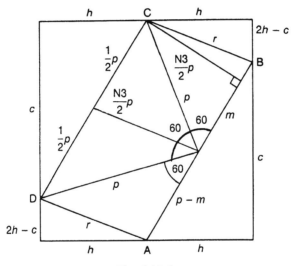

Fig. 5.10.5.

Area of square $= 4h^2$ which is also the area of an equilateral triangle of side $2p$.

So area of inner equilateral triangle of side $p = \frac{4h^2}{4} = h^2$

Height of this inner triangle $= \sqrt{(p^2 - \frac{1}{2}p^2)} = \frac{\sqrt{3}}{2}p$

And its area $= 1/2$ base $\times$ height:

Giving $h^2 = \frac{1}{2}p \times \frac{\sqrt{3}}{2}p$

From which $p^2 = \frac{4h^2}{\sqrt{3}}$ and $p = \frac{2h}{\sqrt[4]{3}}$

$$c = \sqrt{(p^2 - h^2)} = \sqrt{\left(\frac{4h^2}{\sqrt{3}} - h^2\right)} = 1.1442906h$$

So $2h - c = 0.8557094h$

Shorter side of parallelogram $= r$

$$r = \sqrt{(2h - c)^2 + h^2} = \sqrt{(0.8557094h)^2 + h^2} = 1.3161453h$$

Position of inner triangle apex $= m$

$$m = \frac{1}{2}p + \sqrt{r^2 - \left(\frac{\sqrt{3}}{2}p\right)^2} = \frac{h}{\sqrt[4]{3}} + \sqrt{r^2 - (\sqrt[4]{3}h)^2}$$
$$= 0.7598356h + \sqrt{(1.7322384 - 1.7320508)}h$$
$$= 0.7735323h$$

Now that the relationships of c and m to h have been calculated you can mark out the square on wood or cardboard and cut it up. So, for a square of side 15 cm, h is 7.5 cm, c is 8.6 cm and m is 5.8 cm.

Another possibility is to confirm that the inner triangle is indeed equilateral by using the values of c and m when $h = 1$. To explore the dissection further, how about examining the conditions that produce an isosceles triangle from the four pieces? Do you have to start with a square?

Reference

H. Dudeney, *The Canterbury Puzzles* (1907).

5.11

Two Squares from One

BRIAN BOLT

Professor Yoshio Kimura, a friend in Kobe, challenged me to investigate the problem of cutting a square into four pieces which could be rearranged into two squares. He sent me the solutions he had found with his students, four of which are shown below.

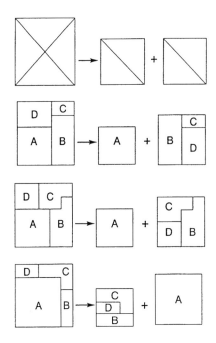

These solutions were a good starting point, and I was soon finding similar solutions of my own. But these were all based on squares in the ratio 3:4:5. Will all solutions

First published in *Mathematics in School* **30**, 5 (November 2001), pp. 30–31.

be linked to squares based on Pythagorean triples? Apart from the first solution above, using the diagonals of a square, is it inevitable that all other solutions will form one of the new squares from three pieces? How many solutions are there for squares in a given ratio?

Two of my early solutions follow which begin to show the possibilities, but certainly would only lead to a finite number of solutions.

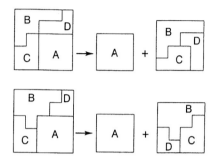

At this point I sent the problem to Professor Derek Holton in New Zealand who initially found further solutions based on rectangular shapes and using the 3:4:5 ratio. But then he came up with the following solution involving the diagonal line *XY*.

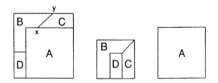

This soon led us to see how to create an infinite number of solutions. Instead of the straight cut *XY*, any cut from *X* to *Y* which has bilateral symmetry about the mediator of *XY* would allow shapes B and C to fit together as in the above solution. Three examples of possibilities for shape C are shown below.

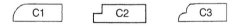

At some point I sent the problem to Robert Goodland, a teacher in Somerset, and he has sent me a solution based on the Pythagorean triple (5, 12, 13). The likelihood is that solutions exist for all Pythagorean triples, but what about other ratios? Take for example a 2×2 square and cut off a 1×1 square from it. Can you find a way of dissecting the remaining L-shaped piece into three pieces to form a square?

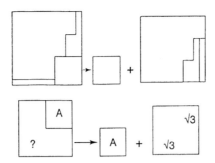

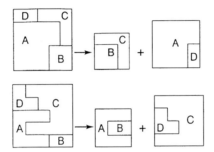

More recently both Derek and Robert have sent me solutions where each of the smaller squares are formed from two pieces of the original, which surprised me as I had come to the conclusion they didn't exist! Two of them are shown below.

The investigation continues. All ideas and solutions welcome!

5.12

Half-Squares, Tessellations and Quilting: Variations on a Transformational Theme

TONY ORTON

There are many ways of dividing a square into two shapes of equal area, and a well known school investigation consists of finding as many different divisions as possible.

However, cut-out squares which are based on one such division and appropriately coloured can be used to create tessellations. The activity suggests many tiling and quilting possibilities. The pictures which follow have been generated on an Archimedes computer using the 'Draw' application.

The basic tile

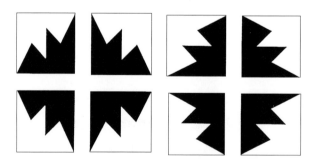

The eight transforms

First published in *Mathematics in School* **23**, 1 (January 1994), pp. 25–28.

An elementary tessellation:

Cats

And another simple one:

Mountains

Slightly more complicated:

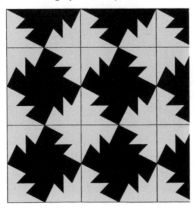

Cogs

And the very beautiful:

Explosion

Tiling has so far produced these tessellating shapes:

and the more complicated:

but other tessellating shapes can be found,
for example:

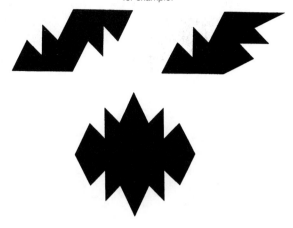

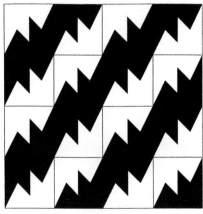

Other attractive designs also exist, for example:

Lightning

which suggests another tessellating shape:

Another attractive design is:

Fire

There are other possibilities with the same basic tile, but here is one based on a different tile.

This produces a tessellation
of two shapes.

Bats

5.13

From Tessellations to Fractals

TONY ORTON

A simple tessellating shape based on the equilateral triangle is shown in Figure 5.13.1.

We can predict that this shape is likely to tessellate because whatever is cut away is replaced elsewhere in a systematic and regular way (see Figure 5.13.2).

This procedure suggests that it may be continued. Each edge of the shape in Figure 5.13.1 may now be replaced in the same way as the original sides of the triangle. Thus, if the equilateral triangle is the Level 0 shape, and Figure 5.13.1 shows the Level 1 shape, then the Level 2 shape will be as in Figure 5.13.3 and this shape will also tessellate, as we shall see.

A new "triangular" shape made from sixteen Level 1 shapes is shown in Figure 5.13.4.

If we apply the same rotation procedure as was demonstrated in Figure 5.13.2 to the shape in Figure 5.13.4 then the shape in Figure 5.13.5 is produced, and this is the Level 2 shape. Thus the Level 2 shape may be seen to be a tessellation of smaller Level 1 shapes, just as the Level 1 shape is a tessellation of smaller Level 0 shapes. So sixteen Level 2 shapes may be assembled to form the Level 3 shape in Figure 5.13.6.

Thus it is possible to investigate this infinite sequence by cutting out simple shapes, fitting them together and reducing them in size (or looking at them from afar). Use of colours will show up the many patterns and relationships very effectively.

Other shapes which tessellate and which may be investigated from the point of view of generating similar infinite sequences of shapes include those shown in Figure 5.13.7. How many Level X shapes will be needed to make a Level $(X + 1)$ shape for each of these?

First published in *Mathematics in School* **20**, 2 (March 1991), pp. 30–31.

Fig. 5.13.1.

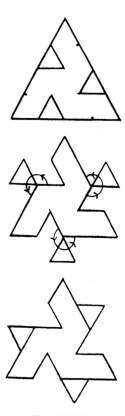

Fig. 5.13.2.

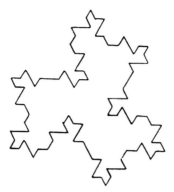

Fig. 5.13.3.

Fig. 5.13.4.

Tony Orton

Fig. 5.13.5.

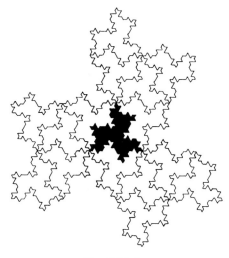

Fig. 5.13.6.

Fig. 5.13.7.

5.14

Paper Patterns with Circles

WILLIAM GIBBS

There are a delightful variety of patterns that can be created from paper circles either by overlapping and interleaving them or by folding and colouring.

Use paper circles of about 8 to 10 cm diameter cut from coloured paper. Draw around a tin to give the circle and cut out with scissors or if you want circles with clean edges use a circle cutter which can be bought at most good stationers.

1. Overlapping Circles

Overlapping the circles in rows creates a striking tessellation. To position the circles in the correct position it helps to draw lines which are the radius of the circles apart.

Fig. 5.14.1(a). Pattern with overlapping circles.

Folding a circle in half will help here.

First published in *Mathematics in School* **19**, 4 (September 1990), pp. 2–8.

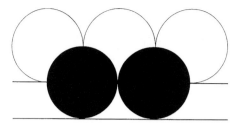

Fig. 5.14.1(b). How to arrange the circles.

Fig. 5.14.1(c).

Variations on this basic pattern can be created by grouping circles of the same colour. For example, grouping together four circles of the same colour will create a new tiling unit;

Fig. 5.14.2(a). Overlapping circles.

Fig. 5.14.2(b).

2. Interleaving Circles Four at a Point

The basic unit for patterns of this type is created by interleaving four circles so that they meet at a point. To help in the positioning of the four circles, fold one of them in half and in half again to give diameters that intersect at right angles. Place the circle that overlaps this so that the creases are tangential.

Fig. 5.14.3(a).

Once the first four circles have been placed accurately further circles can be added to create a tessellation;

Fig. 5.14.3(b).

Arranging circles in different ways or by grouping circles of the same colour together will generate other patterns.

Four circles of the same colour can be grouped to give two new tessellations;

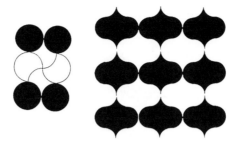

Fig. 5.14.4(a).

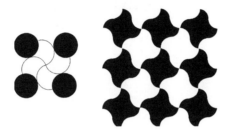

Fig. 5.14.4(b).

If four circles, two of each colour, are interleaved and glued to give this basic unit then by combining several units tessellation patterns based on each of the following shapes can be created;

Fig. 5.14.5.

Here is a pattern created using circles grouped in threes. You will need circles of three different colours.

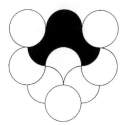

Fig. 5.14.6(a).

Fig. 5.14.6(b).

This pattern is created using just two colours and a slight modification of the original interleaving pattern;

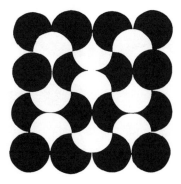

Fig. 5.14.7.

Starting with these three different basic units of four circles each;

Fig. 5.14.8.

and interleaving, this tessellation can be created;

Fig. 5.14.9.

3. Circles at a Point

If the circles are arranged so that three circles meet at a point then this tessellation can be created;

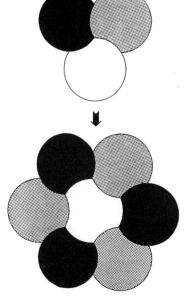

Fig. 5.14.10.

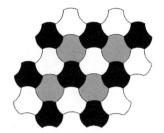

Fig. 5.14.11.

Fig. 5.14.11(a).

4. Six Circles at a Point

To arrange six circles at a point the centre of each circle needs to be marked. Fold in the edge of each circle so that it lies on the centre of the circle.

Fig. 5.14.12.

Then each circle is placed on top of the last so that its circumference lies on the centre of the circle below and the folded edges meet.

Fig. 5.14.13.

Using three different colours creates a basic pattern like this;

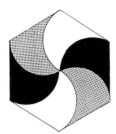

Fig. 5.14.14.

and combining many such units creates this tessellation;

Fig. 5.14.15(a).

Fig. 5.14.15(b).

5. Folding Circles

Folding the edge of the circle in twice creates a new shape with interesting possibilities. For example work with circles of three different colours and combine pairs of folded circles to create rhombi.

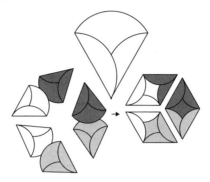

Fig. 5.14.16.

Then combine the three rhombi to make a hexagon which when tessellated creates a new and striking pattern.

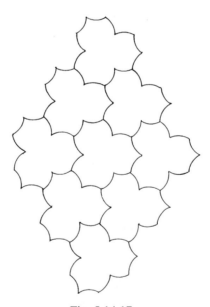

Fig. 5.14.17.

6. Folding and Colouring

Folding the edge of the circle into the centre three times will create an equilateral triangle.

Fig. 5.14.18(a).

These folds can be coloured and interleaved in a variety of ways and a wide range of patterns created. Here are a couple of examples.

Fold the triangle so that just one segment is folded forward and colour to give an equal number of triangles like these:

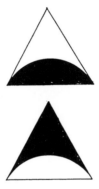

Fig. 5.14.18(b).

Here are two of the many patterns that can be created from these triangles;

Fig. 5.14.19(a).

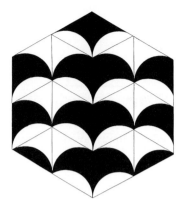

Fig. 5.14.19(b).

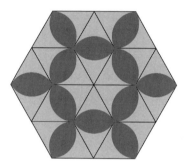

Fig. 5.14.19(c). Single flaps coloured.

Fold and colour pairs of triangles like these.

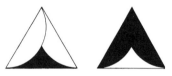

Fig. 5.14.20.

Combine them to make this unit which can then be tessellated;

Fig. 5.14.21(a).

Fig. 5.14.21(b).

If the folds are each of a different colour and interleaved then a triangle like this is created;

Fig. 5.14.22.

These triangles can be combined to make this tessellation;

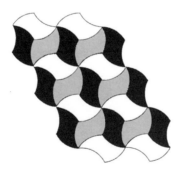

Fig. 5.14.23(a).

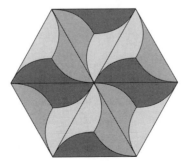

Fig. 5.14.23(b).

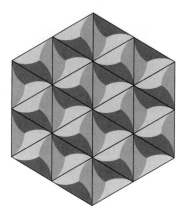

Fig. 5.14.23(c).

Interleave the folds when making the triangle and colour to give two sets of triangles like these;

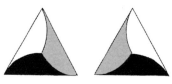

Fig. 5.14.24.

then combine the triangles;

Fig. 5.14.25(a).

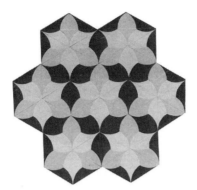

Fig. 5.14.25(b).

7. Folding Thin Paper

As a variation, cut the circles from very thin coloured paper. Fold them into triangles and create patterns. Then fix the final pattern to a window. Very delicate and attractive patterns emerge.

The circle can also be folded into other shapes; a hexagon, a square, a kite and a rhombus.

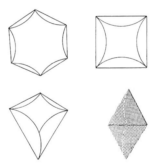

Fig. 5.14.26.

To make the rhombus, for example, the edge is folded to the centre twice as for the equilateral triangle but then on the next two folds the ends of the chords are folded to the centre;

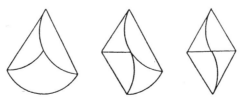

Fig. 5.14.27.

These shapes provide further starting points for investigation in terms of the tessellations and patterns they can create.

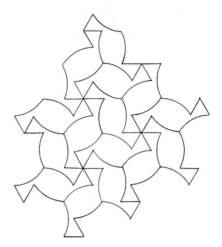

Fig. 5.14.28. Tessellation of interleaved rhombi.

5.15

Tessellations with Pentagons
[with Related Correspondence]

J. A. DUNN

Every triangle can be used to tile or tessellate a plane because the three angles add up to 180° and so can be arranged to form a straight line. In a similar way every quadrilateral will tessellate because the four angles can be grouped round a point.

The 5 angles of a pentagon add up to 540° which can be thought of as $180° + 360°$. So, a tessellation will be possible when two of the angles add up to 180° and the other three to 360°. There are basically only two ways that this can happen. Either (a) the two supplementary angles (x and x') are adjacent – Figure 5.15.1(a) – or (b) they are separated by one of the other angles – Figure 5.15.1(b).

(a) If they are adjacent then two sides of the pentagon are parallel and a tessellation will result from $\frac{1}{2}$-turns about the points marked in Figure 5.15.2. These are the mid-points of the line segments AB, BC, CD, DE, where $DM = BC$.

The tessellation is shown in Figure 5.15.3 where the angles x and x' join together to make straight lines and the other three group round a point.

The well-known Pentiamond tessellation in Figure 5.15.4 is an example of this type with $x = 120°$ and all the sides equal.

(b) If x and x' are separated by one of the other angles then attempts to form tessellations lead to certain constraints on the relative lengths of the sides. This is best shown by an example. In Figure 5.15.5, $AE = BC$ and $AB = ED$ and the resulting tessellation is shown.

This is a strange pattern of interlocking hexagons, each hexagon made up of four pentagons, two of which are mirror-images of the other two.

Finally, if the sides are all equal and $x = x' = 90°$, the tessellation in Figure 5.15.5 becomes Figure 5.15.6 which is shown in Cundy and Rollett and is a favourite street-tiling in Cairo. The geometry of this basic pentagon is shown in Figure 5.15.7.

Published in *Mathematical Gazette* **55** (December 1971), pp. 366–369; **56** (December 1972), pp. 333–335; **67** (June 1983), pp. 139–140.

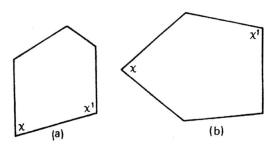

(a)   (b)

Fig. 5.15.1.

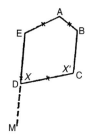

Fig. 5.15.2.

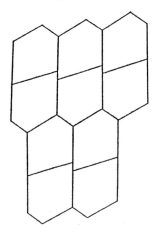

Fig. 5.15.3.

PENTIAMOND

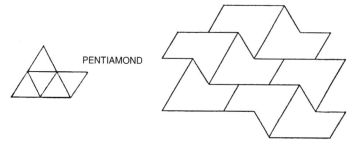

Fig. 5.15.4.

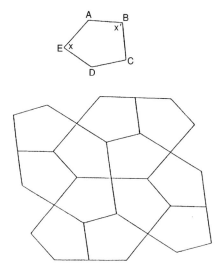

Fig. 5.15.5.

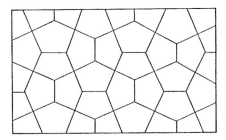

Fig. 5.15.6.

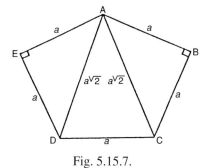

Fig. 5.15.7.

It is possible for this basic (sufficient) condition to be reversed, i.e. to have two of the angles adding up to 360° and the other three to 180°. Such a pentagon will have a reflex angle and tessellations are possible.

Are there other possibilities?

To the Editor, *The Mathematical Gazette*

TESSELLATIONS WITH PENTAGONS

DEAR SIR. – I have just read an article by Mr. J. A. Dunn on "Tessellations with Pentagons", which finishes with the query: "Are there other possibilities?" This can be answered by referring the author to an article by Mr. R. B. Kershner of the Johns Hopkins University, published in the October 1968 issue (vol. 75 No. 8) of the *American Mathematical Monthly*, pp. 839–844, entitled "On paving the plane", where all tessellations with congruent polygons having more than four sides are covered by the following two theorems, which are demonstrated in the article:

Theorem 1. A convex hexagon can pave the plane if and only if it is of one of the following three types:

$$1: A + B + C = 2\pi, \quad a = d;$$
$$2: A + B + D = 2\pi, \quad a = d, c = e;$$
$$3: A = C = E = \tfrac{2}{3}\pi, \quad a = b, c = e, e = f.$$

Theorem 2. A convex pentagon can pave the plane if and only if it is of the following eight types.

$$1: A + B + C = 2\pi;$$
$$2: A + B + D = 2\pi, a = d;$$
$$3: A = C = D = \tfrac{2}{3}\pi, a = b, d = c + e;$$
$$4: A = C = \tfrac{1}{2}\pi, a = b, c = d;$$
$$5: A = \tfrac{1}{3}\pi, C = \tfrac{2}{3}\pi, a = b, c = d;$$
$$6: A + B + D = 2\pi, A = 2C, a = b = e, c = d;$$
$$7: 2B + C = 2D + A = 2\pi, a = b = c = d;$$
$$8: 2A + B = 2D + C = 2\pi, a = b = c = d.$$

It will be seen that not only are Mr. Kershner's results more precise and definite than Mr. Dunn's, but that he also gives two types of pentagon (his types 7 and 8) which are not included in Mr. Dunn's possibilities.

Mr. Dunn's first example is, of course, Mr. Kershner's type 1, and the former's second example is a special case of the latter's type 2. Mr. Kershner's types 3 to 6 agree with Mr. Dunn's general idea of three angles totalling 360° and the other two totalling 180°, but the pentagons are arranged in other ways, in one case (type 5) even *six* pentagons concurring at a vertex. In types 7 and 8 we have vertices where four pentagons meet without any two angles adding up to 180°.

The possibility of non-convex pentagons, mentioned by Mr. Dunn, would remain open.

Yours faithfully,
M. M. Risueño

Buenos Aires

To the Editor, *The Mathematical Gazette*

DEAR SIR. – The point made by Mr. Dunn at the end of his article "Tessellations with Pentagons" about the reversal of the basic condition for a pentagon to be a tessellating cell by having "2 of the angles adding up to 360° and the other 3 to 180°" is interesting. Outlined below is a method of constructing such a pentagon, which will necessarily be re-entrant.

Draw any re-entrant par-hexagon $ABCDEF$ (that is, a hexagon with three pairs of parallel sides), and produce any pair of parallel sides, EF and BC say, by equal distances inward to points P, Q respectively, as in Figure 5.15.8. Join PQ.

Because a par-hexagon enjoys point symmetry about its centre O, we have here that $PO = OQ$. We have thus divided the hexagon into 2 congruent pentagons (the one can be obtained from the other by a half-turn about O), which can now be used to tessellate a plane. It will be noticed that because EP and BQ are parallel,

$$\angle FPQ = \angle PQC;$$

therefore, in the pentagon $ABQPF$,

$$\text{reflex } \angle FPQ + \angle PQB = 360°,$$

and it is easily seen that the other three angles of the pentagon sum to 180°. (It is well-known that any par-hexagon can be used to tessellate a plane surface.)

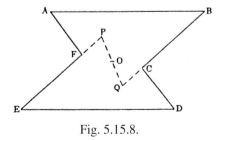

Fig. 5.15.8.

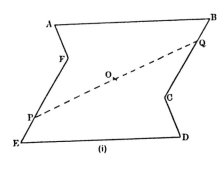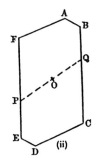

Fig. 5.15.9.

All that the above construction really does is to divide the parallel sides EF and BC *externally* in the *same* ratio. It is interesting to note that if we divide the parallel sides of any par-hexagon *internally* in the same ratio, as in Figure 5.15.9, we have the first type of pentagons, with 2 adjacent angles supplementary, here angles FPQ and BQP. The pentagons so formed are (i) re-entrant or (ii) convex according as the par-hexagon is re-entrant or convex.

It should also be noted that as P and Q travel round the sides of the par-hexagon, provided always that the line segment PQ is wholly contained within the hexagon, a whole class of pentagonal cells is generated, each of which can be used for tessellation.

Yours faithfully
P. Nsanda Eba

N. W. Cameroon

[Mr. Eba also draws attention to Mr. Dunn's second class of tessellating pentagons, of which a specimen is reproduced below. In the original article it is stated that "each hexagon is made up of four pentagons, two of which are mirror images of the other two". Mr. Dunn points out that he here uses the description "mirror images" to describe pentagons related by an opposite isometry, but not of course by a single reflection. His point is that this tessellation includes congruent non-regular asymmetrical pentagons (I and III in the figure) and the same pentagons "turned over" (II and IV), not just rotated in the plane. He remarks "I think this is fairly unusual". Mr Eba adds that I and III are obtained from each other by half-turn about the centre of the hexagon (marked with a cross in the figure), as are II and IV. The construction of this tessellation can be recommended as an instructive and entertaining exercise.

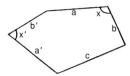

The fundamental pentagon: $a = a'$, $b = b'$ and $x + x' = 180°$.

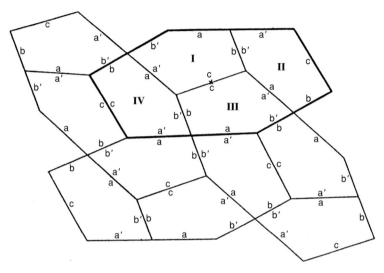

Tessellation of pentagon forming a pattern of interlocking hexagons.

Finally, we are grateful to Mr. Eba for pointing out that the statement on p. 421, that the pentiamond has "all the sides equal" is, of course, incorrect; actually, three of the sides are I unit in length and the other two are 2 units each. D.A.Q.[1]

---

[1] Douglas Quadling, the then Editor of the *Mathematical Gazette*

5.16

Universal Games

HELEN MORRIS

Games have been played for thousands of years, for power, for money, for amusement, all over the world. They began very simply, using the resources available at the time – sand, pebbles, coloured sticks, shells. Most of the games that we play today have descended from these simple ancient pastimes, although we now have access to superior resources, including computers programmed to beat us, and commercially manufactured packages. We cannot always afford to purchase the more sophisticated offerings, but we can always find items around that will still be able to provide children with the experiences and excitement enjoyed around the world for the past 3000 years.

Alquerque (Africa)

A game that has been played for centuries in North Africa and the Middle East. On the West Bank of the River Nile in Egypt there is a temple at Al-Qurna, where an engraving of an Alquerque board has been found, dating back to 1400 BC. The game then spread, by the Moors of Africa, to Spain.

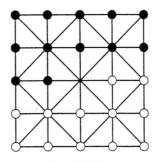

Fig. 5.16.1.

First published in *Mathematics in School* **26**, 4 (September 1997), pp. 35–40.

Two players each have 12 pieces, starting in the positions shown. Pieces can move along a line to an empty point. Pieces can be captured by being jumped over onto an empty point. More than one capture can be made in one move, and the direction of movement can also be changed. If a player misses a chance to capture an opponent's piece, then the offending piece can be removed from the board. The winner is the first person to capture all of the opponent's pieces.

Ba-Awa (Africa)

A Mancala game, played to the same rules as Wari, with one exception. Two players have 6 cups containing 4 seeds in each. The first player lifts the seeds from one of their cups and sowing one into each cup in an anti-clockwise direction. Where the last seed falls, the same player lifts and sows those seeds. This continues until the last seed falls into an empty cup. The second player now starts from any of their own cups. From now on, if any seed makes a cup up to 4 seeds, the owner of the cup immediately transfers them to their store. If the last seed makes up a 4 the turn ends. The player with the most seeds wins.

Chatatha (Native America)

Originally played by Native Americans using sea shells and a shallow basket. The players toss six shells in the basket, and win sticks according to their score. Three shells facing up and three down score 1 stick, six shells up or six down score 2 sticks. The winner is the first person to score 8 sticks.

Dara (Nigeria)

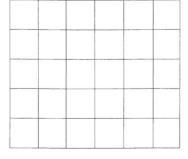

Played by the Dakarkari people in Nigeria using stones, pieces of pottery or shaped sticks. The board consists of 5 rows of 6 holes. Each player has 12 pieces, which are placed, in turn, into the holes. Once all of the pieces have been placed moves are made. A piece can be moved into an adjacent empty hole (not diagonally). The aim is to make a row of three pieces, not diagonally. When a line of 3 is formed the player removes one of the opponent's pieces from the board. The game ends when a player is unable to make a line of 3 pieces.

Exchange Kono (Korea)

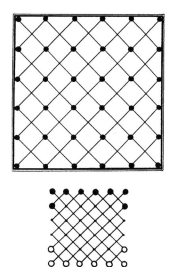

Fig. 5.16.2.

Each player has 8 pieces, with the starting position as shown. The players take turns to move a counter one space diagonally onto a black spot. The aim is to be the first to occupy the opponent's starting positions. There are no jumps or captures.

Fox and Geese (Iceland)

This game was played by the Vikings in Iceland. There are 13 geese and 1 fox. The geese start in the positions shown; the fox starts on any empty spot. Geese move first, along a line. The fox kills a goose by jumping over it to a vacant point. The geese win if they surround the fox. The fox wins if there are so many geese killed that it cannot be surrounded.

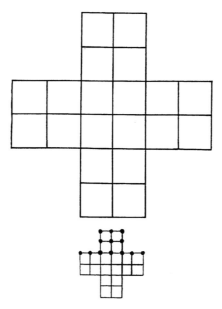

Fig. 5.16.3.

Go Bang (Japan)

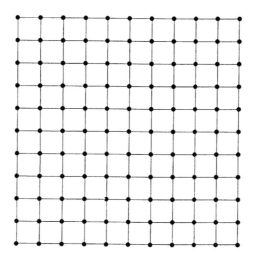

In Japan the most popular game is Go, with professional players earning a lot of money from the game. Go Bang is a simpler version of Go, arriving in England in 1885. Counters are placed alternately on the intersections of a 10×10 square board. The aim is to form 5 counters in a row in any direction.

A game that has been played for centuries by the Baggara tribe in North Africa. The Baggara tribe were nomadic and the game portrays the perils of nomadic life.

The game can be played with 2 to 6 players. Each player has a counter which represents their mother. There is also a bowl with 50 counters (*Taba beans*). Scoring is made with 3 coins. One head – take 1 Taba from the bowl. Two heads – move 2 places and the go ends. Three heads – move 3 places. No heads – move 6 places. A go only ends when two heads have been thrown.

You start with one Taba. A mother must reach the well with an exact throw, or make up the difference with Tabas. She then pays 4 Tabas to wash her clothes and starts back again. If she is short of Tabas she must throw them, but she can count up her scores while she is waiting and use them all at once when she starts back. The first mother back (does not have to be an exact throw) pays 2 Tabas and releases the Hyena, which doubles its coin scores. At the well it needs 10 Tabas for a drink and then it races back, eating any mothers it overtakes. So a mother can win, lose or be eaten.

Ise-Ozin-Egbe (Nigeria)

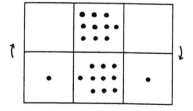

Fig. 5.16.4.

The player starts by lifting the 10 beans in the bottom row and sowing one at a time. The next moves start from the hole where the last sowing finished. The aim is to arrive back to the original position.

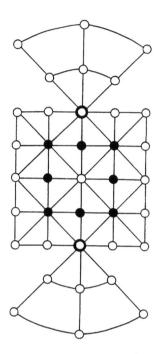

A battle game between 2 Princes and 24 Lamas (the priests, not the animals!). The Princes and 8 Lamas are placed as shown. The first player (the Prince) can move a Prince one space or capture a Lama by jumping over it to an empty space beyond. The second player (the Lama) plays by placing a Lama on the board until all 24 have been used. Then the second player continues play by moving Lamas on the board. The Prince wins if only 8 Lamas remain. The Lama wins if the Princes are trapped. Lamas can force a capture in order to help trap a Prince. Multiple captures are allowed.

Ludus Latrunculorum (Italy)

Popular with Roman soldiers, this game was used when they were resting.

An 8 × 8 chessboard is used. Each player needs 16 men, plus a Dux or King (special counter). The players take it in turns to place their men, two at a time, on the squares of the board. The Dux is placed last. After all pieces have been placed, the second player goes first, moving any piece one square (not diagonally). If an enemy piece is sandwiched between two opposing pieces, it is captured and removed from the board. The moving piece can continue with another move, after a capture. Corner pieces can be captured by being surrounded on either side. A piece can be moved in between two enemy pieces in safety. If an opponent cannot move then a player must move again. The Dux can also jump over any single piece. If this ends up in a sandwich the sandwiched man is captured. The Dux can be captured. The player with the most men left at the end is the winner.

Mu Torere (New Zealand)

A blockade board game played in New Zealand by the Ngati Porou people. It is the only native Maori board game known. The board would have been marked on the ground with twigs or stones used as counters.

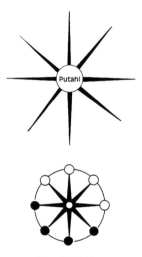

Fig. 5.16.5.

Two players have 4 pieces (known as *perepere*) each placed on adjacent points of the star (called the *kewai*). The aim is to block the opponent from moving. The centre space is called the *putahi*. Moves can be made (a) from one kewai to an adjacent empty kewai, (b) from the putahi to a kewai, (c) from a kewai to the putahi as long as either one or both of the adjacent points is occupied by an enemy piece. Only one piece can occupy the same place at the same time. Jumping is not allowed.

1. Nyout (Korea)

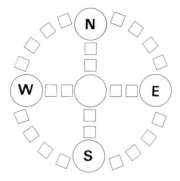

A racing game from Korea, for 2 to 4 players, probably from before 1100 BC. It is still a very popular gambling game. These games made their way to America by 800 AD.

The board consists of points made into a cross and a circle. Each player can use four counters (*Mal*) which stand for horses. The players take turns to toss a dice (called *Pam-Ny-out*) and move their horses anti-clockwise around the board, starting at N, aiming to finish at N. If a player's horse lands on E, S or W, they may take a short cut through the middle of the board to the exit or *Ch'ut* (N). If a horse lands on the same square as another of it's own horses they can join together and move as one horse. Three or four horses can join together. If a player's horse lands on a square occupied by an opponent, then the opponent is sent back to the start, but not if the opponent has a joined horse there greater than the landing horse. If a score of 4 or 5 are thrown, the player has another throw; the scores are added together and can be used for one horse or split between two or more horses. The winner is the first player to land on the Ch'ut.

Ou-moul-ko-no (Korea)

The Korean name for Pong Hau K'i.

Pong Hau K'i (China)

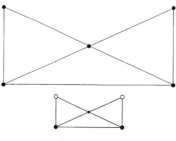

Fig. 5.16.6.

Traditionally played by children in Canton, usually with scraps of paper.

A game for two players, one with two white counters, the other with two black counters. Players take it in turns to move one counter along a line to an empty space. The aim is to block the opponent from being able to move.

Quirkat (Arabia)

The original Arabian name given to Alquerque.

Rithmomachia

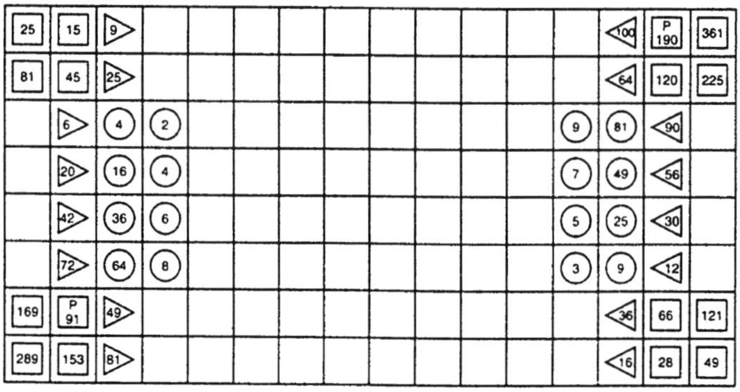

Fig. 5.16.7.

Popular game played by intellectuals in the Middle Ages. It is played on a board of 8×16 squares. Each player has 'round', 'triangle', 'square' pieces in their colour, and one 'pyramid' piece. Players take turns and move one piece at a time. A 'round' can move to any adjacent, empty space. A 'triangle' can move 3 empty spaces in any direction. A 'square' can move 4 empty spaces in any direction. A 'pyramid' can move in any of the previous 3 ways.

The aim is to capture the opponent's pieces, which can be done in 4 different ways.

(1) By landing on an occupied square.
(2) By surrounding a piece on 4 sides.
(3) By moving to a position so that the number on the piece multiplied by the number of vacant squares between this piece and an opponent's piece is equal to the number on the opponent's piece.
(4) By moving pieces to either side of an opponent's piece so that the sum of the pieces is equal to the opponent's number.

There are several ways of winning and players need to agree in advance the Victory method that they are playing to. Some common Victory rules follow.

(1) Capturing a certain number of pieces.
(2) Capturing pieces to a total value of, say, 160.
(3) Capturing pieces to a total value but with a specified number of pieces.

For the more advanced player, Victory can be gained on achieving any of the following outcomes (made up of own and at least one captured piece).

(1) Having 3 pieces which form an arithmetic progression (eg 3,6,9) or a geometric progression (eg 2,4,8) or a harmonic progression (eg 2,3,6 as the reciprocals form an arithmetic progression).
(2) Having 4 pieces which form two different progressions (eg 2,3,4,6 gives an A.P. with 2,3,4 and an H.P. with 2,3,6).
(3) Having 4 pieces which form 3 different progressions (eg 4,6,9,12 gives 6,9,12 as an A.P., 4,6,9 as a G.P., and 4,6,12 as an H.P.)

Shap Luk Kon Tseung Kwan (China)

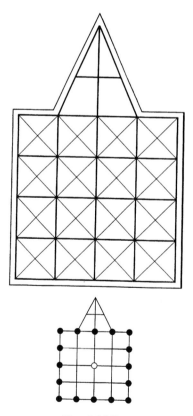

Fig. 5.16.8.

The name means 'sixteen pursue the general'. It is played on an Alquerque board with a triangular extension at the top. One player is the general and the other controls the 16 soldiers. They can all move one step along any line in any direction. The general can enter the triangle but the soldiers cannot. The general and the soldiers can capture. The general can capture two soldiers by moving to an empty point between them. Both soldiers are then captured and removed from the board. If the soldiers can position themselves so that they are directly beside the general on the same line the general is captured and loses. If the general is trapped inside the triangle he is captured and loses.

Tabula (Italy)

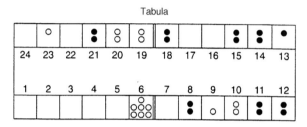

Fig. 5.16.9.

Played by the Emperor Claudius (50 AD) and Zeno (450 AD).

Each player has 15 counters of their colour. All pieces go onto the board at 1 and leave at 24, moving anti-clockwise. Three dice are thrown, the total of which can be used to move one piece, or used separately to move 3 pieces, or two of the dice can be combined to move two pieces. If a player has two or more pieces on a square the opponent cannot land on that square. If a player lands on a square occupied by only one opponent's piece, that piece is sent back to the beginning. The winner is the first person to get all 15 counters safely round and off the board, with an exact throw needed to leave the board.

Ur (Iraq)

The Royal Game of Ur is sometimes called the Sumerian Game, and dates from 3000 BC. A board and pieces were found in the Royal Cemetery at Ur.

Each player has 7 counters. 3 coins are used. No heads scores 0. One head scores 1. Two heads scores 4. Three heads scores 5. When 3 heads are thrown,

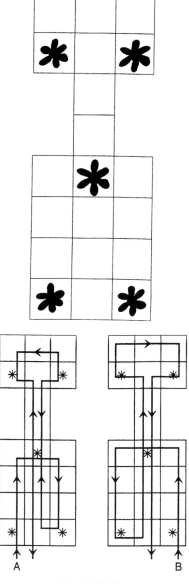

Fig. 5.16.10.

have another go. Throw 1 to enter the board. One player follows route A, the other route B. Two or more men on a square are safe, but single men can be hit and sent back to the start. Rosettes can be shared by opposing players, and are safe. The exact score must be thrown to leave the board.

Vultures and Crows (India)

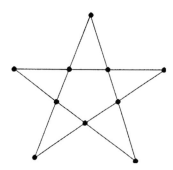

Also called Kaooa, it uses a Pentalpha board, and involves one vulture and seven crows. A crow is placed, then the vulture, then another crow, then the vulture can move one space, until all crows have been placed on the board. The vulture captures the crows by jumping over them to a vacant spot. The crows can pen in the vulture to stop it from moving.

Wari (Arabia)

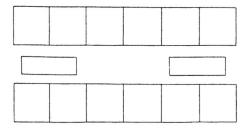

Also known as Mancala, and has become the general name for all 'transferring' games where counters (or seeds) are moved from one cup to another. Two players have 6 cups containing 4 seeds in each. Players take turns, lifting the seeds from one of their cups and sowing one into each hole in an anti-clockwise direction. If the last seed drops into an enemy cup to make a final total of 2 or 3, the seeds are captured, along with the seeds of any unbroken sequence of 2s or 3s on the opponents side adjacent to and behind the plundered hole. The player with the most seeds at the end is the winner.

Yote (West Africa)

Players take turns to place twelve counters of their own colour. Once all of their counters are placed the pieces can be moved one space up or down. A capture is made by jumping over an enemy piece to a space directly beyond. Any player who makes a capture can also remove another counter belonging to the opponent. Multiple jumping is not allowed.

Zamma (Sahara)

This game is sometimes called 'quadruple Alquerque', with similar rules. Each player has 40 pieces – moves can be made directly or diagonally forward. When a piece reaches the opposite side of the board it becomes a *mullah*, and can be 'crowned' with another piece. A mullah can move any amount of spaces in any direction, and can jump across empty spaces between itself and an opponent's piece to an empty space immediately on the other side.

References

R. Bell, & M. Cornelius, *Board Games Round the World* (Cambridge: Cambridge University Press, 1988).

C. Cornelius, & A. Parr, *What's Your Game?* (Cambridge: Cambridge University Press, 1991).

C. Irons, & J. Burnett, *Mathematics from Many Cultures* (London: Kingscourt Publishing, 1995).

M. C. Krause, *Multicultural Mathematics Materials* (Virginia: The National Council of Teachers of Mathematics, 1983).

R. Sheppard, & J. Wilkinson, *The Strategy Games File* (Diss: Tarquin Publications, 1989).

Desert Island Theorems

Group E: Geometrical Physics

E1

Euler's Identity

KEITH DEVLIN'S CHOICE

In 1748, the Swiss mathematician Leonhard Euler discovered the amazing identity $e^{i\pi} + 1 = 0$ which relates the five most significant and most ubiquitous constants in mathematics: e, the base of the natural logarithms, i, the square root of -1, π, the ratio of the circumference of a circle to its diameter, 1, the identity for multiplication, and 0, the identity for addition.

Another way to write the identity is $e^{i\pi} = -1$ which highlights the fact that taking an irrational number and raising it to an irrational imaginary power gives (on this occasion) an integer answer. As an arithmetical identity, therefore, the result seems highly mysterious. But that's because it isn't really a theorem of arithmetic. It's a result of geometry. I choose it as my Desert Island Theorem of geometry because it demonstrates the incredible power of the geometric approach.

The trick is to view multiplication as a geometric process that takes place in a two-dimensional, Euclidean plane. At elementary school we all learned how to add, subtract, multiply, and divide whole and fractional numbers, and to think of those processes in terms of points on the number line. Adding, we learn, corresponds to counting further along the line. For example, to add 3 to 5 we start at the point 5 and count on three further integer points to the right. Subtraction corresponds to counting backwards, i.e., to the left. Multiplying a number X, i.e., a point on the number line, by a whole number N corresponds to repeating the "X-step" (the line segment from 0 to X) N times. Dividing X by N corresponds to cutting up the line segment from 0 to X into N equal subsegments. And so forth.

Arithmetic of complex numbers can likewise be viewed geometrically, and although it is never (as far as I know) taught at elementary school, and only rarely at high school, it is not significantly more difficult. The main difference is that complex numbers are essentially two-dimensional entities, and thus must be viewed

Keith Devlin, Mathematician and Information Scientist, Stanford University. *Guardian* Columnist and Author of *The New Golden Age* and *The Math Gene*.

geometrically as the points in a two-dimensional plane (the Euclidean plane). Instead of numbers being represented as (the end-points of) line segments that start at the origin and run left or right along the (real) number line, we have to think of them as (the end-points of) line segments that start at the origin and run in any direction whatever in the plane. The length of the line segment represents the size (or "modulus") of the complex number. Its direction can be specified by measuring the angle through which the positive Real-axis has to be rotated counterclockwise to point in that direction. For example, the imaginary number i has Real-coordinate 0 and Imaginary coordinate 1; so its geometric coordinates are $(0,1)$. Thus, its modulus is 1 and you have to rotate the Real-axis a counterclockwise angle of $\pi/2$ to get it to point toward i. This angle of direction is called the "argument" of the complex number.

When complex numbers are thought of as points in the plane in this way, addition of two complex numbers becomes vector addition. That is to say, you complete the parallelogram determined by the two line segments that represent the given numbers, and then the diagonal of that parallelogram represents the sum of the two numbers. This is illustrated below:

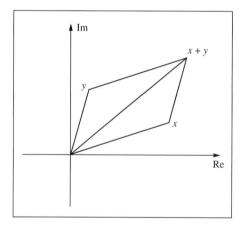

Multiplication is a bit more complicated. To obtain the product xy of two complex numbers x and y, you multiply their moduli, to give the modulus of the product, and then you add together their arguments to give the argument of the product. The second step here corresponds to performing the two rotations one after the other, first a rotation through the argument of x, then an additional rotation equal to the argument of y. Visually, then, multiplication of one complex number x by another complex number y involves a rotation of x together with a change in its length – the rotation is by the argument of y, the change in length is by a factor equal to the modulus of y.

Now comes Euler's key insight. As the real number x varies from 0 up to 2π, the complex number e^{ix} traces out a circle of radius 1, whose centre is at the origin.

This is illustrated below:

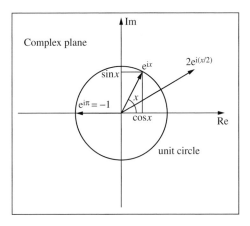

By elementary trigonometry, this implies that

$$e^{ix} = \cos x + i \sin x.$$

This is also illustrated in the diagram. So too is the result of multiplying e^{ix} by $2e^{-ix/2}$.

   In particular, therefore, taking $x = \pi$, we see that $e^{i\pi}$ is the point you reach when you start on the positive Real-axis and go around the unit circle anticlockwise through an angle π. But that takes you exactly half way round the circle, so you end up at the point -1. Hence $e^{i\pi} = -1$. In other words, Euler's seemingly mysterious identity is a fairly simple geometric observation. At least, it is once you know that the formula e^{ix} traces out the unit circle. That result, which provides a geometric explanation of e^{ix}, is the part that required the genius of Euler.

E2

Clifford Parallels

MICHAEL ATIYAH'S CHOICE

As we all know "parallel lines never meet", and we can picture the (x,y) plane filled out by the lines parallel to say the x-axis. Similarly three-dimensional space with coordinates (x,y,z) can also be filled out by parallels to the x-axis. This is essentially the content of Euclid's famous axiom which played such an important part in the history of geometry. But the brilliant mathematician, William Kingdom Clifford (1845–1879), discovered a 'twisted version' of this story which turned out to be of profound significance. The Clifford 'lines' are actually circles of varying radii r (including lines as a limiting case when r $\to \infty$) which 'never meet' and again fill out three-dimensional space. But now they have the remarkable new property that **any two are linked**. A picture of these Clifford circles is given on the next page.

Just as stereographic projection from the North Pole, N, onto the plane sends circles to circles (or lines if the circle goes through N), so an analogous projection identifies the 3-dimensional sphere (with its North Pole removed) with 3-dimensional Euclidean space. Clifford's circles are just the images of a family of 'parallel' circles on the 3-dimensional sphere. If we take this sphere as given by the equation

$$x_1^2 + x_2^2 + x_3^2 + x_4^2 = 1;$$

then, introducing two complex variables, $z_1 = x_1 + ix_2$; $z_2 = x_3 + ix_4$, we can write it as

$$|z_1|^2 + |z_2|^2 = 1.$$

For each value λ (including ∞) of the complex ratio z_1/z_2 we get a circle C_λ and these are Clifford's circles. It is clear that they are disjoint (i.e. 'parallel') and fill out space.

Michael Atiyah, Geometer, University of Edinburgh. Former Master of Trinity College, Cambridge. Fields Medalist, 1966.

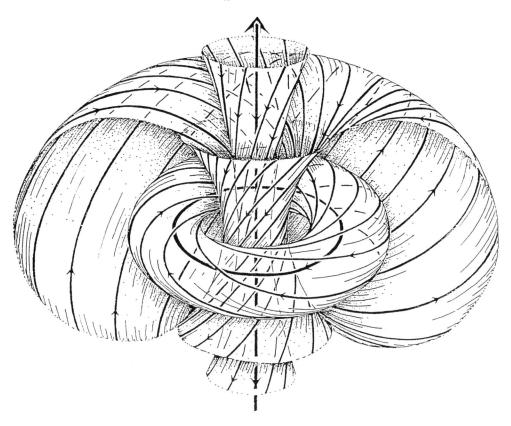

It is not hard to see that any two circles are linked. In fact, by continuity it is enough to check it for the two values $\lambda = 0$ (i.e. $z_1 = 0$) and $\lambda = \infty$ ($z_2 = 0$), and this is a good exercise for the reader!

The fascination of this theorem of Clifford lies partly in its inherent beauty but even more in its deep significance. It is the first example of such twisted fibrations (as they are now called) which dominate modern geometry. It lies behind the theory of birational transformations which enable us to remove singularities of algebraic curves. Finally, and most surprisingly, it is the key ingredient in Dirac's explanation of why all fundamental particles in the world have an electric charge which is an integral multiple of that of the electron. This explanation requires quantum mechanics which only came fifty years after the death of Clifford.

Clifford's discovery was a portent of the future.

E3

The Tait Conjectures

RUTH LAWRENCE'S CHOICE

> Here we may remark that it is obvious that when the crossings are alternately +
> and − no reduction is possible, unless there be essentially nugatory crossings.

So wrote Peter Guthrie Tait in 1877 in his communication to the Royal Society of Edinburgh entitled 'On knots' [1]. It took over a century before a rigorous proof of this 'obvious' fact appeared in 1987 ([2],[3],[4]).

What is a knot? It is a mathematical idealisation of a loop of string without end – a zero-thickness curve. However, we do not care exactly how this loop hangs. If we can continuously deform a loop into another one without cutting and gluing then we want to consider the associated knots as identical. We picture knots by thinking of them lying over a piece of paper with the over- and under-crossings marked. Different depictions of the same knot may have different numbers of crossings; one with the least number of crossings is said to be a *reduced diagram* of the knot.

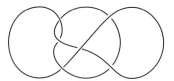 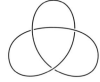

Fig. E3.1. Equivalent non-reduced and reduced diagrams of a knot.

The simplest knot is the *unknot*, a simple loop. There are no non-trivial knots with one or two crossings. Then comes the trefoil with a minimum of three crossings. The trefoil comes in two types, mirror images of each other. Think of laying the knot on a mirror. The image will have the same shape except that the orientation of each crossing will have been switched. Next comes Listing's knot (the so-called figure-8 knot) which is the same as its mirror image. Such knots are said to be

Ruth Lawrence, Topologist, Hebrew University of Jerusalem.

amphichiral (or *amphicheiral*), meaning 'either handed'. They can be deformed into their mirror images. (That Listing's knot is amphichiral is not immediately obvious – try it with a piece of string!)

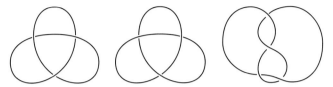

Fig. E3.2. Two trefoils and the amphichiral Listing's knot.

As with many subjects in mathematics, knot theory was first seriously investigated by Carl Friedrich Gauss. His student, Johann Benedict Listing, wrote a treatise on the subject in the 1840s [5]. This work was largely forgotten until Sir William Thomson (later Lord Kelvin, of thermodynamic fame) proposed his 'vortex atom' theory of matter, in an attempt to explain the then recently discovered array of atomic elements and their spectra. This theory proposed that empty space (vacuum) should be thought of as filled by æther and matter should be considered as a vortex in the flow of the æther locally described by a knot. Just as atoms can be combined into molecules, so can knots be combined to make more complicated ones, by tying them one after the other on the piece of string before closing the ends together.

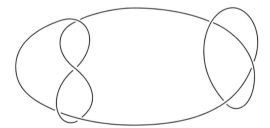

Fig. E3.3. A composite knot formed by combining Listing's knot and a trefoil.

A prime or *simple* knot is one that cannot be decomposed into other (non-trivial) knots, and they were to correspond to atoms. There is however an infinite number of simple knots; so in Tait's words, 'why have we not a much greater number of elements than those already known to us?'. The answer was to be some theory of kinetic stability.

But in any case, there is the purely mathematical question as to how many different simple knots there are with a given small number of crossings. James Clerk Maxwell, Tait and Thomson collaborated on this project and Tait took on the task of enumerating knots of a given number of crossings. This problem splits into two: to draw all combinations of projections with that number of crossings and then to identify which ones are really just different forms of the same knot (or of simpler knots with fewer crossings). Indeed some quite complicated knots can be reduced to simpler ones, although there are examples of knots, such as Millett's unknot, which have to be made more complicated before they can be simplified.

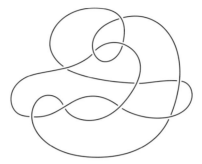

Fig. E3.4. Millett's unknot.

In general this is a difficult task, but the Tait conjectures enable it to be carried out easily for a special class of knots.

Pick a point on any of the knots drawn above and follow around the knot from there, eventually returning back. Whenever you encounter a crossing, it will either be by following the upper path (an overcrossing $+$) or the lower path (an under-crossing $-$). A knot diagram with the property that these signs alternate as one traces it, is called an *alternating knot diagram*. At first sight you may think that a non-alternating diagram must be reducible. This is not true.

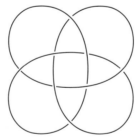

Fig. E3.5. An 8-crossing non-alternating reduced diagram.

On the other hand, the statement we began with (while not exactly obvious) is true! Alternating knots, at least, are automatically irreducible unless there is a *nugatory* crossing. (Crossings were dubbed 'nugatory' by Tait if, when removed by cutting the two strands, they could be reconnected to produce two loops which are physically separable.) Another of Tait's conjectures specifies exactly when two alternating diagrams, without nugatory crossings, can represent the same knot; namely if they are related by a series of *flypes*. (The Scots word 'flype' means to turn inside-out.)

Fig. E3.6. The effect of flyping.

These conjectures together enable a relatively easy classification of at least all the simple alternating knots of a given number of crossings ([6], [7]). Thus there are 19536 prime alternating knots with 14 crossings resulting from the distinct types coming from 170,692 reduced diagrams of prime alternating knots with 14 crossings.

We will here indicate the method of proof for the first conjecture. The idea is to use some *invariant* of knots, that is something readily calculable from a description of a knot (say a knot diagram) which is unchanged when the knot is deformed. This is hard to come by. For example, the smallest number of crossings in a diagram is an invariant but almost incalculable. Several such invariants are known. In 1984/5, Vaughan Jones discovered a new invariant for knots, the values of which are polynomials in an indeterminate t, coming out of completely different work he was doing on subfactors of von Neumann algebras. The original definition [8] is rather technical. Here is a combinatorial prescription. The procedure for calculation is to recursively relate values on a link with those on links differing only at one crossing, by either removing it (which leads to a diagram with fewer crossings), or using the mirror image crossing (which, if the crossing is carefully chosen, gives a reducible diagram). The procedure terminates when we reach the simplest knot or link, with no crossings. Here is the prescription and a sample calculation.

Prescription:

$$\langle \rangle = t^{-4}\langle \rangle = \left(t^{-1}-t^{-3}\right)\langle \rangle$$

$$\langle d \text{ unlinked and unknotted circles}\rangle = \left(t+t^{-1}\right)^{d-1}$$

Calculation:

$$\langle \rangle = t^{-4}\langle \rangle + \left(t^{-1}-t^{-3}\right)\langle \underset{unknot}{} \rangle$$

$$= t^{-4}\left(t+t^{-1}\right) + \left(t^{-1}-t^{-3}\right) = t^{-1}+t^{-5} \quad span = (-1)-(-5) = 4$$

$$\langle \rangle = t^{-4}\langle \underset{unknot}{} \rangle + \left(t^{-1}-t^{-3}\right)\langle \rangle$$

$$= t^{-4} \qquad + \left(t^{-1}-t^{-3}\right)\left(t^{-1}+t^{-5}\right)$$

$$= t^{-2}+t^{-6}-t^{-8} \quad span = (-2)-(-8) = 6$$

It can be shown that the *span* (the difference between the highest and lowest exponents of t) of the bracket polynomial of an alternating knot (without nugatory crossings) is twice the number of crossings in the knot, and that for any knot this is the maximum span attainable. Since the polynomial is an invariant, this means that any reformation of an alternating knot must have at least as many crossings and gives a proof of Tait's first conjecture.

In conclusion, it is worth noting that Tait's prediction that ' ... the theory of knots ... is likely soon to become an important branch of mathematics' indeed came true, though it took over a century for it to really flower [9]. Although the real impetus for the development of knot theory came from Thomson's vortex theory – a proposal which has long been dismissed as a valid physical theory – it is perhaps ironic that it has found new applications in physics in the last couple of decades, once again in the pursuit of the understanding of elementary particles, but this time in the guise of string theory and quantum field theory.

References

1. P. G. Tait, On knots, *Trans. Roy. Soc. Edin.* **28** (1877), 145–190. Tait wrote two further papers with the same title and a number of others on the topology of knots. The three papers 'On knots' are in the *Scientific Papers of Peter Guthrie Tait*, Vol. I (Cambridge University Press, 1898), 273–347. A fuller list is given by Chris Pritchard, in *Provisional Bibliography of Peter Guthrie Tait*, at www.maths.ed.ac.uk/~aar/knots/taitbib.htm
2. L. H. Kauffman, State models and the Jones polynomial, *Topology* **26** (1987) 395–407.
3. K. Murasugi, Jones polynomials and classical conjectures in knot theory, *Topology* **26** (1987), 187–194.
4. M. B. Thistlethwaite, A spanning tree expansion of the Jones polynomial, *Topology* **26**, 3 (1987), 297–309.
5. J. B. Listing, *Vorstudien zur Topologie (Introduction to Topology)*, Abgedruckt aus der Göttingen studien, Göttingen (1847). The word 'topology' was first used in this book. Listing's more definitive study is *Der Census räumlicher Complexe (The Census of Spatial Complexes)*, Göttingen (1862).
6. W. Menasco & M. B. Thistlethwaite, The classification of alternating links, *Ann. of Math.* **138** (1993), 113–171.
7. B. Arnold, M. Au, C. Candy, K. Erdener, J. Fan, R. Flynn, J. Hoste, R. J. Muir & D. Wu, Tabulating alternating knots through 14 crossings, *J. Knot Theory Ramifications* **3**, 4 (1994), 433–437.
8. V. F. R. Jones, A polynomial invariant for knots via von Neumann algebras, *Bull. Amer. Math. Soc.* (New Series) **12**, 1 (1985), 103–111.
9. J. H. Przytycki, History of knot theory from Vandermonde to Jones, XXIVth National Congress of the Mexican Mathematical Society (Spanish) (Oaxtepec, 1991), 173–185, *Aportaciones Mat. Comun.* **11**, Soc. Mat. Mexicana, Mexico (1992).

E4

Kelvin's Circulation Theorem

KEITH MOFFATT'S CHOICE

Kelvin's Theorem (1869) is one of the cornerstones of modern fluid dynamics. The "circulation" round any closed curve C in a fluid is the integral round the curve of the tangential component of velocity on C. The curve can be marked by the fluid particles that lie on it; as these particles move with the flow, so the curve C is transported by the flow also; it is then described as a *material* curve. Kelvin showed that in an ideal fluid, subject only to conservative forces, the circulation round every material closed curve is constant. The theorem is intimately related with an earlier result of Helmholtz that 'vortex lines are transported with the flow'. (The diagram shows the vorticity field ω trapped within C.) It implies that topological properties of vortex lines (e.g. their links and knots) are conserved. Indeed it was Kelvin's Theorem that stimulated P. G. Tait to initiate his famous classification of knots in the 1870s, from which the discipline of Topology subsequently developed.

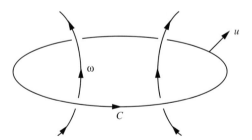

I imagine that from a mountain peak on my desert island, I will be able to observe the circulation of the currents around the island, and to consider the consequences of Kelvin's Theorem in the island context.

Keith Moffatt, Fluid Dynamicist, Isaac Newton Institute for Mathematical Sciences, University of Cambridge.

E5

Noether's Theorem

LEON LEDERMAN'S AND CHRIS HILL'S CHOICE

Noether's Theorem: For every continuous symmetry of the laws of physics, there must exist a conservation law. For every conservation law, there must exist a continuous symmetry.

Conservation laws, like the conservation of energy, momentum and angular momentum (these are the most famous), are studied in high school. We now see from Emmy Noether's Theorem that they emerge from symmetry concepts far deeper than Newton's laws.

It is an experimental fact that the laws of physics are invariant under spatial translations. This is a strong statement. For example, if space had the structure of a crystal, then moving the origin of coordinates from a nucleus to a void would change the laws of nature within the crystal. The hypothesis that space is translationally invariant is equivalent to the statement that one point in space is equivalent to any other point, i.e. the symmetry is such that translations of any system or, equivalently, the translation of the coordinate system, does not change the laws of nature. Equivalently, the laws (and equations that express these laws) are invariant to translations (translational symmetry). In this case, Noether's Theorem states:

The conservation law corresponding to space translational symmetry is the Law of Conservation of Momentum.

So, we learn in senior physics class that the total momentum of an isolated system remains constant. The ith element of the system has a momentum in Newtonian physics of the form $p_i = mv_i$ and the total momentum is just the sum of all of the elements, $P_{total} = p_1 + p_2 + \cdots + p_N$ for a system of N elements. Noether's Theorem states that P is conserved, i.e., it does not change in time, no matter how

Leon Lederman, Particle Physicist, Fermi National Accelerator Laboratory, Batavia, Illinois. Nobel Laureate, 1988.
Chris Hill, Particle Physicist, Fermi National Accelerator Laboratory, Batavia, Illinois.

the various particles interact, because the interactions are determined by laws that don't depend upon where the whole system is located in space!

Turning it around, the validity of the Law of Conservation of Momentum as an observational fact, via Noether's Theorem, supports the hypothesis that space is homogeneous, i.e., possessing translational symmetry. The more we verify the law of conservation of momentum, and it has been tested literally trillions of times in laboratories all over the world, at all distance scales, the more we verify the idea that space is homogeneous!

The experimental evidence also favours very strongly the homogeneity of time, i.e. any point on the time axis is as good as any other point, or the laws of physics are invariant under translations in time. What conservation law then follows by Noether's Theorem? Surprise! It is nothing less than the law of conservation of energy:

The conservation law corresponding to time translational symmetry is the Law of Conservation of Energy.

Since the constancy of the total energy of a system is extremely well tested experimentally, this tells us that nature's laws are invariant under time translations.

We also live in a world where the laws of physics are rotationally invariant:

The conservation law corresponding to rotational symmetry is the Law of Conservation of Angular Momentum.

Conservation of angular momentum is often demonstrated in lectures by what is usually called 'the 3 dumbbell experiment'. The instructor stands on a rotating table, his hands outstretched, with a heavy dumbbell in each hand. (Who is the third dumbbell?) A student starts the lecturer rotating on his table. He turns slowly, then brings his hands (and dumbbells) close to his body. His rotation speed (angular velocity, ω) increases substantially. What is kept constant is the angular momentum, J, the product of I, the moment of inertia, and the angular velocity ω. Hence, $J = I\omega$. By bringing his dumbbells in close to his body, the moment of inertia I, is decreased but J, the angular momentum, must be conserved, so ω must increase. Skaters do this trick all the time.

Atoms, elementary particles, etc., all have angular momentum and in any reaction, the final angular momentum must be equal to the initial angular momentum. Like our planet earth, particles spin and execute orbits and both motions have associated angular momentum. Data over the past 70 or so years confirms conservation of this quantity on the macroscopic scale of people and their machines and on the microscopic scale of particles. And now, (thanks to Emmy) we learn that these data imply that space is isotropic – there is no preferred direction. All directions in space are equivalent.

Editor's Notes

1. The full version of the text of 'Noether's Theorem' can be found at www.emmynoether. com a website set up and maintained by Leon Lederman and Chris Hill, and devoted to 'Teaching Symmetry in the Introductory Physics Curriculum'. It is a good source of information about Emmy Noether and her groundbreaking work at the interface of physics and geometry.

2. Emmy Noether was born in 1882 in Erlangen where her father Max Noether was already a distinguished mathematician at the local university. Her doctoral thesis on invariants associated with ternary biquadratic forms was successfully defended in 1907 and, without ever receiving the formal status of her male peers, she stayed on at Erlangen's Mathematical Institute for the next eight years. During this period she collaborated with Hermann Minkowski, Felix Klein and David Hilbert and it was Hilbert who invited her to join a team at the Mathematical Institute in Göttingen in 1915. Its task was to resolve a number of troublesome aspects of the general theory of relativity, most notably that energy is not conserved locally as in classical field theories. Noether's two theorems of 1918 – Leon Lederman and Chris Hill are concerned with the first of them above – enabled her to clarify, quantify and prove what Hilbert could only adumbrate. The 1920s were fertile years for her as she wrestled with the foundations of abstract algebra and developed the theory of ideals, and she was in great demand. But with the rise of National Socialism in the early 1930s there was little future and no small danger for her as a female, Jewish pacifist and she moved to Bryn Mawr in the United States, where she died suddenly in 1935. Noether's contemporaries considered her to be an outstanding mathematician. In a letter to the *New York Times* a little after her death (4 May 1935), Einstein wrote of her, 'In the judgement of the most competent living mathematicians, Fraulein Noether was the most significant creative mathematical genius thus far produced since the higher education of women began.' For a more detailed biography of Noether, see Lynn M. Olsen, *Women in Mathematics* (Cambridge MA: MIT Press, 1994).

3. A discussion of how Emmy Noether came to discover the theorem named after her was given by Nina Byers at the symposium *The Heritage of Emmy Noether in Algebra, Geometry and Physics*, held at Bar Ilan University, Tel Aviv in December 1996. Its published form is:

   Nina Byers, E. Noether's Discovery of the Deep Connection Between Symmetries and Conservation Laws, Israel Mathematical Conference Proceedings **12** (1999).

4. The paper in which the theorem was unveiled was read by Felix Klein at a meeting of the Göttingen Academy of Sciences, held on 16 July 1918. It is available in the original (German) and in English translation in both paper and electronic form:

   E. Noether, Invariante Variationsprobleme, *Nachr. d. König. Gesellsch. d. Wiss. zu Göttingen, Math-phys. Klasse* (1918), 235–257; English translation by M. A. Tavel, in *Transport Theory and Statistical Physics* **1**, 3 (1971), 183–207 and www.physics.ucla. edu/~cwp/pubs/noether.trans/english/mort186.html

E6

Kepler's Packing Theorem

SIMON SINGH'S CHOICE

As somebody who is interested in engaging the public in mathematics, I have chosen Kepler's Packing Theorem. When I write about a mathematical theorem, it helps if the theorem is important, has a long history, has at least one hero associated with it, and is not too abstract and esoteric. Kepler's Packing Theorem wins on all counts.

The theorem was born in 1611, when Johannes Kepler wrote a paper entitled 'On the Six-Cornered Snowflake'. He explained why each snowflake has a unique hexagonal structure by suggesting that every snowflake begins with a hexagonally symmetric seed, which grows as it falls through the atmosphere. Continually changing conditions of wind, temperature and moisture ensure that each snowflake is different, and yet the seed is so small that the conditions determining the pattern growth will be identical on all of the six sides, ensuring that symmetry is maintained.

His snowflake research led him to another question about the arrangement of particles, namely, which is the most efficient way to pack spheres such that they occupy the least possible volume? One of the first arrangements examined by Kepler is now known as the face-centred cubic lattice. This can be constructed by first creating a bottom layer of spheres such that each sphere is surrounded by six other spheres. The second layer is generated by positioning spheres in the 'dimples' of the first layer, and subsequent layers are created in the same way. This arrangement is identical to the one used by grocers stacking pyramids of oranges, and has an efficiency of 74%.

Kepler studied a variety of configurations but they were all less efficient than 74%, so he came to a conclusion which he included in his paper about snowflakes, namely that it was the face-centred cubic lattice for which 'the packing will be the tightest possible'. The challenge for mathematicians has been to prove Kepler's conjecture, even though grocers and the rest of the world assumed that centuries of failure to find a better arrangement was proof enough. The British sphere-packing expert

Simon Singh, Populariser of Mathematics. Author of *Fermat's Last Theorem* and *The Code Book*.

C. A. Rogers commented that Kepler's conjecture is one that "most mathematicians believe, and all physicists know."

Eventually, in 1998, Professor Thomas Hales at the University of Michigan announced a solution to the problem that had plagued mathematicians for almost four centuries. Hales had spent ten years developing a complex 250-page argument, which in turn relies on 3 gigabytes of computer files. His approach is based on analysing a single equation consisting of 150 variables, which describes every conceivable arrangement. The challenge for Hales has been to check that there is no combination of variables which leads to a packing efficiency that is higher than that of the face-centred cubic lattice. Optimising such an equation required a major computing effort. Checking Hales's proof has been difficult, because the software has had to be examined, as well as the mathematical argument. During the refereeing process, there has been no shortage of grocers declaring, "I told you so!"

Editor's Notes

1. Simon Singh's selection is reminiscent of the story he told so engagingly in *Fermat's Last Theorem*. Fermat's marginal note was probably made in 1637 but it wasn't until late in 1994, some eight years after he started, that Andrew Wiles was able to provide a definitive proof that for $n > 2$ it is impossible to find numbers x, y and z satisfying the relationship $x^n + y^n = z^n$.
2. It is a measure of Simon Singh's fascination with such long struggles by the mathematical community to solve 'old chestnuts' that to accompany his Desert Island Theorem, he has included at his website, http://www.simonsingh.net a series of links to articles and pages about Kepler Packing Theorem.

VI

The Teaching of Geometry

6.1

Introductory Essay: A Century of School Geometry Teaching: From Euclid to the 'Subject Which Dare Not Speak Its Name'?

MICHAEL PRICE

EUCLID'S ELEMENTS. By AN ANCIENT GEOMETER
FAREWELL! old Euclid: loved of yore, and may be loved again,
When our beatific vision sees thy plane surface, plain;
Unfettered now, we range without thy limited confines;
Thy concept had no breadth at all. We must have broader lines.

(S.C. 1907 [1])

The place of geometry in the curriculum
We believe that geometry has declined in status within the English mathematics curriculum and that this needs to be redressed. It should not be the *'subject which dare not speak its name'*.

(Royal Society 2001 [2, p. 5])

1. Introduction

The two opening quotations span nearly a century of geometry teaching in England which is the focus for this chapter. They also capture *the* central twentieth-century transformation in the conception of geometry for a general secondary education: from largely unadulterated Euclid, through various forms of practical and deductive geometry for schools, to the latest National Curriculum specification for 'shape, space and measures' from the Department for Education and Employment (DfEE) and the Qualifications and Curriculum Authority (QCA). The Royal Society even go so far as to recommend that the shorter title 'geometry' be readopted for the twenty-first century!

In mathematics curriculum history, the progress of geometry has arguably been more controversial and complex than the other branches of mathematics in education [3]. To help unravel the complexity and clarify the controversies, this essay is accompanied by reprints from the *Mathematical Gazette* of a number of key contributions to the history of geometry teaching [4–10]. Some of these articles survey developments over major sub-periods and others contribute to the debates at the time. Cross-references will be provided throughout the largely chronological survey in this introductory essay.

As well as contributing to the debates and pedagogical literature, leading individuals such as Charles Godfrey, Arthur Warry Siddons and Clement Vavasor Durell also made an important impact on classroom practice through their geometry textbooks, and played prominent roles in professional association, particularly but not exclusively through The Mathematical Association (MA). Individuals, professional organizations, government bodies, examination boards, educational publishers and suppliers have all contributed to the shaping of twentieth-century school geometry.

In writing about geometry teaching it is important to distinguish between 'intended', 'implemented' and 'attained' curricula [11, p. 268]. These categories capture in turn what is recommended within an educational system, what is thought and taught by teachers, and what pupils actually learn. Primary and secondary historical sources, including those chosen for this present publication, typically provide more evidence for intended as opposed to implemented and attained curricula in any period. This limitation in historiography has at least been recognised.

In addition to *descriptions* about geometry teaching, in principle or practice, it is also important to frame some *explanations* for change, resistance to change, progress, or relative stability in any period. Key questions to be investigated here include:

1. In what educational settings or sectors is geometry being taught?
2. For whom is geometry intended?
3. What is the scope of the geometry being taught?
4. What purposes does the geometry teaching serve?
5. What associated pedagogy and teaching and assessment methods are involved?
6. What major constraints are at work, such as examinations and mathematics teacher supply?

This is a challenging list of questions for the curriculum historian but it will serve to focus the necessarily brief discussion of major developments in each of the chosen historical periods which will now be examined.

2. 'Farewell! Old Euclid'

The extraordinary nineteenth-century durability of editions of Euclid's early geometrical books as the staple for public and other secondary school geometry – restricted to the plane and predominantly for boys only – has been fittingly captured by T. A. A. Broadbent, a former editor of the *Mathematical Gazette* [12, p. 186]:

The Elements, with its orderly system, its brilliant logical triumphs in the theory of parallels and the theory of proportion . . . was a peak of Greek culture. How quite it came to be inferred from this that the book was one pre-eminently suited for the education of small boys in Victorian England, how it came to pass that the small boy learned his Euclid as he learned his list of the Kings of Israel and Judah, how it became accepted as incontrovertible that immense mental and spiritual benefits must result from these exercises, forms a curious chapter in the history of education in this country.

This geometry curriculum was largely restricted to the learning and reproduction of Euclidean proofs – 'bookwork' – the sequence of propositions following Euclid's order. Potential links with arithmetic, measurement, algebra and trigonometry were all largely neglected. Geometry was pure deductive geometry, and practical geometry, involving the drawing and use of simple geometrical instruments, was not involved: Euclid's constructions were accommodated in theory but not in practice! Furthermore, the use and application of geometrical knowledge was also neglected: problem solving – the working of 'riders' – was generally not a priority. The justification for such a curriculum was grounded in mental and moral discipline and the cultural ideals of a classical education.

The MA's predecessor, the Association for the Improvement of Geometry Teaching (AIGT) from 1871, struggled hard against this dominant Euclidean tradition but with little success. A thumbnail sketch of the *status quo* up to the turn of the century is provided by Siddons [8, pp. 161–162] and a full assessment of the AIGT's nineteenth-century work is provided in my history of the MA [3, ch. 2]. It required a wide-ranging movement for reform, located outside the confines of the Association, to achieve a breakthrough in the 1900s.

The reform movement was associated with the name of John Perry, an engineer and educational reformer who promulgated ideals for a new mathematics curriculum, based on grounds of utility and a 'practical' pedagogy, emphasizing the learner's interests, motivation and involvement through experiment and practical activity. In the case of geometry, a separate branch – practical plane and solid geometry – had developed in the context of technical and vocational education. But this teaching of geometry with instruments was divorced both from Euclid and the classical ideals of the influential public schools. Perry's conception of geometry within his scheme of 'elementary practical mathematics' was very radical in 1900 [13, pp. 319–320]:

Geometry.- Dividing lines into parts in given proportions, and other illustrations of the 6th Book of Euclid. Measurement of angles in degrees and radians. The definitions of sine, cosine and tangent of an angle; determination of their values by drawing and measurement; setting out of angles by means of a protractor when they are given in degrees or radians, also when the value of the sine, cosine or tangent is given. Use of tables of sines, cosines and tangents. The solution of a right angled triangle by calculation and by drawing to scale. The construction of a triangle from given data; determination of the area of a triangle. The more important propositions of Euclid may be illustrated by actual drawing; if the proposition is about angles, these may be measured by means of a protractor; or if it refers to the equality of lines, areas or ratios, lengths may be measured by a scale and the necessary calculations made arithmetically. This combination of drawing and arithmetical calculation may be freely used to *illustrate* the truth of a proposition.

Perry's scheme also included measurement of areas and volumes, uses of squared paper, some co-ordinate geometry in two and three dimensions, two-dimensional representation of three-dimensional figures, and simple vectors. Perry's broad interpretation of geometry has many connections with the specifications for 'shape, space and measures' from the DfEE [14, pp. 36–39, 49–52, 65–68] a century later! But conspicuous by its absence from Perry's scheme is any reference to systematic deductive geometry.

Perry's attack on Euclid came from *outside* the mathematical establishment and embodied the views of the practical users of geometrical knowledge. But Euclid was also coming under attack from *within* the mathematical establishment and, here, Perry had a surprising ally in Bertrand Russell, a young Cambridge mathematician working in logic and the foundations of mathematics. By this time, Euclidean geometry was one branch of pure geometry, alongside non-Euclidean and projective geometries. Furthermore, the development of formal axiomatics and new canons of rigour revealed logical imperfections in Euclid's foundations. The stylish critique of Russell [4] embodies this thrust with particular reference to proofs in Euclid Book I. However, in the debates concerning Euclid for schools it was pedagogical and not mathematical or philosophical arguments which largely won the day.

Inevitably, for the public and grammar schools in the early twentieth century, what the Perry movement actually achieved in geometrical reform was a *compromise* which combined the two streams of practical and theoretical plane – but not solid – geometry in a 'new geometry' for secondary schools.

Siddons's sketch [8, pp.162–3] points to the pivotal role of Cambridge University in examination reform, particularly affecting the public schools. The new schedule [15, p. 179] for the general entrance examination stated:

(1) In demonstrative geometry, Euclid's Elements shall be optional as a text-book, and the sequence of Euclid shall not be enforced. The examiners will accept any proof of a proposition which they are satisfied forms part of a systematic treatment of the subject.

(2) Practical geometry is to be introduced, along with deductive geometry, and questions will be set requiring careful draughtsmanship and the use of efficient drawing instruments.

The reform movement prompted the MA to establish its first Teaching Committee in 1902 and a first report on geometry soon followed [5]. Notable here are the introduction of a geometrical foundation involving numerical, practical and experimental work; the parallel development of a course of theorems and geometrical constructions; and the support for rider work. But the shadow of Euclid still dominated the Committee's proposals for deductive geometry based on Books I–IV and VI. Siddons and Godfrey were both members of the Committee and the latter subsequently confessed [16, p. 20]:

On the eve of our liberation the M.A. published a report on Geometry teaching, a very conservative report, as it was considered impracticable to secure the abolition of the sequence. This report became obsolete in 1902.

A full assessment of the Perry movement, the farewell to Euclid, and the MA's contribution, is provided in my history of the MA [3, ch. 3].

3. 'New Geometry' for Secondary Schools

The major examination reforms affecting secondary school geometry were accompanied by widespread distribution of geometrical drawing instruments and teaching aids, and, by 1905, a remarkable surge in the output of new geometry textbooks. These new books covered preliminary, practical, theoretical or combined treatments of geometry and titles also included other market-capturing labels such as 'experimental', 'heuristic', 'observational', and 'based on experiment and discovery'.

The new thinking concerning teaching methods was also disseminated through professional association and periodicals such as *Nature*, the *School World*, the *Educational Times* and the *Mathematical Gazette*, which collectively accommodated a wealth of discussions, articles and textbook reviews. The general situation in both public and the new grant-aided secondary schools under the government's Board of Education must have been a stimulating but confusing one for many teachers seeking new directions in an unprecedented period of freedom for curriculum development.

Important characteristics of the new geometry were: an emphasis on understanding and problem solving through the working of riders, alongside the bookwork; the linking of geometry with arithmetic, measurement and algebra; and the accommodation of experimental and practical work. Bookwork still involved a selection of propositions and constructions from Euclid Books I–IV and VI, and fifty or more geometrical proofs might still be required for an examination. Furthermore, although pupils' intuitions might be utilized in early work on angles, parallels and congruence, formal proofs of fundamental propositions in Euclid Book I were still being demanded. Learning about proof in mathematics through deductive geometry was still of central importance. The justification for this in a general secondary education was still in terms of mental discipline.

The greatest variation in textbooks and teachers' practices was in the accommodation of practical work and its relations with deductive work. Concern was expressed at an early stage that practical work was being overemphasized, without clear purposes, and at the expense of deductive geometry [17, p. viii]:

It must ... be admitted that the particular type of intellectual discipline obtainable from mathematical study on its formal, systematic, and logical side, is in considerable danger of becoming temporarily sacrificed during a too extreme swing of the pendulum of reform.

Godfrey, who had produced a pioneering practical and theoretical geometry textbook with Siddons, also cautioned against practical excesses [18, p. 256]:

... it was soon realized that this would make the subject invertebrate, that there must be a certain element of severity in every school study, and that for purposes of general education geometry must still stand or fall by the logical training it gives.

Before the first world war, it was the Board of Education and not The Mathematical Association which provided all secondary schools with much-needed clarification and direction concerning the relations between practical and theoretical geometry. Circular 711 (1909) was the work of W. C. Fletcher and drew on his wealth of experience as a mathematics and head teacher, textbook writer, and chief inspector of grant-aided secondary schools. Both Godfrey [6, p. 195], writing in 1910, and Siddons [8, p. 163], looking back some forty-five years later, emphasized the pivotal role of this circular in the history of geometry teaching. Fletcher died fifty years after the circular's publication and, in an obituary, Siddons judged that the impact of the circular was 'very far-reaching and led up to the reforms that have since been made in the teaching of geometry' [19, p. 86].

The circular was seminal in two major respects. Firstly, it introduced a pedagogical model of *three stages* in geometrical education, involving a varying mix of practical and theoretical elements. This model provided for *progression* throughout the secondary school years and clarified purposes at each stage. The three stages, beginning with the fundamental notions and the first use of instruments, are summarized by Siddons [8, p. 163] and by Godfrey [6, p. 200]. Secondly, in the second stage, the circular adopted a progressive line concerning the roles of intuition and experiment in providing a much broader basis for the deductive work in the third stage. The difficulties in deductive geometry for schools were still largely associated with the fundamental propositions and their sequencing in Euclid Book I. The circular advocated an inductive treatment of these propositions in four groups: angles at a point, parallel lines, angles of a triangle and polygon, and congruence of triangles. These groups would then provide a broad base of assumption for the deductive third stage. As Godfrey [20] pointed out; 'This endless controversy as to the best sequence in Book I arises from the impossibility of compromising satisfactorily between the claims of pedagogy and mathematical rigour.' Significantly, he added: 'The Board of Education circular simply cuts the knot, by treating these fundamental theorems as postulates.' Here Fletcher could point to recent developments in the foundations of geometry – as exemplified by Russell's [4] critique of Euclid – to justify his own radical treatment of the foundations for the purposes of *school* geometry.

The circular was notably ahead of examination requirements but it clearly pointed the way forward and had a major impact on geometry textbook writing. The circular was also the focus for a major discussion, led by Siddons, at the first meeting in 1910 of the new London Branch of the MA. The discussion attracted an audience of over two hundred [21], exceeding the attendance for the MA's annual meeting at this time.

4. Inter-War Consolidation and Conservatism

Both institutionally and in terms of curricula, the period between the two world wars was relatively stable. Secondary schooling, as opposed to extensions of elementary schooling, was still the preserve of the minority in independent and grant-aided secondary schools. The needs of girls as well as boys were increasingly recognised but served by a largely common mathematics curriculum, including the new geometry. Constitutional change within the MA reflected institutional developments: in 1912 differentiated interests were first accommodated in separate special committees for public schools for boys, other secondary schools for boys, and girls' schools; in 1923 the needs of public and other secondary schools for boys were merged in one committee for boys' schools.

The geometry curriculum was principally shaped by the requirements of examinations and associated school mathematics textbooks, and through professional association, particularly the communion and publications of the MA. The influence of individual innovators such as Godfrey, Siddons and Fletcher continued to be felt, but one outstanding new figure came to prominence in this period: C. V. Durell. Durell's textbook production, including various titles in geometry, was extraordinarily successful and spanned over half a century. Quadling's [22, p. 121] assessment of his contribution is fitting: 'By the thirties and forties his name was for many pupils almost synonymous with mathematics. . . .'

The role of the MA in developing inter-war geometry was significant but controversial. Looking back in 1920 on the MA's conservative geometry report of 1902, Godfrey [16, p. 20] noted that 'no report on Geometry has emanated from your Teaching Committee since that date'. Godfrey explained:

an Association cannot easily do more than register the average opinion of sound teachers. . . . I do not consider that the time has yet come for the Association to support with its authority any particular method of learning Geometry.

The Board of Education had distributed a follow-up circular 851 on geometry in 1914 and, soon after the war, the Assistant Masters' Association (AMA) also turned its attention to geometry teaching.

In the early inter-war period, concern over the question of sequence for propositions in deductive geometry resurfaced, and the possibility of imposing a new uniformity to replace Euclid was again raised. The AMA's [23] report of 1923, supported by Fletcher in particular, provided no more than a *suggested* schedule of numbered theorems. Furthermore, to lighten the burden of learning proofs and to increase the use of intuition, as recommended by the Board's circulars, a number of formal proofs were classified as not for examination purposes and others as inappropriate for school work. Subsequent developments in mainstream examination syllabuses and textbook production followed these recommendations. The MA was also consistently opposed to uniformity of sequence but, in a new geometry report of 1923, took a very different pedagogical standpoint to that of the Board and the AMA.

The MA's [24, p. 16] report is perhaps best known for its model of stages – following the Board but using letters – A for the Board's first two stages combined, B for the Board's third stage, and C for a proposed new 'systematising stage'. The bulk of the report is devoted to Stage C for 15–17 year old pupils and the tone is clearly elitist:

A course ending with Stage B might be stimulating, but would certainly be ragged and unfinished. With dull boys, probably nothing more can be accomplished; the Committee believes that they will derive more benefit from a frankly preliminary course like this than from an old-fashioned course on Euclidean lines. On the other hand, able boys will feel the need of rounding off and consolidating their study. This is the purpose of Stage C.

Durell [7], who had contributed to the AMA's pragmatic report, was notably critical of aspects of the MA's more speculative contribution at this time. The major bone of contention for Durell was the predominant concern to explore the controversial details of systematisation for teachers' benefit, at the expense of practical pedagogical advice grounded in classroom experience. A valuable recent secondary perspective from Geoffrey Howson [9] also highlights the report's backward-looking stance, both in its preoccupation with foundations and systematisation, and in its neglect of newer twentieth-century possibilities for developing geometrical properties using symmetry and transformations as opposed to congruence and similarity along Euclidean lines. The title of Howson's critique is short and to the point: 'Milestone or millstone?'

As we have seen, for his presidential address to the MA in 1925, G. H. Hardy [25, p. 316] chose the question 'What is geometry?' Significantly, he urged the MA to adopt a forward-looking stance in relation to mainstream *school* geometry:

I am well aware of the very great services which the Association has rendered in the improvement of geometrical teaching. I think that it might well now concentrate its efforts on a general endeavour to widen the horizon of knowledge, recognising as regards niceties of logic, sequence and exposition that the elementary geometry of schools is a fundamentally and inevitably illogical subject, about whose details agreement can never be reached.

The MA returned to geometry teaching in the 1930s and a second report was published in 1938 [26]. Much more emphasis was now placed on pedagogical details to support teachers in Stages A and B. Stage C was now subdivided into two stages, Ca and Cb – the organization of derived and, finally, primitive propositions – and the last sub-stage was now recognized as only of very specialist interest. There was some encouragement to 'widen the horizon' through work in three dimensions but, predictably, no fundamental break with the Euclidean approach to developing geometrical properties. Some of the critical observations of a contemporary German commentator, G. Wolff [27, p. 196], are particularly pertinent here and they anticipate post-war English developments in the use of symmetry and transformations by some twenty years:

There is still a discrepancy in the method of teaching geometry [in England]. On the one hand it is desired to have more concrete and empiric treatment, on the other hand teachers will not give up abstract and pure geometry. It is universally recognised that . . . the methods of motion, changes, transformation, projecting, are as sure a proof as many a deduction.

In relation to subsequent major developments, one other issue addressed in the 1938 report deserves mention: the relative emphasis on reproduction of geometric proofs of standard theorems (bookwork) and work on problem solving (riders), possibly using algebraic or trigonometric methods. A minority, notably led by Fletcher, argued in a dissentient appendix that the report 'goes too far in its insistence on learning proofs of standard theorems and on a particular style of writing out these proofs' [26, p. 188]. The case was argued for a shift towards problem solving using a variety of methods. As Tuckey [28. p. 237] subsequently observed, the minority view was prophetic: 'the ideas set out in Appendix 14 . . . proved the most influential of the whole Report, and led directly to the geometrical part of the modern "alternative syllabus".' It is to the development of new post-war syllabuses that we now turn.

5. Post-War Reconstruction

Following the 1944 Education Act, post-war educational reconstruction proceeded on two major fronts: institutional reorganization and the reform of examinations and curricula [3, 1994, ch. 6]. Universal primary education was now followed by secondary education for all in *some* form within independent or grant-aided secondary grammar, modern or technical schools. From 1951, secondary school examinations based on the single school certificate system were replaced by a multi-choice, single-subject system – the General Certificate of Education (GCE) – which very gradually widened access to mathematical certification for pupils outside the grammar and independent schools.

Following the Jeffery report of 1944, mathematical syllabus reform was still focused on the needs of the minority proceeding to school certificate and, later, GCE mathematics. New syllabuses and papers were offered as an *alternative* to traditional examinations in arithmetic, algebra and geometry, and the consequences for the place and shape of geometry proved to be far-reaching.

The Jeffery syllabus was grounded in the key principles of *unification* or fusion of the separate branches, and *broadening* of the range of mathematics in a general secondary education. Inevitably, drastic pruning and simplification of the traditional syllabus was necessary and major gains here were achieved by substantially reducing the list of required geometrical proofs to around ten key theorems. For geometry, unification involved the formalization of links with algebra, co-ordinate geometry and trigonometry, and broadening extended the work to *three* dimensions and wide-ranging practical applications [29, p. iii]:

The whole syllabus is inspired by the desire to bring mathematics more closely into relation with the life and experience of the pupil. The flat geometry of the blackboard should be linked with the solid geometry of the real world by the study of plans and elevations and the consideration of the simpler solid figures. The simple geometry of the surface of a sphere has an obvious interest for the earth-dweller. The interests of the pupil in air and sea navigation, and in handicrafts, should be used as opportunities to demonstrate the power of elementary mathematics as an aid to understanding.

This was a radical blueprint for the post-war mathematics curriculum but, as an alternative, it was only gradually adopted by schools. Textbook production for some authors involved little more than 'cut and paste' of chapters from existing textbooks on the separate branches. Nevertheless, a report of AMA [30, p. 12] judged that the Jeffery syllabus had 'a far-reaching influence on the deformalization of geometry and the integration of all the branches of mathematics'.

By about 1950, Tuckey [28, p. 236] could point to a major historical trend in geometry teaching:

away from the formal setting-out of a series of propositions in logical sequence as in Euclid's *Elements* and towards the somewhat haphazard discussion of various geometrical properties arrived at by methods which permit the use of algebra, trigonometry and perhaps accurate drawing . . . at the start of the century a pupil was expected to learn proofs of 100 theorems; about 1925 teachers were content to exact proofs of about 50 theorems, while in 1951 they consider themselves fortunate if their pupils master a dozen.

In parallel with this trend was the major shift from an emphasis on Euclidean-style rigour to methods based on intuition and experiment as embodied in the stages model.

In 1953 the MA [31] produced its third and last major twentieth-century report on geometry teaching. The focus was now on 'higher geometry', broadly conceived but, as in the 1923 report, largely presented to support and enhance teachers' own subject knowledge. The 1953 report opened up the whole question of progression in geometry for the future specialist, from spatial geometry, based initially on physical observation and measurement in Stage A, to abstract geometry, whether pure or analytical, based on formal and rigorous deductive methods. Here two twentieth-century discontinuities were identified. The first was a consequence of the tendency to move towards the spirit of the Jeffery syllabus and away from systematic Stage C geometry. The second was a consequence of the development of abstract geometry in undergraduate courses. The report – a substantial one of over a hundred pages – aimed to provide a bridge for teachers between spatial geometry and abstract, mainly algebraic, Cartesian and projective geometry, in the interests of smoothing the passage of high attaining pupils through a difficult field.

In his presidential address to the MA in 1956, G. L. Parsons [32, p. 3] reflected on half a century of reform:

All revolutions lead to a period of chaos in which all sorts of ideas about new freedoms and the breaking of old bondages naturally come to the fore. Perhaps the chaotic period resulting from the revolution in teaching geometry was to some extent prolonged by the intervention of two wars but it may not be out of place to express the hope that it may soon be ended.

Given the impending major curricular upheaval associated with 'modern mathematics' for schools, Parsons's hope proved to be a forlorn one.

6. Modernizing School Geometry

There are clear and continuous lines of development in school geometry from Euclid, through the Perry movement, to the Jeffery syllabus. However, the modernization of school mathematics curricula from around 1960 destabilized the place of geometry and introduced alternative new paradigms for school geometry, initially through radical new mathematics syllabuses and textbooks for GCE mathematics. The background and progress of the modern mathematics reform movement as a whole, including its international dimension, is surveyed in my history of the MA [3, ch. 7]. Major curriculum development projects, involving teams of authors as opposed to individuals or pairs, included the School Mathematics Project (SMP),

Midlands Mathematical Experiment (MME), Contemporary School Mathematics (CSM), Mathematics in Education and Industry (MEI), and, across the border, the Scottish Mathematics Group (SMG).

There was no uniformity in the place and shape of geometry across these projects. The 1960s was a period of major curricular upheaval and experimentation comparable to the post-Euclid reforms of the 1900s. The range of new questions to be addressed in the modernization of school geometry has been well summarized by Quadling [33, p. 257] in a special issue of the *Mathematical Gazette* devoted to 'the perennial problem of geometry':

Is there still a place for the vestiges of the Euclidean tradition?
Is the answer to re-structure the subject matter in terms of transformations or vectors?...
Is geometry any more than a subset of algebra?...
Where does topology come in?

Projects' responses to these questions were varied and there was no consensus in the 1960s and 1970s. But, in the case of the Euclidean tradition, a number of projects did at least agree with the influential SMP's [34, p. 308] view that:

for the majority of pupils, formal geometry offers little training in logical reasoning and emphasizes, instead, practice in memorizing of theorems and proofs of no particular worth.

For such reasons, the SMP and other UK projects rejected a formal axiomatic treatment of geometry, whether based on Euclidean or modern foundations.

The central concern for the geometry curriculum was now the wide choice of sometimes competing tools with which to develop school geometrical knowledge: symmetries, transformations, groups, vectors and matrices, alongside the Euclidean tools of parallelism and congruence. The perspective of the Royal Society's [2, p. viii] report of 2001 is pertinent here:

We might refer to this, not unwelcome, problem as one of **abundance**.... At one extreme there is the danger of choosing eclectically from this abundance in a way that leads to the teaching of a lot of apparently unconnected 'bits'. At the other extreme there is the danger of developing a tightly organised body of knowledge which addresses only a very small part of geometry. Our challenge has been to combine breadth with both educational and mathematical coherence – a problem we refer to as **coherence**.

A most valuable survey of some of the leading new projects' responses to these curricular challenges is provided by Willson [35]. His analysis is based on a broad taxonomy of five mathematical approaches which are characterised as:

A. Durell, working in the Euclidean tradition
B. Klein, using transformations as a means to develop properties of geometrical figures

C. Dieudonné, using vectors as the means

D. Jeger, focusing on the properties of transformations *per se*

E. Coordinate, using matrices.

Willson's [35, p. 21] major conclusion concerning these projects' choices among the five approaches, including the treatments of SMP, MME and SMG, was that 'authors and teachers usually provide a mixture of some or even all of them'. Willson also identified a number of other major consequences for school geometry of the modernization movement.

1) Modern school geometry was part of a unified course of mathematics but its place as a whole was not marginalized, in spite of competition from other modern topics such as statistics and probability. For SMP O-level, taking geometry 'to include topology and work on coordinates, but excluding trigonometry and the bits of graphical work which might be called "pre-calculus"' Willson [35, p. 23] found about forty-five per cent of the course was devoted to geometry.

2) A modern common core for two-dimensional work is identifiable [35, p. 57]:
   (i) Measurement and calculation of lengths and of areas of figures formed by straight lines and circles.
   (ii) Measurement and calculation of angles. Parallelism. Angle properties of polygons.
   (iii) Classification and symmetry properties of triangles, quadrilaterals and circles (including tangents).
   (iv) Pythagoras' theorem.
   (v) Angle properties of circles. (Angles at the centre, angles in the same segment and angles in a cyclic quadrilateral.)
   However, what was not common was the weighting of each topic and the choice of geometric tools for the development.

3) Following the Jeffery syllabus, modern school geometry included three-dimensional work and a common core is again identifiable [35, p. 72].
   (i) Measurement and calculation of lengths, areas and volumes of figures formed by lines and planes, the cylinder, cone and sphere.
   (ii) Measurement and calculation of angles.
   (iii) Classification and symmetry properties of polyhedra, cylinder, cone and sphere.
   (iv) Application of Pythagoras' theorem in three dimensions.
   (v) The geometry of the sphere.

4) 'Practical geometry', interpreted as ruler and compass constructions, was a major 'casualty' of the reform movement [35, p. 51]: 'they are not included in any of the modern mathematics "O" Level syllabuses, and some of the texts ignore them altogether'. Geometrical drawing might still be a part of the curriculum but as a separate subject, known as 'technical drawing' and 'taught by different teachers, and perhaps confined to boys or to less academic pupils' [35, pp. 55–56]. There are striking historical links here with the nineteenth-century divorce between practical geometry and Euclid for educational purposes.

5) 'Proof' as part of Stage B geometry was another major casualty of the reform movement [35, p. 120]:

> the new approaches which have been used in geometry have generally been accompanied by a drastic reduction in the amount of attention paid to deduction in this subject (in some cases amounting to its virtual elimination).

The consequences for the place of proof attracted some early critical reaction, particularly in relation to the needs of higher attaining pupils and future mathematics specialists. Armitage [10] was a notably outspoken critic from higher education, in his 1973 contribution to the *Mathematical Gazette*, where he made a powerful case for re-establishing geometry – broadly conceived and on a systematic basis – in the school curriculum. Admittedly, Armitage's principal concern was 'from the top down', for the needs of the top five to ten per cent of the ability range, but his thrust anticipates many views most recently broadcast by the Royal Society [2]. Armitage [10, pp. 275–277] also highlighted the post-war erosion of the place of geometry in *university* mathematics curricula and the inevitable consequences for mathematics teacher capability and geometry's status and development in schools, particularly in the upper years.

6) Alongside geometry syllabus reform, the modernization movement also promoted the development and dissemination of the use of visual and tactile aids for two- and three-dimensional work [35, ch. 11]. On this front the work of the Association for Teaching Aids in Mathematics (ATAM) from 1952 was seminal. As Willson points out: 'The majority of teaching aids and materials for mathematics, especially in the secondary school, are essentially geometrical in nature' [35, p. 173]. The ATAM became the Association of Teachers of Mathematics (ATM) in 1962 and also played an important part in the development of modern mathematics syllabuses and textbooks [3, 1994, ch. 6–7].

In 1965 the MA abandoned its long-term strategy of publishing major reports on the separate branches of school mathematics for 'academic' secondary and sixth-form work. As in the 1900s, the period of major curricular upheaval in the 1960s was a difficult one for the MA in terms of providing professional leadership in school mathematics. Unification in the mathematics curriculum was now reflected in the MA's decision to focus on mathematics as a whole for the 11–16 age range in a major report which was eventually published in 1974 [36].

Geometry was separately accommodated in a single chapter, significantly entitled 'space'. Modern approaches to school geometry are discussed and exemplified, and the potential of an eclectic treatment is recognized, partly on grounds of unification [36, p. 103]:

the Euclidean method . . . sets geometry apart from other branches of mathematics as a separate study with its own axioms, procedures, and objectives. Vector and transformation approaches, on the other hand, continually stress the interplay between geometry and algebra, analysis and application of mathematics to physical science.

There is no explicit mention of Stages A, B and C but some general acknowledgement of the need for progression through stages, from intuition and experiment to rigour and systematisation, depending upon the ability of the pupil. This report recognized the need for some *differentiation* of curricula to serve the differing needs of pupils in grammar, secondary modern, and, increasingly, comprehensive schools. The question: 'What geometry should we provide for the multitude?' was now pressing.

7. Geometry for the Multitude?

After the 1944 Education Act, secondary modern schools catered for the majority of secondary pupils and their mathematics curriculum was based on the principle of broadening from a foundation in arithmetic to include elements of 'practical mathematics' in the spirit of the Perry movement: some practical plane and, possibly, solid geometry, measurements and, possibly, simple numerical trigonometry. However, for the 'non-academic' pupil, formal deductive geometry was not an aspiration. Utility and not mental discipline was the principal ground for geometry and contexts for developing geometrical work included surveying, navigation, shape in the environment, and technical drawing. Teaching methods based on intuition and experiment and the use of visual and tactile aids were influenced by the work of the ATAM.

Given its roots in the public and grammar schools, the MA was not in a position to report in a major way on mathematics for secondary modern schools until 1959 [37]. This report distilled much of the best thinking and practice up to that time but it was soon to be overtaken by other developments affecting mathematics for the multitude: the modern mathematics movement, the development of the Certificate of Secondary Education (CSE) from 1962, alongside the GCE, and the spread of comprehensive schooling.

Increasingly in the 1960s and 1970s, through the dual system of single-subject GCE and CSE examinations and associated textbooks, elements of modern school geometry were disseminated to a much wider range of pupils in comprehensive schools. Essentially, this was a top-down approach to the modernization of mathematics for the multitude, and the influential Cockcroft Report of 1982 underlined some generally worrying features [38, p. 130]:

We believe ... that the changes in the examination system and in the organisation of secondary schools which have taken place in recent years have influenced the teaching of mathematics in ways which have been neither intended nor sufficiently realised. At the present time up to 80 per cent of pupils in secondary schools are following courses leading to examinations whose syllabuses are comparable in extent and conceptual difficulty with those which twenty years ago were followed by only about 25 per cent of pupils. Because ... it is the content of O-level syllabuses which exerts the greatest influence, it is the pupils whose attainment is average or below who have been most greatly disadvantaged.

Inadequate *differentiation* in the mathematics curriculum – including too much breadth, mismatching of cognitive demands, and insufficient regard for educational purposes – was the major bone of contention. On the credit side in geometry, the Cockcroft Committee [38, p. 82] welcomed the introduction of modern geometrical work on symmetry, reflection, rotation and co-ordinates for the majority of pupils.

To tackle the essential problem of differentiation, the Cockcroft Committee boldly advocated the turning of the top-down approach on its head [38, p. 133]:

We believe that this is the wrong approach and that development should be 'from the bottom upwards' by considering the range of work which is appropriate for lower-attaining pupils and extending this range as the level of attainment of pupils increases.

The aim was to reverse the historical trend [38, p. 130]:

We have . . . moved from a situation in which, twenty years ago, there was in our view too great a difference between the mathematics syllabuses followed by those who attempted O-level and those who did not to one in which, at the present time, there is far too little difference in the mathematics syllabuses which are followed by pupils of different levels of attainment.

A key guiding principle in the design of differentiated courses was also laid down [38, p.133]:

We believe it should be a fundamental principle that no topic should be included unless it can be developed sufficiently for it to be applied in ways which the pupil can understand.

Matrix multiplication without application e.g. to geometric transformations was a case in point here.

The Cockcroft Committee also went so far as to propose a 'foundation list' as a basis for the mathematics curriculum at the bottom i.e. for the lowest forty per cent of the attainment range. Cockcroft's basic list covers geometry under the heading 'spatial concepts' [38, pp. 139–140]: vocabulary and properties of plane and solid shapes; angles, parallels and bearings; simple geometric drawing, including representation of three-dimensional figures in two dimensions; measurement of lengths, areas and volumes, including circles; co-ordinates, similarity and scale; and simple mechanical movements and linkages. For progression, isolated additional topics such as Pythagoras's theorem and simple trigonometry are mentioned [38, p. 144], but the committee was not motivated to expand on the content and approach to geometry for more able pupils.

Significantly, the Cockcroft Report's index includes no references to geometry, deduction or deductive work, reasoning, logic or proof. A single entry for 'shape and space' refers only to work for the primary years. The committee's rationale for mathematics in a general education is based on its usefulness as a means of communication and its wide-ranging applications. By contrast, the mental discipline ground is far from strongly held [38, p. 2]:

It is often suggested that mathematics should be studied in order to develop powers of logical thinking, accuracy and spatial awareness. The study of mathematics can certainly contribute to these ends but the extent to which it does so depends on the way in which mathematics is taught. Nor is its contribution unique; many other activities and the study of a number of other subjects can develop these powers as well. We therefore believe that the need to develop these powers does not itself constitute a sufficient reason for studying mathematics rather than other things.

It is fair to conclude that, in the post-Cockcroft era, the overriding claims of utility and the needs of pupils from the bottom up have served further to undermine the place of deductive geometry in the mathematics curriculum for *all* pupils. From the 1980s, the overall shape of the geometry curriculum and the means of differentiation were also fundamentally affected by the development of the General Certificate of Secondary Education (GCSE) examinations and the introduction of the National Curriculum for England.

8. Government Control and National Curricula

To replace the dual O-level and CSE examination system, a common system of examinations at sixteen plus was introduced from 1985. Building on the Cockcroft proposals for differentiation, two-year syllabuses were now defined at three levels: 'foundation' for low attainers, 'intermediate' for average attainers, and 'higher' for high attainers. This model for differentiation has fittingly been referred to by Howson [39, p. 10] as 'nesting . . . in the manner of Russian dolls'. The range of topics and demands in geometry largely followed the main lines identified in Willson's [35] survey.

Growing central government involvement in the details of curricula is evident through the introduction of national criteria for the differentiated syllabuses. Again, following the Cockcroft proposals, opportunities for assessed coursework were now generally provided. However, geometrical topics and methods for investigation, including possible deductive elements, were not prescribed as part of the common core syllabuses.

Following the 1988 Education Act, more radical approaches to differentiation and central government control were implemented in England and Wales through the introduction of the National Curriculum, which covered both mathematics syllabus and assessment requirements for the full 5–16 age range. As Howson [39, p. 10] frankly points out, the model of differentiation is 'a more basic form of nesting':

'It is the "piece of spaghetti" model – one swallows as much as one can in the time available.' Furthermore, the model 'makes a remarkable assumption' [39, p. 33]:

that curricula can be independent of [i.e. defined without reference to] the student's age and ability. Differentiation through depth of treatment or expected outcome is largely ignored. Students of all abilities are expected to follow the same path – the only accepted variable is rate of progress.

By comparison with the models for national curricula adopted in thirteen other countries, Howson [39, p. 217] could claim: 'The structure devised for the curriculum and testing may be best described as "unique".' External testing was now imposed nationally for seven-, eleven-, and fourteen-year-old pupils, alongside sixteen-plus examinations through modified GCSE arrangements.

Reshaping school mathematics for all abilities, including geometry, to fit the National Curriculum model would have been a major challenge for curriculum developers under *any* circumstances. But the working group charged with this task was much driven by political as opposed to professional imperatives, as Margaret Brown [40, p. 269], a group member, emphasized in her 1991 MA presidential address:

the haste with which the subject groups (especially those in mathematics and science) had to report, and the management structure chosen, prevented any serious collection or discussion of essential information, such as comparative information from other countries, proper theoretical underpinnings of a curriculum framework, or research results;
consultation was not taken seriously, either in the way it was conducted or in the regard for the results;
essential trialling and piloting of the attainment targets and programmes of study did not take place, and the trialling of the assessment was disrupted by policy changes;
the implementation in schools was premature, before teachers had had sufficient information, teaching resources, specialised in-service input, or planning time.

Progression for all pupils over the age range 5–16 had now to be defined at ten levels. Despite Brown's major caveats, the final 1988 report of the working group [41] set the general pattern for geometry teaching through the 1990s. Geometry was now titled 'shape and space', to include three essential ingredients – shape, position and movement [41, p. 32] – alongside work on measures [41, pp. 30–31]. Progression in shape and space was defined under two broad headings: 'recognise and use the properties of two-dimensional and three-dimensional shapes' [41, pp. 33–34]; and 'recognise location and use transformations in the study of space' [41, pp. 35–36].

Elements of both traditional and modern school geometry were accommodated, and the language adopted for the level descriptions is indicative of the cognitive demands on all pupils: 'know', 'identify', 'recognise', 'use', 'understand', 'determine' and 'calculate'. References to explanation and proof in geometry are conspicuous by their absence, bar a single entry at level 5: 'explain and use angle properties associated with intersecting and parallel lines and triangles' [41, p. 33]. However, the need to progress from experimental to deductive thinking involving reasoning and proof was still generally acknowledged: 'the experimental approach needs to be complemented by due attention to intellectual rigour as the pupil progresses' [41, p. 32].

This progression in reasoning and proof is nowhere to be found in the National Curriculum specification for geometry itself, but is separately accommodated under the generic heading 'using and applying mathematics'. Here are to be found a collection of reasoning elements: 'making and testing generalisations and simple hypotheses'; 'defining'; 'following a chain of mathematical reasoning'; 'using "if ... then"'; 'proving and disproving and using counter-examples'; 'recognising and using necessary and sufficient conditions'; and 'constructing a proof including [oddly?] proof by contradiction' [39, pp. 224–229]. The laudable aim was to promote the development across the mathematics curriculum of coursework, including pupils' skill development, teachers' support for pupils' learning, and assessment. But, in the case of reasoning and proof in geometry, the National Curriculum model served only to marginalize this aspect and its progression up to GCSE higher level.

By the mid-1990s there was mounting concern about the place of reasoning and proof in the school mathematics curriculum. The younger sister to the *Mathematical Gazette, Mathematics in School*, published a 'proof special' in 1994, in which Porteous [42, p. 3] in particular pointed to 'truly an impoverished role' for proof, following the major 1960s' reactions to its relatively secure place within the Euclidean tradition.

In 1995, concerted reaction from the higher education lobby was embodied in a well-publicized joint report, from the London Mathematical Society, the Royal Statistical Society and the Institute of Mathematics and its Applications, provocatively entitled *Tackling the mathematics problem*. Inadequate attention to precision and proof in school mathematics and consequent difficulties in transition to higher education were identified as serious parts of the 'problem' [43, p. 6].

In 1996, the ATM's *Mathematics Teaching* [44] devoted a special issue largely to proof, including many and varied reactions to the 1995 report. The MA has since published two contrasting handbooks for teachers, to support the development of reasoning and proof, including many geometrical examples, throughout the school years [45, 46]. One of the authors has, most recently, pointed to a remarkable turn-of-the-century transformation in the centrally controlled secondary curriculum: 'Proof is back!' [47]. The circumstances surrounding this surprising judgement and the implications for school geometry remain to be considered.

9. School Geometry for the Twenty-First Century?

In sharp contrast to the Board of Education's approach through published suggestions for geometrical reform in the 1900s, the 1990s has seen the power of central government *prescriptions* to effect radical and relatively rapid curriculum change in the sphere of school geometry. The contrasts between the 1989 and 1999 National Curriculum specifications for geometry are particularly striking.

For Key Stage 3 (11–14 age range), the latest version of the National Curriculum for England [14, pp. 36–39] integrates generic skills of using and applying mathematics in the detailed 'programme of study' (or syllabus) for 'shape, space and measures'. Introductory elements of proof are now made explicit in a separate section devoted to 'geometrical reasoning' about angles, polygons, circles, and areas. Work involving geometrical drawing instruments is now made explicit in a separate section devoted to 'measures and construction'. 'Modern' school geometry is now restricted to some work, without proofs, on 'transformations and coordinates'.

Countering Howson's earlier critique, for assessment purposes a minimalist approach is adopted for the geometry attainment targets which are now summarized in single paragraphs at eight levels plus a top level of 'exceptional performance'.

For Key Stage 4 (14–16 age range), major differentiation is built into the detailed programmes of study at two levels: 'foundation' and 'higher'. The two specifications for shape, space and measures follow the structure for Key Stage 3, including significant differentiation in the reasoning and proof demands between the foundation and higher levels [14, pp. 49–52, 65–68]. Proofs using congruence, including justifications for some geometrical constructions and the group of circle theorems are now prescribed for higher ability pupils. Significantly, the work specified for transformations and coordinates, including use of vectors, makes comparatively little demand in terms of reasoning and proof, where Euclidean methods have held sway. These new National Curriculum requirements are now embodied in new specifications for GCSE mathematics.

For Key Stage 3, another layer of government prescription affecting school geometry has most recently been introduced through the extension of the National Numeracy Strategy from primary schools to include a DfEE *Framework for teaching mathematics: Years 7, 8 and 9* [48]. This massive publication of nearly three hundred pages builds upon and exemplifies in great detail the latest National Curriculum for mathematics, including shape, space and measures. Here two major features of geometry for Key Stage 3 are identified [48, p. 17]:

- developing **geometrical reasoning** and **construction** skills, and an appreciation of logical deduction;
- developing **visualisation** and **sketching** skills, including a dynamic approach to geometry, making use of ICT and other visual aids.

The reference to ICT is significant and highlights another potentially powerful visual tool for developing geometrical intuition and experiment. LOGO programming and dynamic geometry software are both suggested as major ICT tools, and detailed exemplification of their uses is provided by the DfEE to help teachers maximize this potential in the new millennium.

The renewed concern for the place and shape of geometry in schools is also evidenced by the most recent attentions of both the Royal Society [2] and, in this publication, the MA. The geometry report from the Royal Society and Joint Mathematical Council was published in 2001. The report details the most recent national developments in England and exposes the lack of consensus worldwide about the place and shape of geometry in a general education, including the role and potential contributions of ICT and of proof.

The report generally supports the main thrust of the most recent government-driven reforms in 11-16 geometry, but also points to the marginalization of geometry in post-16 school mathematics [2, p. viii]:

the working group . . . is far less sanguine about the state of geometry in 16–19 education. The geometrical content of the current AS/A-level specifications in pure mathematics is very small and offers little by way of progression from what has come before.

The report identifies this as a major problem of **progression**, which also recurs in geometry at the transition from post-16 to higher education.

In response to the curricular challenges of **abundance** and **coherence** for 11–16 geometry (see the quotation on p. 474 above), the report [2, pp. 75–80] goes so far as to include a detailed planning framework based on the National Curriculum and four main themes: **Euclidean plane geometry** [my stress], coordinate geometry, 3D geometry, and symmetry, transformations and vectors. Opportunities are specified throughout the framework for the use of intuition and experiment, the use of ICT, deduction and proof, and links with relevant contexts and applications.

To echo Waring [47], 'Proof is back!', and, what is more, Euclidean plane geometry is now starting to enjoy a new lease of life. The Royal Society gathered wide-ranging views from the worlds of mathematics and education, and concluded [2, p. 27]:

The most widely held view was that basic plane geometry, including such things as proofs of angle and circle theorems, should be reinstated as a major part of the high school mathematics curriculum.

This is hardly a 'subject which dare not speak its name' [2, p. 5] at the start of this century. One can only speculate on the prospects for Euclid in school geometry by the end of the twenty-first century!

Acknowledgement

I would like to thank Mary Walmsley for her helpful feedback on a first draft of this essay.

References

1. S. C., Euclid's elements. By an ancient geometer, *Journal of Education* **450** (January 1907) 32.
2. Royal Society, *Teaching and learning geometry 11–19. Report of a Royal Society/Joint Mathematical Council working group*, Royal Society (2001).
3. Many of the historical judgements in this chapter are drawn from research detailed in the following two works:
   M. H. Price, *The Reform of English Mathematical Education in the Late Nineteenth and Early Twentieth Centuries*, unpublished PhD thesis, Leicester (1981).
   M. H. Price, *Mathematics for the Multitude? A History of the Mathematical Association* (Leicester: Mathematical Association, 1994).
4. B. Russell, The teaching of Euclid, *Math. Gaz.* **2** (May 1902), 165–167.
5. Report of the M. A. committee on geometry, with introductory comments by W. J. Greenstreet, *Math. Gaz.* **2** (May 1902), 167–172.
6. C. Godfrey, The Board of Education circular on the teaching of geometry, *Math. Gaz.* **5** (March 1910), 195–200.
7. C. V. Durell, The teaching of geometry in schools, *Math. Gaz.* **12** (January 1925), 274–276.
8. A. W. Siddons, Fifty years of change, *Math. Gaz.* **40** (October 1956), 161–169.
9. A. G. Howson, Milestone or Millstone?, *Math. Gaz.* **57** (December 1973), 258–266.
10. J. V. Armitage, The place of geometry in a mathematical education, *Math. Gaz.* **57** (December 1973), 267–278.
11. Institut für Didaktik der Mathematik, *Comparative studies of mathematics curricula: change and stability 1960–1980*, University of Bielefeld (1980).
12. T. A. A. Broadbent, *The Mathematical Gazette*: our history and aims, *Math. Gaz.* **30** (October 1946), 186–194.
13. J. Perry, The teaching of mathematics, *Nature* **62** (August 1900), 317–320.
14. Department of Education and Employment and Qualifications and Curriculum Authority, *The National Curriculum for England: Mathematics* (HMSO and QCA, 1999).
15. Mathematical reform at Cambridge, *Nature* **68** (June 1903), 178–179.
16. C. Godfrey, Geometry teaching: The next step, *Math. Gaz.* **10** (March 1920), 20–24.
17. B. Branford, *A Study of Mathematical Education* (Oxford: Clarendon Press, 1908).
18. C. Godfrey, The teaching of mathematics in English public schools for boys, *Math. Gaz.* **4** (May 1908), 250–259.
19. A. W. Siddons, Obituary: W. C. Fletcher, *Math. Gaz.* **43** (May 1959), 85–87.
20. C. Godfrey, The question of sequence in geometry, *School World* **14** (September 1912), 357.
21. Local branches: The first meeting of the London branch, *Math. Gaz.* **5** (July 1910), 289–299.
22. D. Quadling, A century of textbooks, *Math. Gaz.* **80** (March 1996), 119–126. Belatedly, Durell will be included in a new entry in the *New Dictionary of National Biography*: M. Price, in *NewDNB* (Oxford University Press, 2004) forthcoming.

23. Assistant Masters' Association, *The Teaching of Elementary Geometry* (Oxford University Press, 1923).

24. Mathematical Association, *The Teaching of Geometry in Schools* (G. Bell and Sons, 1923).

25. G. H. Hardy, What is geometry? *Math. Gaz.* **12** (March 1925), 309–316. Reprinted as the first article in this book.

26. Mathematical Association, *A Second Report on the Teaching of Geometry in Schools* (G. Bell and Sons, 1938).

27. G. Wolff, The second report on the teaching of geometry, *Math. Gaz.* **23** (May 1939), 185–197.

28. C. O. Tuckey, The geometry reports, *Math. Gaz.* **35** (December 1951), 236–238.

29. Cambridge Local Examinations Syndicate, *School Certificate Mathematics*. [Jeffery] MA reprint (1944).

30. Assistant Masters' Association, *The Teaching of Mathematics in Secondary Schools* (Cambridge: Cambridge University Press, 1973).

31. Mathematical Association, *The Teaching of Higher Geometry in Schools* (G. Bell and Sons, 1953).

32. G. L. Parsons, Teaching the teacher, *Math. Gaz.* **41** (February 1957), 1–8.

33. D. A. Quadling, The perennial problem of geometry, *Math. Gaz.* **57** (December 1973), 257–258.

34. Quoted in A. G. Howson, *Geometry in Great Britain in Recent Years* [11, 304–25].

35. W. W. Willson, *The Mathematics Curriculum: Geometry* (Blackie and Schools Council, 1977).

36. Mathematical Association, *Mathematics Eleven to Sixteen* (G. Bell and Sons, 1974).

37. Mathematical Association, *Mathematics in Secondary Modern Schools* (G. Bell and Sons, 1959).

38. Committee of Inquiry into the Teaching of Mathematics in Schools, *Mathematics Counts*. [Cockcroft] (HMSO, 1982).

39. A. G. Howson, *National Curricula in Mathematics* (The Mathematical Association, 1991).

40. M. Brown, The second iteration, *Math. Gaz.* **75** (October 1991), 263–274.

41. Department of Education and Science and the Welsh Office, *Mathematics for Ages 5 to 16* (HMSO, 1988).

42. K. Porteous, When a truth is seen to be necessary, *Mathematics in School* **23** (November 1994), 2–5.

43. Cited in *Mathematics Teaching* **155** (June 1996), 6.

44. See [43]. For a very recent special issue on proof see *Mathematics Teaching* **177** (December 2001).

45. D. French and C. Stripp (Eds.), *Are you Sure? Learning About Proof* (The Mathematical Association, 1999).

46. S. Waring, *Can You Prove it? Developing Concepts of Proof in Primary and Secondary Schools*, MA (2000).

47. S. Waring, Proof is back!, *Mathematics in School* **30** (January 2001), 4–8.

48. Department for Education and Employment, *Framework for Teaching Mathematics: Years 7, 8 and 9* (DfEE Publications, 2001).

6.2

The Teaching of Euclid

BERTRAND RUSSELL

It has been customary when Euclid, considered as a text-book, is attacked for his verbosity or his obscurity or his pedantry, to defend him on the ground that his logical excellence is transcendent, and affords an invaluable training to the youthful powers of reasoning. This claim, however, vanishes on a close inspection. His definitions do not always define, his axioms are not always indemonstrable, his demonstrations require many axioms of which he is quite unconscious. A valid proof retains its demonstrative force when no figure is drawn, but very many of Euclid's earlier proofs fail before this test.

The first proposition assumes that the circles used in the construction intersect – an assumption not noticed by Euclid because of the dangerous habit of using a figure. We require as a lemma, before the construction can be known to succeed, the following: If A and B be any two given points, there is at least one point C whose distances from A and B are both equal to AB. This lemma may be derived from an axiom of continuity. The fact that in elliptic space it is not always possible to construct an equilateral triangle on a given base, shows also that Euclid has assumed the straight line to be not a closed curve – an assumption which certainly is not made explicit. When these facts are taken account of, it will be found that the first proposition has a rather long proof, and presupposes the fourth. We require the axiom: on any straight line there is at least one point whose distance from a given point on or off the line exceeds a given distance.

The fourth proposition is a tissue of nonsense. Superposition is a logically worthless device; for if our triangles are spatial, not material, there is a logical contradiction in the notion of moving them, while if they are material, they cannot be perfectly rigid, and when superposed they are certain to be slightly deformed from the shape they had before. What is presupposed, if anything analogous to Euclid's proof is to be retained, is the following very complicated axiom: Given a triangle

First published in *Mathematical Gazette* **2** (May 1902), pp. 165–167.

ABC and a straight line DE, there are two triangles, one on either side of DE, having their vertices at D, and one side along DE, and equal in all respects to the triangle ABC. (This axiom presupposes the definition of the two sides of a line, for which see below.) When the existence of a triangle thus equal in all respects to ABC is assured, we can prove that the triangle considered in the fourth proposition is this triangle.

The sixth proposition requires an axiom which may be stated as follows: If OAA', OBB', OCC' be three lines in a plane, meeting two transversals in A, B, C, A', B', C' respectively; and if O be not between A and A', nor B and B', nor C and C', or be between in all three cases; then, if B be between A and C, B', is between A' and C'. This axiom is the basis of the measurement of angles by distances, and is required for proving that if D be on AB, and BD be less than BA, the triangle DBC is less than the triangle ABC.

The seventh proposition is so thoroughly fallacious that Euclid would have done better not to attempt a proof. In the first place, it uses an undefined term in the enunciation, namely, *on the same side*. The definition requires an axiom, and may be set forth as follows: Given a line AB and a point C, with regard to any point D in the plane ABC, three cases may arise; (1) the straight line CD does not meet AB; (2) CD meets AB, produced if necessary, in a point not between C and D; (3) CD meets AB in a point between C and D. In cases (1) and (2), C and D are said to be on the same side of AB; in case (3), on opposite sides. The above very complicated axiom is better replaced by the following two: (1) Given three points A, B, C, a point D between B and C, and a point G between A and D, BG produced meets AC in a point between A and C; (2) A, B, C, D being as before, and E being between A and C, AD, and BE meet in a point between A and D and also between B and E.[1] (The definition of *between* is long, and I omit it here for want of space.) The proof of I. 7; further assumes that if C and D be on the same side of AB, then if CB is between CA and CD, DA is between DC and DB; while if CB is between CD and AC produced, then AD produced is between DC and DB. This is a very complicated assumption, of which Euclid is to all appearance completely ignorant. The assumption may be stated more simply as follows: Of three lines in a plane starting from a point, either there is one which is between the other two, or else any one of them produced is between the other two. But in this statement, the meaning of *between* has to be very carefully defined.

I. 8 involves the same fallacy as I. 4, and requires the same axiom as to the existence of congruent triangles in different places. In the following propositions, we require the equality of all right angles, which is not a true axiom, since it is demonstrable.[2] I. 12 involves the assumption that a circle meets a line in two points

---

[1] Cf. Pasch, *Vorlesungen über neuere Geometrie*, Leipzig, 1882; Peano, *I Principii di Geometria*, Turin, 1889.
[2] Cf. Hilbert, *Grundlagen der Geometrie*, Leipzig, 1899, p. 16.

or in none, which has not been in any way demonstrated. Its demonstration requires an axiom of continuity, by the help of which the circle can be dispensed with as an independent figure.

I. 16 is false in elliptic space, although Euclid does not explicitly employ any assumption which fails for that space. Implicitly, he uses the following: If ABC be a triangle, and E the middle point of AC; and if BE be produced to F so that $BE = EF$, then CF is between CA and BC produced. In spaces where the straight line is not a closed series, this follows from the axioms mentioned in connection with I. 6 and I. 7. No other points of interest, except that I. 26 involves the same fallacy as I. 4 and I. 8, arise until we come to parallels; and the treatment of parallels in Euclid is, so far as I know, wholly free from logical defects.

Many more general criticisms might be passed on Euclid's methods, and on his conception of Geometry; but the above definite fallacies seem sufficient to show that the value of his work as a masterpiece of logic has been very grossly exaggerated.

6.3

The Board of Education Circular on the Teaching of Geometry

CHARLES GODFREY

There are few things more distasteful to the people of this country than a Government Department. It is supposed that from such a place we must expect neither light nor leading. How far this is a just view I need not now discuss: it is equally true that when a Government department *does* give a strong lead, it is condemned for outrunning public opinion. If not King Log, then King Stork.

A case in point is the Board of Education circular on Geometry, a strong pronouncement which has been received by some people with alarm. The Board circular is inspired by clear and constructive ideas. It attempts to crystallize in a new shape the notions that have been set in solution by the discussions of the last few years. Strong solvents were needed before the modern schoolboy could be freed from the system framed for Greek University students. The Board found the process of solution in an advanced state; it has now given a much-needed lead to the process of reconstruction.

Why do we teach geometry? The Board avoids this question, though it lays down that a training in rigid deduction is an essential element in school geometry. *An* essential element: apparently then not the only essential element.

Is it not, in fact, unphilosophical to seek a single *raison d'être* for each subject in the curriculum? If subjects could be justified on such simple grounds education would be an easier thing than we find it. But human affairs are complex. We teach geometry because it has been taught for a long time; and *vis inertiae* decrees that we shall go on teaching it. Probably I shall be putting this view in a more acceptable form if I say the reasons for our present curriculum are historical.

We cannot label each subject, as if it were a drug, with the name of its particular virtue or efficacy. Geometry has long been recommended for the sake of the logical training it imparts. But this claim needs a little examination. Do we really believe that a person who has not studied geometry is likely to be deficient in reasoning

First published in *Mathematical Gazette* **5** (March 1910), pp. 195–200.

powers. It may be so, but I cannot say that I take this view. Must we not limit the expression "reasoning powers" before we confer the monopoly upon geometry? Will a person talk grammatically who has not studied grammar? Will a man make a good speech if he has not studied rhetoric? Very possibly – Yes. At the same time, if he has studied grammar and rhetoric, his conversation and his oratory will be on more orthodox lines. Logic was a school study in the middle ages, and produced the type of reasoning used by Shakespeare's clowns. But perhaps at the same time it had a disciplinary effect on the mind. A man who has no geometry may reason correctly. But a course of geometry would have trained him in the use of a particular logical form – the nearest approach to formal logic that our schools now teach.

What is the peculiar value of this formal training? What chastening disciplining effect has it on the mind? I should be one of the last to suggest that there is no such good effect. But to me this is a subject of great obscurity. What is the effect of drill on a soldier? Cannot a man walk without learning the goose-step? At first sight, ceremonial drill would seem to be a clumsy way of teaching a man to perform simple actions stiffly. But the experience of all military men forbids us to assume that drill is useless to the soldier. It is formal training.

I can distinguish at least three separate currents in educational ideals: the formal ideal, the Herbartian ideal, and the utilitarian ideal. With regard to the utilitarian ideal, I shall only say here that this is the ideal that appeals most strongly to boys; and, that as long as this is so, we shall do well not to ignore it.

Then there is the formal ideal, the ideal of mental gymnastic: this has affinities with the classical as opposed to the romantic ideal in literature and art. The learner is subjected to drill in some stereotyped form; the Euclidean drill, the grammatical drill, the military drill. The subject matter of drill may be remote from real life; a logical process seldom used, a dead grammar, and so forth. But this is not held to impair the fortifying value of the exercise. One accepts this ideal loyally and piously, as the educational light of centuries; loyally, but without enthusiasm. And yet it has its enthusiasts. From this ideal flows the cult of formal geometry, a cult which still sways the great majority of the best and most up-to-date teachers of mathematics. We have still to regard training in the use of a logical form as the unique element in geometry teaching, though we must beware of claiming for geometry any monopoly among educational subjects in the development of reasoning power in the wider sense.

A second educational ideal is what I shall, subject to correction, call the Herbartian ideal. By this I understand the ideal that seeks to implant in the mind fruitful ideas – ideas which will find a congenial soil in the boy's existing knowledge and interests. This ideal has been at work powerfully in the recasting of geometry teaching. By its belief in the fertility of ideas, it has emphasized the value of geometrical knowledge as opposed to the study of logical form. Let the boy be thoroughly at home with a new fact or property before he begins to apply formal logic to it. To attain this familiarity, do not reject at any stage the help of experiment, and the recourse to common objects and experience. Geometrical experiment may use models, frameworks, machines; but there is a limit to the amount of apparatus that is convenient. We rely, therefore, in the main upon figures; freehand sketches, where a sketch will reveal the fact that we are looking for; accurate figures, where eye and hand alone are not clever enough. Hence the amalgamation of geometrical drawing with geometrical theory, subjects once divorced, to the great loss of both.

There is a certain educational value in measurement pure and simple, but this is soon exhausted. After the earliest stage we do not make boys draw or measure for the sake of drawing or measuring, but for the sake of the geometrical truths that we are trying to discover. I cannot emphasize this point too strongly. Time has been wasted by setting boys to draw and measure without an adequate aim in view; and I should like to turn aside; for a moment to deal with this matter.

Drawing is legitimate as an experimental exercise in leading up to a new theorem, though it must be remembered that a freehand sketch may sometimes be as useful as an accurate figure, and boys should not be allowed to consider themselves helpless without instruments. Again, accurate drawing is generally desirable in solving a problem of construction. The whole point of Euclidean constructions is that they are to be performed with ruler and compass, and nothing else. Unless the performer has to carry out the work with these actual instruments, he cannot grasp their possibilities and their limitations.

On the other hand, to draw an accurately finished figure in writing out a proposition is sheer waste of time. The figure must be large and neat; straight lines must look straight; parallels parallel and right angles right. But all this can be done freehand and quickly, with the help of the compass perhaps for a circle.

To draw a figure simply for the sake of measuring it and getting marks *may* be waste of time. But let us here do a little clear thinking. Say that the master has given numerical data for a certain construction, numerical in order that all the class may be at work on exactly the same problem. The boys have drawn the figure accurately. How is the master to check it and mark it? He may have an accurately drawn figure on tracing paper, and test with this. But this is probably a sinful waste of master's time. Generally, there can be found some test line in the figure. If the boy measures this, the master may estimate the accuracy of the whole figure well enough by comparing the measurement with the correct answer. Measurements of this kind are not aimless, but a real convenience; and a good many people have quarrelled with such measurement problems as this without giving enough consideration to the matter.

Drawing is not to be abandoned at a definite epoch in the geometry course; practice and theory should advance hand in hand. There are a certain number of teachers who would have a preliminary practical course, all drawing and no thinking; and a subsequent theoretical course, all thinking and no drawing. I admit that this is an unfair way of stating their position; but I cannot agree with this school of thought, and I am glad to find myself confirmed by the Board of Education. The circular says: "The importance of practical work varies from point to point, rising highest where a new idea has to be effectively assimilated. Thus, when the conceptions of locus, envelope, ratio, similarity are first introduced, the practical work should expand. Apart from this, once the earliest stages of the subject are passed, the practical work is of less value, and in the first instance at least attention should be chiefly directed to the development of geometrical and logical power."

The expression "geometrical and logical power" brings me back to the distinction between the formal and the Herbartian ideals from which I started, and for which I quite expect to be taken to task. We have to recognise that "Geometrical power" is not the same as "Logical power." Geometrical power is the power we exercise when we solve a rider. Solving riders was to the average Euclid-trained schoolboy of the past almost as high a flight as turning out a copy of verses was to myself. Why was this so? He had the logical power but had not developed geometrical power.

Riders cannot be solved by logic alone. And unless a boy knows that he can tackle a fairly easy rider with good chance of success, there is no zest in his work. He will not enjoy the sense of mastery which is the true reward of schoolwork done faithfully. To my mind this has always been the keystone of the whole problem – How to make a boy do his work with zest. If you can get him to work with zest, then he can digest your formal training. But formal training, with or without marks and prizes, will not produce zest in the average boy.

There is no subject which can more readily be made exciting to a boy than geometry, if one goes to work in the right way. There must be a good foundation of practical work, and recourse to practical and experimental illustration wherever this can be introduced naturally into the later theoretical course. Only in this way can the average boy develop what I will call the geometrical "eye": the power of seeing geometrical properties detach themselves from a figure. A French geometrician has given us an account of how he makes his geometrical discoveries. He always works without figures; he sees the figure in his head; he finds that the best way of solving a problem is to go to a concert, or take a seat on the top of a bus, and close his eyes. Well, this is the way of the born geometrician. The average English boy is not a born geometrician, and is perhaps less capable of dealing with abstractions than other boys. He wants all the material help that he can get; given this, most English boys can become very fair, or at any rate, very interested geometricians. Perhaps there is something particularly material and concrete in the English intellect; hand and eye must co-operate with brain in order to produce.

A friend of mine who had not taught mathematics for some years past took a mathematical class for a term recently, and his criticisms were interesting as showing the result of the new methods of teaching. He noticed that the boys could see much more in a figure than he used to find formerly. They attacked deductions with more confidence, and with fair success. On the other hand, he found that they were not so good at writing out propositions.

This last criticism did not surprise me. Boys used to be very ready at writing out propositions; and no wonder, for they did little else. The weak candidate in responsions always chose to take geometry in preference to algebra. However hopeless he was at mathematics he could always make sure of getting his propositions by heart. The whole thing was overdone: perhaps nowadays it is underdone.

The list of essential propositions is now very moderate: the Cambridge Little-Go schedule requires only about 40 theorems in the first three books; most of these easy. There is no reason why boys should not by the age of sixteen be able to write these out. If the advice of the Board of Education is taken, the list will be further reduced by the omission from the list of several formal theorems of Book I.

The theorems that the Board proposes to arrive at by induction rather than deduction fall into four groups, referring respectively to angles at a point, parallel lines, angles of triangle and polygon, congruence of triangles.

We are recommended to arrive at these facts, not by deduction from two geometrical axioms, but inductively from experience. For instance, how shall we justify the statement that two triangles are congruent if three sides of the one are equal to three sides of the other? Take three white sticks of different lengths; and three green sticks respectively equal to the white sticks. Make a triangle of the three white sticks, and make another triangle of the three green sticks. Who can doubt that these triangles are congruent? But don't be too sure about it. Make a quadrilateral of four white sticks; and another quadrilateral of four equal green sticks. You won't find these two quadrilaterals necessarily congruent. Why then were the triangles congruent? Essentially because three sticks make a rigid framework, and four sticks do not. If you look at Euclid's original proof you will find that all it amounts to is that three sticks make a rigid framework.

"These fundamental propositions are those on which all the subsequent deduction depends, and the essential thing in regard to them is not to analyse them and reduce them to the minimum number of axioms, or, rather, postulates (which is Euclid's method), but to present them in such a way that their truth is as obvious and real to the pupil as the difference between white and black, or between his right hand and his left. Any process which interferes with this directness of vision and apprehension is vicious, whatever claim it may have to logical value, and avenges itself in gross mistakes in subsequent work, due to haziness or lack of grasp of the fundamental facts which have been so laboriously 'proved'.

"With beginners, then, Euclidean proofs of these propositions are out of place, and attention must be concentrated not on formal proofs but on vivid presentation, and accurate, firm apprehension of the propositions themselves."

After these facts have been grasped, the Board contemplate that subsequent work shall be on the well-established lines familiar to all good teachers. "Henceforward, though intuition and experience should be largely used to discover propositions, rigid deductive proof on the basis of the fundamental propositions defined above must be insisted upon."

In effect, the Board argues as follows. We want boys to know geometry and we want then to build it up deductively. But we can please ourselves as to what are the foundations of this deductive building. Euclid's foundations consisted of certain axioms two of which were geometrical: his geometrical foundation was therefore very narrow. We advocate a *broad* geometrical foundation nothing less, in fact, than the four groups of facts mentioned above.

What will be the effect of this circular? Presumably it will have an immediate effect on the schools controlled by the Board. But the effect in the long run will probably depend on the attitude of the teaching profession towards the proposals put forward. And the point that teachers will want to satisfy themselves about is whether the adoption of these proposals will impair the value of geometry as a training in deductive logic.

It may be admitted at once that the number of propositions to which deductive logic is applied will be diminished. But this in itself will not be a serious objection; we want quality rather than quantity. And even as to quantity, we may gain at one end as much as we lose at the other. Oxford and Cambridge require the substance of three books of Euclid for responsions and Little-Go, and this amount has seemed no small matter to many candidates at the age of 18. But the Board says that, in their experience, some schools find it possible to cover effectively in a single year the substance of Euclid Books I. and III. However this may be, few teachers will doubt that any boy of 16 ought to have mastered this amount of geometry with ease, if treated as the Board suggests.

The question then reduces itself to the following. Will the training derived from a strict deductive treatment of the remaining propositions be impaired by the inductive character of the arguments used to establish the fundamental theorems? Now who can suppose that this is a real danger? Was the Euclidean training neutralised by the fact that the boy had never reasoned deductively before he began Euclid? Surely there is no antagonism between inductive and deductive reasoning, as if between an acid and an alkali. The two nodes of thought are complementary, and must be *combined*, not isolated, for any fruitful purpose.

One Euclidean ideal has been sacrificed, and we may regret it: the aesthetic ideal of developing the great structure of geometry by pure logic from the minute germ of Euclid's axioms. A great deal of work has been done during the last half century in examining the foundation of geometry. It is work of the greatest philosophical difficulty, and there is as yet no agreement as to what are the fundamental axioms. But there is agreement on one point – that Euclid's axioms are not the true fundamental axioms of geometry. The Euclidean ideal then has been destroyed, not by the Board of Education, but by the labours of pure mathematicians.

A practical advantage of the new system will be the short circuiting of all difficulties as to sequence of propositions. These difficulties are confined almost entirely to the sequence of the fundamental propositions. The controversy is never-ending, for the reason that no sequence that can be proposed is free from some objection, practical or theoretical; and each person has to judge for himself which is the least of the evils before him. There is certainly some inconvenience in the existence of various sequences, though I believe that the inconvenience has been exaggerated. On the other hand, it would be an educational disaster should the outcry about sequence lead to the stereotyping of a new sequence. Well, the trouble melts away if we agree to establish the fundamental propositions inductively, referring each of them straight back to experience and intuition.

To recapitulate, it is recommended that early geometry teaching shall be in three Stages. *The First Stage* aims at instilling the primary concepts, and the meaning of geometrical terms. It is not intended to give accuracy in the use of instruments; here instruments are used to help ideas. The meaning of geometrical terms is not taught by definition but by properly planned series of experiments and questions bearing on everyday objects.

The Second Stage establishes informally, four groups of fundamental facts, these facts to serve as a foundation for the deductive course that is to follow. By the end of this stage, the pupil must be perfectly familiar with these facts; but he need not have them in any definite order, as each fact is referred straight back to experience. He should be able to quote each fact in good set terms; either in the terms of the book, or in an intelligent variation of his own.

Incidentally, this stage is the time to teach accurate drawing; practice being obtained by drawing triangles, etc., to data; or by problems on heights and distances.

The Third Stage is essentially Euclid revised. We arrive at Euclid's goal, but not by Euclid's road.

6.4

The Teaching of Geometry in Schools: A Report
Prepared for the Mathematical Association

C. V. DURELL

There is little doubt that many teachers are seriously bewildered by this Report. It is not an easy document to read and frequently it is difficult to ascertain the precise nature and application of its recommendations. It suffers from attempting to cover too much ground: the reader is confused by the double purpose it tries to serve, (i) instruction on the axiomatic basis of geometry for teachers who have made no special study of the logical foundations of the subject, (ii) pronouncements on practical procedure in teaching and geometrical method.

No one can doubt the importance of the first of these two objects and numerous teachers will welcome gratefully an intelligible account of the scrutiny modern research has made of the foundations of geometry. But such an account is not a matter of opinion or experience, but of scientific fact. It should be issued as a *tract*,[1] not as a *report*. What modifications should be made in elementary teaching in the light of modern research must, however, be determined by the school teacher. The reports of the Association have in the past been confined to methods of teaching, syllabus, etc., and their influence has been due to the fact that they have summarised teaching experience: they have not hitherto been used to provoke discussion, but to urge particular methods which have been tested by experience.

This distinction between a tract and a report is not trivial for the following reasons:

(i) A tract is the work of an expert: a committee may suggest minor improvements by pointing out ambiguities of expression or difficulties the ordinary reader is likely to encounter. But criticism as to fact does not arise. A report is the work of a committee: it attempts to express the best informed experience, its recommendations are matters of opinion, not of fact.

(ii) The smaller the limits to which a report is confined, the more effectively a committee can handle it and mould it into a workmanlike shape. If, however, numerous statements of fact are woven into expressions of opinion, the document

First published in *Mathematical Gazette* **12** (January 1925), pp. 274–276.
[1] Cf. "Cambridge Tracts in Mathematics."

may become so unwieldy that committee-criticism is impracticable. The dimensions of this Report make it possible to doubt (a) whether the Report represents the *considered judgement* of all members of the sub-committee and (b) whether the approval given by the Teaching Committee was much more than a mere matter of form.

(iii) When a method is condemned, it may not be clear whether the objection is based on grounds of logic (the criticisms of the expert) or experience (the criticism of the practical teacher).

(iv) The qualifications for presenting an authoritative scientific statement such as a tract must contain are not necessarily associated with the elementary teaching experience essential for an authoritative pronouncement on methods of procedure and without which it is difficult to know the kind of practical details for which the ordinary teacher will look. In the present case it is certain that a great many teachers will be left in doubt as to what practical procedure the committee recommends and what precisely are the important changes they are asked to make. There is a real danger that geometrical teaching may be injured by a misunderstanding and mis-applications of the views expressed.

If these reasons carry conviction, it appears that much of the matter of the Report is out of place, valuable though it is. The following remarks refer only to those sections which appear to make recommendations affecting the practical teacher.

There are two main themes in the Report, (a) Congruence, (b) Parallelism. In both cases it is not always clear whether the Report is calling attention to an ideally scientific method or to a method considered feasible for class-room use.

1. Congruence

The "Principle of Congruence" as set out in the Report will strike terror into many teachers' hearts. Compare it with the concrete practical method of the Board of Education's circular. Fundamentally they agree. In the circular the pupil is asked to say what measurements must be made in order to copy a given triangle. The Report quotes this with approval, but sets out in a formal statement the underlying principle (pp. 29–31). Is this simply instruction for the teacher? Probably not, for the Report appears to advise (pp. 30, 31) that the statement of the principle, analysed into axioms, and the deduction from it of the formal proofs of I. 4 and I. 26 should be taught to pupils of ages 15–17. Such advice may do grave harm. It calls for exceptional powers of exposition from the teacher and a high order of intelligence on the part of the pupil. In the absence of either, this method would be disastrous. Moreover the subtleties of language involved (cf. the distinction between constructing a figure and instructions that direct attention to a figure, p. 32) are beyond the capacity of all ordinary pupils.

By all means let superposition proofs be cut out from every syllabus and every examination. But may we not then adopt the method of the circular for obtaining the results of I. 4 and I. 26 and thereafter treat these as part of our present axiomatic basis? To substitute the proofs of this Report will condemn most boys to committing to memory a series of phrases and sentences to which they attach no meaning.

2. Parallelism and Similarity

It is generally accepted that the concept of similarity should be introduced early: this view is endorsed by the Report. But the controversial paragraphs deal with the application of the "Principle of Similarity" to the organising stage (ages 15–17).

A previous article in the *Gazette* has criticised one of the suggested proofs. But apart from any question of vigour, the real obstacle is the intrinsic difficulty of the proposed method. It is unlikely that many attempts will be made to follow it out unless a specially written text-book is provided. It therefore seems needless to examine the reasons given in support of the recommendation to substitute the "Principle of Similarity" for Euclid's method or one of the modern equivalents. The report then proceeds to comment on the use of the idea of direction in the theory of parallels. It is not easy to discover precisely what these comments mean. No one will dispute the statement that a logical proof of the fundamental parallel theorems cannot be based on the idea of direction or the idea of rotation. Nor does the Board's circular say it can be. It does say that these two ideas can profitably be used to make the pupil understand what these theorems mean and to link up his everyday experience with his geometrical notions. If our axiomatic basis is to be so broad as to include the fundamental parallel theorems, this seems the most practicable way of approaching the subject. The danger is that this specific criticism of the Board's circular may lead some teachers to think that the use of such illustrations in building up the axiomatic basis is condemned by the Report.

It is, of course, desirable to consider how far (if at all) an attempt should be made in the *organising stage* to narrow the axiomatic basis. This is discussed on page 20, etc. [What by the way is the meaning of the statement, p. 20: "the first group of theorems (angles at a point) should certainly be proved here"? Can anything more be said at this stage than has been said long ago in any geometrical course? What sort of formal proof is wanted ?] The Report recommends that the axiomatic basis should be narrowed as follows: (i) the deductions of I. 4 and I. 26, from the Principle of Congruence, (ii) the treatment of parallelism and similarity *either* by Euclid *or* by the Principle of Similarity.

This problem of the narrowing of the basis is of fundamental importance and demands a wide interchange of opinion among teachers: the definite recommendation of the Report seems premature. It cannot be denied that the enumeration or

exposition of all the axioms required for a scientific study of geometry is specialist work: school geometry must be built on a broader basis than will satisfy the
philosopher. Compromise is necessary and it rests with teachers to discover and
decide what compromise will give the best results. All our axioms must clearly
be in accord with a boy's experience and it is equally obvious that proofs must be
given as soon as results occur which are not self-evident. But subject to these two
limitations, the foundation should be determined by what experience shows is the
best-jumping off point for school work and best adapted for extensive reading and
applications. The more time is spent on narrowing the basis, the less time remains
for developments. This at any rate is the question at issue, and it is doubtful whether
the time is yet ripe for a dogmatic pronouncement.

The Report deals with many other matters of great interest, too numerous to discuss within the bounds of a single article. The writer however, disagrees so seriously
with the section on the treatment of limits that he hopes to be allowed to express
his views on this subject on a later occasion.

6.5

Fifty Years of Change

A.W. SIDDONS

There must be many teachers of mathematics today who do not realise the great changes, both in what was taught and in methods of teaching, that were made in the early days of this century and the part that the Mathematical Association took in making them.

To understand what those changes are and to appreciate the benefits that they have conferred on teachers and pupils, it is desirable to know something of the history of the Mathematical Association.

In 1871 the Association for the Improvement of Geometrical Teaching was founded at the suggestion of my old mathematical master, Rawdon Levett. In 1897 its name was changed to the Mathematical Association.

The A.I.G.T. and the M.A. up to about 1912 may be said to have consisted of Public and Grammar School masters with a fair sprinkling of more advanced teachers who were interested in what was taught in the schools concerned. In spite of the resultant narrowness it had considerable influence on the work of other types of schools. In 1902 the first Teaching Committee was formed and reports were published on Geometry, Arithmetic and Algebra. In 1907 a report was published on mathematical teaching in Preparatory Schools. In 1912 committees were formed for dealing with mathematics in (i) Public Schools, (ii) Other Secondary Schools for boys, (iii) Girls' Schools. In recent years the scope of the work has been increased to include mathematics in all types of schools.

The first number of the *Mathematical Gazette* appeared in 1894. On the cover of that number it said, "We intend to keep strictly to Elementary Mathematics: while not excluding Differential and Integral Calculus, our columns will, as a rule, be devoted to such school subjects as Arithmetic, Algebra, Geometry, Trigonometry and Mechanics." This limitation gradually disappeared and the Gazette developed a great reputation for its reviews of books, both elementary and advanced, and

First published in *Mathematical Gazette* **40** (October 1956), pp. 161–169.

articles on advanced subjects have figured largely in its pages. In recent years the number of articles dealing with elementary topics has been comparatively small, in spite of frequent appeals by the Editor; but some items of interest to the main body of teachers have appeared in the form of Notes. At the moment steps are being taken in hopes of providing many more articles that deal with elementary topics and that will be of interest to the main body of teachers in all types of schools.

The main object of the A.I.G.T. was to free teachers and pupils from the bondage of Euclid, whose treatise was intended for the use of university students and is quite unsuitable for introducing school children to geometry. In the early days of the A.I.G.T., Euclid's proofs were required in examinations and no other proofs were allowed. First of all the A.I.G.T., after very careful work, produced a syllabus of geometry and in 1886 a textbook of geometry was published. Even these would hardly be considered suitable for beginners today. Both the Syllabus and the Textbook were very favourably received, but Cambridge would not move from its position until 1888 when it announced that proofs other than Euclid's would be accepted *so long as they did not violate Euclid's order.*

At that time and even into the early days of the present century, in many schools, all the geometry that was done was the learning of Euclid's proofs. In many cases it was done unintelligently. To take an extreme example here is a story told me years ago by the then Master of Jesus College, Cambridge. Up till 1850 no man was allowed to take the classical tripos at Cambridge until he had passed the mathematical tripos. A brilliant classic who was weak at mathematics went to a coach and asked him to choose 20 propositions that were likely to be set in the mathematical tripos and he said "I will learn them by heart". Later the number was reduced to 10. After the examination the delighted undergraduate rushed up to his coach's room and said "I am through; I got 8 of the propositions you chose and I got them all right to a comma". As an afterthought he added "I am not sure that I put the right letters at the right corners, but I suppose that does not matter".

I was a boy at King Edward's School, Birmingham, at which the mathematical teaching was probably as good as at any school in the country. We did many riders from the start, but there were very few riders in our textbook and we generally had to take them down when they were set.

When I started teaching at Harrow in 1899, in the middle divisions of the upper school I found many boys who professed never to have done a rider – that does not mean that there were boys in higher divisions who had not done many riders, but it shows how low was the standard of the majority.

At the beginning of the present century Professor John Perry started an agitation against existing methods of teaching mathematics and the matter that was being taught. At the Glasgow meeting of the British Association in 1901 he opened a discussion on this subject and put forward a schedule of work. 15 or 16 people spoke after Professor Perry – only one of them a working schoolmaster. After the meeting, the British Association appointed a committee to go into the matter; on that committee, if I remember rightly, there was only one working schoolmaster and he taught science, not mathematics.

However, Professor A. R. Forsyth, who was chairman, invited Charles Godfrey, the senior mathematical master at Winchester, to write a letter to the committee suggesting reforms that could be made. This letter was signed by 23 schoolmasters and was reprinted in *Nature* and in the *Mathematical Gazette* of January 1902.

In January 1902 the M.A. appointed its first Teaching Committee. The committee included several men who had been active members of the A.I.G.T. in its later days; it also included 4 or 5 schoolmasters who were under 30 years of age, and later other youngish schoolmasters were coopted. In less than a year reports were published on Geometry, Arithmetic and Algebra. These reports recommended moderate changes with which it was hoped most teachers would agree, so that examining bodies might be almost forced to make the changes desired.

In December 1902 Cambridge appointed a Syndicate to report on University pass examinations and, in particular, what concerned school teachers most, the "Little go". This Syndicate recommended that Euclid's order should no longer be required, but that "Any proof of a proposition should be accepted which appeared to form part of a systematic treatment of the subject". The report was accepted by the Senate and Euclid's order no longer dominated the teaching of Geometry. The first examination under the new regulations was held in March 1904.

The importance of this change can hardly be exaggerated, because the requirements of the University entrance examinations dominated the teaching of Geometry in the schools, and, until the Universities moved, the schools could do little. It was probably the greatest reform ever made in the teaching of any subject; it was the forerunner of many reforms in mathematics and in other subjects.

After this historical sketch, I will now consider in detail some of the changes that have been made. As my main experience has been in Public Schools, I will deal with those, but most of what I shall say will apply equally well to other schools, both boys' and girls'. I shall quote freely from the letter of the 23 schoolmasters which shows fairly well the aims of the reformers.

That letter said:

"As regards Geometry, we are of opinion that the most practical direction for reform is towards a wide extension of accurate drawing and measuring in the Geometry lesson. This work is found to be easy and to interest boys; while many teachers believe that it leads to a logical habit of mind more gently and naturally than does a sudden introduction of a deductive system."

As to this work of drawing and measuring, very little was done until after the beginning of the present century. In his presidential address to the M.A. in 1947,

Mr. W. F. Bushell said that in his last term at Charterhouse, in 1903, he was introduced to a protractor for the first time. As soon as Euclid was dethroned, this work was taken up in most schools; but the spirit of the reformers was often misunderstood, in many cases it was overdone and did not lead to anything useful. However, in 1909 the Board of Education issued the famous circular 711 (mainly the work of Mr. W. C. Fletcher) which was the first authoritative statement on the subject and put out the aims of the reformers in a very clear form.

Circular 711 divided the teaching of geometry into three stages. The following explains what those stages are, but not necessarily in the words of the circular.

First stage. To gain familiarity with and clearness of perception of fundamental geometrical concepts by means of observation of the common facts of life and practical work.

The concepts to be dealt with are solid, surface, line, point, volume, area, length, direction, angle, parallelism.

Definitions should be avoided, but emphasis should be laid on the right use of words.

Second stage. Discovery of the fundamental facts of geometry, by experiment and intuition; including the facts relating to angles at a point, parallels, angles of a triangle and polygon, congruent triangles.

When discovered each fact or group of facts should be followed by numerical examples and easy riders intended to illustrate and drive home the facts discovered. It is well worth while memorising in words the facts discovered.

In the course of this stage, the pupil, besides becoming familiar with the fundamental facts, should learn the accurate use of instruments and the elementary ideas of logical argument.

Though circular 711 does not mention it, there is much to be said for including in this stage a course on similar figures. Pupils, from scale drawing and maps, already have a good deal of subconscious knowledge about similar figures, and there are easy riders on the subject.

Third stage. A logical course of deductive geometry, based on the fundamental concepts and facts developed in the previous stages.

No attempt should be made at proving the fundamental facts about angles at a point, parallels and the cases of congruence: these should be regarded as axioms on which the deductive geometry is based. Today in most examinations proofs of these theorems are not required.

As has been said above, the old work in geometry consisted largely of learning the proofs of Euclid's propositions. Today it may be said that at first the teaching aims at bringing into the conscious plane many facts of which subconsciously the pupil is already aware; then developing these ideas and gradually giving geometrical power and some power of intuition. A knowledge of the theorems should be led up to by doing suitable riders, and by experiment and measurement whenever these help. Then, as a final stage, there is the development of a systematic course.

In the systematic course related facts should be grouped together. The one thing that the pupil has to remember is what is the fundamental fact in each group.

An able teacher will find today that geometry is not a difficult subject: in fact it can be attractive to most pupils.

It has been said that "Geometry is the most powerful weapon in Mathematics, but it requires the skill of a Newton to use it". No doubt the author of this epigram had in mind Newton's great book "the Principia" which was largely geometrical, but modern methods do not bear out this description of its difficulty today, at all events as far as the elementary stages are concerned.

To sum up it may be said that 50 or 60 years ago mathematics was presented to pupils in the cold, systematic, logical form in which the academic mathematician might ultimately pigeon-hole it in his mind. This form was quite unsuitable for beginners and, though later it might appeal to the real mathematician, it would never appeal to the vast majority of pupils; certainly it is true to say that the education of 95 per cent. of pupils was being spoilt for the very doubtful benefit of the remaining 5 per cent.

In so short an essay it has been impossible to do more than indicate some of the immense changes, not only in what is taught in the Mathematical classroom, but also in the method of presentation. I do not delude myself into the belief that we have now reached finality. Probably we are still on the upward path and learning slowly to teach better. It is Izaak Walton, in the *Compleat Angler*, who says that "Angling may be said to be so like the mathematics that it can never be fully learnt". Probably the best method of presentation can never be fully learnt, but I do claim that there has been an enormous advance during my lifetime. It may well be that a member of our Association, writing in the year 2000, will look back with pitying scorn on all he had to endure in his schooldays in the middle of this century. Indeed I hope this may be so, thus indicating yet another advance. But I do claim that my generation has done something for the improvement of the teaching of Mathematics.

6.6

Milestone or Millstone?

GEOFFREY HOWSON

1973 sees the jubilee of the Mathematical Association's first report on *The teaching of geometry in schools* and the centenary of the birth of one of its most influential authors, Charles Godfrey. It would seem appropriate, therefore, to look back at that report, at its origins and its effects, and to pay especial attention to the part played by Godfrey in the development of the teaching of geometry.

The position in 1922 when the General Teaching Committee of the Mathematical Association appointed a sub-committee to prepare the geometry report was not unlike that today. For, despite those modern innovators who believe that curriculum innovation began in the 1960s [1], great changes in the mathematical curriculum had taken place in schools early this century. Thus by 1912, Godfrey [2] could say of mathematics teaching in England that

"The use of graphical methods in elementary algebra is universal and entirely a 20th-century development. Other aspects of the same movement are the adoption of descriptive geometry by the mathematicians, the use of handy 4-figure tables [3], and of graphical methods in statistics."

D. E. Smith [4] was also able to report that

"about 90% of the schools (in England) state that the graphical study of statistics is given . . . for a new subject it has made rapid progress having both the encouragement of the mathematicians and an abundant opportunity for application. The graphical representation of functions is taught in all public schools. . . . The use of vectors is found in a large majority of schools, in connection with mechanics (velocities, accelerations, forces) . . . (and) . . . in some schools . . . in connection with complex numbers."

Changes then were being made, but nowhere was this more apparent than in the teaching of geometry. In 1903 Cambridge ruled that in its examinations "Any proof of a proposition will be accepted that appears to the examiners to form part of a

First published in *Mathematical Gazette* **57** (December 1973), pp. 258–266.

systematic treatment of the subject" [5]. What effect this had on the examiners, faced with what appeared to be a well-nigh hopeless task, we do not know, but Godfrey [6] recounts how this "released a volume of pent-up energy. Every enthusiast felt that he might now teach geometry in his own way. Numerous textbooks were produced, embodying every shade of opinion. As was to be expected, many of the new developments were extravagant and had no permanent effect."

One book, though, had a far from temporary effect – Godfrey and Siddons' *Elementary Geometry*. This book was published in 1903 during the aftermath of the famous 1901 meeting of the British Association – the meeting which initiated the 'Perry movement' and prompted Oliver Heaviside's famous 'fudge and fiddlesticks' paper [7]. Both Godfrey and Siddons were extremely active in the reform movement at that time and early in 1902 *Nature* had printed a letter composed by them and signed by some twenty other schoolmasters which proposed that the whole of the school mathematical curriculum should be drastically revised [8]. *Elementary Geometry* was an indication of the way in which the two authors saw mathematical education progressing. Thus the book opened with an experimental portion and, to quote E. M. Langley's review [9], "they (the authors) have given a profusion of concrete examples well calculated to lead the young student to recognise there is geometry everywhere, and not merely between the covers of a school-book. Among these we notice the forms of constellations given as an exercise on graphs, and the motion of a point on the connecting rod of a steam engine as an exercise on loci. But while the many and varied illustrations give the book a very different aspect from the pages of the Potts and Todhunter of our youth, there is no decline from Euclidean rigour in the formal proofs."

The period from 1903 was one of especially rapid change and "in those schools which moved with the times the immediate effect was probably rather chaotic" (Godfrey [10]). For example, drawing and measurement were seen by many teachers as ends in themselves. To combat this, G. St. L. Carson, Senior Mathematical Master of Tonbridge School, sought to help teachers distinguish between the distinct aims and purposes of 'experiment' and 'intuition'. He objected [11] to the way children were "instructed to draw two straight lines, measure the vertically opposite angles, and state what they observe. It is hard to see what good can be derived from such exercises and they may do much harm. If it be true that the education of the child should follow the development of the race [12] they are condemned at once, for it is inconceivable that these ideas were suggested by such processes. . . . They can only be regarded as intuitions from rough experience; a person in whom they are wanting is ignorant of space. . . . A mind which has ranged over all its experience and has made these intuitions has gained a sense of power and accepted truth which cannot be induced to an equal extent by any substitute for the process."

Preliminary experimental work then should develop intuitions and convey a 'knowledge' of space; it should not be used merely to make certain teacher-selected results appear 'reasonable'.

Some of these ideas, but by no means all, were embedded in the 1909 Board of Education Circular on the *Teaching of geometry and graphic algebra*. This was the report which first attempted to break down geometry teaching into stages (although nowadays the credit for this is generally accorded to the 1923 report). Again, Godfrey and Siddons were not slow to act and in 1912 they published *A shorter geometry*. The Preface sets out the plan of the book – divided in accordance with the Board's recommendations:

"*First Stage* Introductory work concerned with the fundamental concepts, and not primarily designed to give facility in using instruments.

Second Stage Discovering of the fundamental facts of geometry, by experiment and intuition.... When discovered and enunciated each fact or group of facts is followed by numerical examples and easy riders.... (Thus) the pupil, besides becoming familiar with the fundamental facts, is taught... the elementary ideas of logical argument as used in strict theoretical geometry.

Third Stage Deductive development of the propositions subsequent to those dealt with by experiment and intuition in the second stage."

The two authors recognised that "a possible disadvantage in the new method of teaching geometry is that the pupil may waste a good deal of time in carrying out experiments without a notion of what he is aiming at. On the other hand, if he knows what he is aiming at, experiment may be a sham. The fact is that many of these heuristic exercises are essentially for class work, the class cooperating with the master in developing a new idea, the master giving enough guidance to focus the whole discussion."

Not surprisingly, Godfrey was very much in favour of the Board's report. He did, however, note [13] that it had studiously avoided the question "Why do we teach geometry?". His own reason: "We teach geometry because it has been taught for a long time; and *vis inertiae* decrees that we shall go on teaching it" is hardly a convincing defence, but, to be fair, he himself did not seem altogether satisfied by this answer.

This then is part of the background to the 1923 report and those who know the report will recognise much of its spirit in the various passages I have quoted. Its first three "stages", as it acknowledges and as we shall later see, have much in common with those reprinted above. It too avoids the question of "Why teach geometry?" and in general appears to work on the belief that Heaven, like Plato's Academy, bears over its gates the inscription "Let no one ignorant of geometry enter within". But there was opposition to this view within the Association, as a paper given in 1917 by C. J. L. Wagstaff [14] shows. Wagstaff's message was forthright and unambiguous:

"Geometry (in the spirit of Euclid) in a Mathematical Syllabus seems to me to occupy much the same position that Greek does in the Classical. The two have similar merits and similar defects, and I am doubtful whether either deserves its place except in specialist classes . . . many believe that classical teaching absorbs too much time. Some assert that while few boys acquire anything of real value from it, many are made dull and listless, and, after suffering from a diet of stones when their desire was for bread, come to regard all school work as distasteful and valueless: eventually without exactly knowing why, they will come to despise their school education. . . . For myself, I must acknowledge, after much sorrowful reflection, that almost everything that 'reformers' say of Greek, I am compelled to say of geometry.

Both are excellent: intellectual delights, humanising influences, mental gymnastics, and so on *ad nauseam*. But only to the few does either appeal at all."

Godfrey and Siddons were not, however, the only teachers with foresight, and indeed others had perhaps outstripped them. Thus Carson in 1912 [11] writes that "the course (of geometry) may also be regarded as an introduction to the idea of a manifold and a function. The importance of these concepts has of late been realised . . ." That is, one aim of elementary geometry teaching could be to provide an early intuitive approach to some of the key concepts of modern mathematics. Not only could one have aims other than those sought by Euclid, but one could use methods unknown to, or ignored by, him: namely, analytical geometry and what is now termed transformation geometry. Analytical geometry made very slow headway in English schools. Godfrey [6] tells how in the early 1900s some teachers made what "amounted to a premature attack on analytical geometry". He and the general run of teachers were more cautious and he tells how boys at "a certain English Public School", although introduced to cartesian coordinates and graph plotting in algebra at the age of 14, did not study the analytical geometry of the straight line and circle until *after* they had studied harmonic ranges and pencils, principle of duality, inversion, pole and polar, cross ratios, etc. The need to overcome this reluctance to use algebraic methods and to link geometry teaching with that of other branches of mathematics was being cogently argued by another Association stalwart, W. I. Dobbs. Of all the reformers of that period, Dobbs gives the impression of being most in advance of his time. One wonders what his contemporaries made of some of his arguments. His thesis [15] was that the object of mathematical education was 'to develop, in combination, the power (i) to apply mathematics successfully to matters of human interest, (ii) to appreciate in some measure that 'realm of order' in which mathematical ideas exist".

To attain these ends he thought it might "be desirable to postpone specialisation in mathematics until after the time of the general school examination – say about the age of sixteen. If so, a school course of mathematics should be written as one subject, not the thing of shreds and patches hitherto insisted upon. . . . Algebra, Geometry, Trigonometry, etc. should not emerge as separate subjects until the specialist stage after the time of the general school examination. . . .The custom of external examinations in setting separate papers on algebra and geometry is a great obstacle to progress on these lines" [16]. Further evidence of the way in which Dobbs foreshadowed later developments lies in his *Gazette* articles [17] and his textbook [18] which attempt to show how school geometry can be based on the study of such plane transformations as reflections, rotations, translations and enlargements.

At the time the 1923 report was written, then, questions were being asked about the place of geometry in the curriculum, its possible merging at an elementary level with other topics, and alternative approaches to that based on congruence and similarity. One would hardly think it from reading the report. The three stages are laid down with their corresponding age levels, the "right time (for Stage C) is at the age of 15 or thereabouts", "Stage C will last until the age of 16 to 17"; the writers hardly acknowledge the existence of algebra; and although encouraging remarks are made about symmetry, one is told that although, for example, it follows immediately from the symmetry of a pair of circles about the line of centres that the common chord is bisected perpendicularly by the line of centres, this consideration is sufficient only for Stage B – "In Stage C we prove the same properties without reference to symmetry".

Yet fifty years later, at the Exeter Congress, Hans Freudenthal [19] described a similar proof by symmetry as "a most natural and lucid proof compared with the artificiality and obscurity of the Euclidean method" and regretted that textbooks still prove the theorem using chains of congruences!

This is but one example, though, of the mathematical imbalance of the report. Other criticisms were quickly made by H. F. Baker and others [20].

The too rigid model of the various stages, which suggested that Stage C, like acne, struck at the age of 15 and, with luck, cleared by 17, was put to right in the *Second report* of 1938 – a document which had much more to say about the classroom and relatively less about logical and illogical niceties. By that time, however, the damage had been done and the general philosophy of the first report had been accepted. Its authoritative nature stifled criticism from the ordinary teacher. Geometry still stood in glorious isolation, and every School Certificate pupil was expected to grapple with the highest achievements of Greek Mathematics – whether it meant anything to him or not. Experimental work on developing alternative approaches had come to an end and little – apart from a reshuffling of syllabus content following the Jeffery Report [16] – was to happen until the 1960s when, as in 1903, pent-up energy was again released with all the same effects – both good and ill – of sixty years before.

From the story as told so far, one might well suspect that Charles Godfrey had a strong influence on the sub-committee responsible for the report, for it follows very much the line one would have expected from those writings of his I have quoted, and indeed would seem to encapsulate his views of around 1912. I must admit that my own thoughts had always run along these lines – it was not until I read a paper of Godfrey's written in 1920 [10] that I realised my error. For Godfrey not only saw one revolution in geometry teaching as being accomplished, he was also beginning to prepare for a second.

Then, as now, many teachers were unhappy with the choice they were being offered and longed to be told what they should do. What was the best sequence for teaching geometry? Godfrey realised the dangers of such a question and warned of the difficulties which would follow "if any attempt were made to impose any one fixed sequence on teachers; happily there seems to be no immediate prospect of any such proposal being accepted". Nevertheless, two years later he was a member of a subcommittee assembled "to consider the effects of diversity of sequences on the teaching of geometry, and the desirability and feasibility of a return to the uniformity that prevailed thirty years ago" [21]. Not a very profitable line to follow it would seem, and one which serves to emphasise that in mathematical education, as in mathematics, the first thing to ensure is that one is asking the right question. The dangers of producing an 'authoritative' report were also realised by Godfrey: "We (Teaching Committee) have felt that during the period of rapid change it was undesirable that the Association should attempt to crystallise teaching. Until matters have settled down, it is for individuals to urge new ideas; an Association cannot easily do more than register the average opinion of sound teachers; and at the present moment a report . . . would delay rather than advance the movement of reform. I do not consider that the time has yet come for the Association to support with its authority any particular method of teaching Geometry" – sentiments that would seem equally valid today.

Nor had Godfrey's view on geometry teaching remained constant. He was now unhappy with the three stages of the Board of Education's Circular: "Is there any natural break in the development of a child's mind which justifies a sudden step from the quite informal methods of Stage 2 to the severely formal methods in Stage 3? I say that a boy of 13 is not ripe for the full rigour of the game: I am sure he is not ready for it before the age of 15, and I expect that even 15 is too early. I should like to recast the work of Stage 3 and to relegate the strict Euclidean method to a Stage 4, which should certainly not be taken up before 15 or 16. The natural place for Stage 4 would be after School Certificate age . . ."

Now this is a most important statement, especially in the light of the criticisms of geometry teaching which were to mount in the 1950s and 1960s. For, in the 1923 report we are told (p. 14) that the Committee would amalgamate the first two of the Board's Stages into Stage A and would subdivide the third. That is, the report's Stages B and C are Godfrey's Stages 3 and 4 respectively. Nor was Godfrey alone in doubting the ability of the non-specialist, pre-school Certificate pupil to cope with formal Stage C work. T. P. Nunn (who, with E. H. Neville, was largely responsible for writing the 1923 report) agreed with Godfrey, saying [22] "Like Mr Godfrey I urged that Stage 4 (a stage of 'strict' geometry based upon a minimum number of carefully examined postulates) should be postponed until after the age of 16". It seems then that both Nunn and Godfrey did not wish to see Stage C work attempted before the age of 16. Why was this not made clear in the report? One cannot tell for certain, but a hint is conveyed in the way in which Godfrey completed the sentence left unfinished above: "; but", he continued, "until the examining bodies have been converted – that is the next campaign – it may be necessary to attack Stage 4 a year or so before the School Certificate is taken".

That two of the world's leading mathematical educators were willing, temporarily at least, to sink their principles and to encourage teachers to attempt something they did not consider possible, seems quite astounding and clearly demonstrates the power wielded at that time by the external examining bodies. Nevertheless, it is clear from Godfrey's words that he was preparing to do battle on this point, but alas he was not to be given the opportunity. As a result, the 1923 report perpetuated the myth that such work was both desirable and possible with the average secondary school pupil. Moreover, it surprisingly avoided mention of the misgivings held by Godfrey and Nunn.

With such rigorous work postponed to the sixth form, what did Godfrey envisage as suitable work for the new Stage 3 (Stage B)?

"Now I want to be free at this stage to adopt whatever methods seem appropriate in each instance. There are many propositions which can perfectly well be dealt with by Euclidean methods in this stage: an instance is the group of propositions dealing with the angle properties of a circle. . . . On the other hand, there are properties of such a character that the boy does not at this age appreciate the need for formal treatment. Of these I instance two classes: properties connected with Symmetry and properties connected with Similarity. . . . I would abolish all these congruence exercises. I would present to boys once and for all the two types of plane symmetry – Symmetry about a line and Symmetry about a centre – and after that I would use the principle of symmetry fearlessly. . . . Traditionally, similarity comes rather late in the book, and the reason for this is historical . . . there is every reason for putting similarity quite early in the course. A boy is appreciating the existence of similar figures whenever he makes a drawing from nature, whenever he takes a photograph, whenever he goes to a cinema. . . .

Stage 3, then, is to cover the whole of elementary Plane Geometry (and some Solid Geometry, I hope) by eclectic methods – sometimes a formal proof, sometimes an appeal to intuition."

This sounds remarkably like some of the talk one heard in the 1960s, except that in those days the value of a mix was not always recognised – witness the horrific proofs of the angle properties to be found on pp. 164–5 of SMP Book T4. Yet as D. A. Quadling writes [23], this approach of "Can you find a proof of . . . by transformations ?" was soon abandoned by the SMP in favour of one embracing a variety of methods.

Godfrey then was willing to base his main-school course on an eclectic approach and in this he was once again supported by Nunn [22] who, however, disagreed slightly with Godfrey over the status of proofs by symmetry. He believed that questions demanding written proofs based on symmetry could be set at School Certificate level and referred to 'the excellent models in several standard French textbooks'.

Both Godfrey and Nunn appear, therefore, to have reached a state of thinking well in advance of that set out in the 1923 report. One can only regret that the doubts and the ideas which they shared could not have been spelled out in that report. Certainly it is unlikely that the sub-committee would have wished to recommend sweeping changes, for wholesale revision would have had little chance of success. Yet, by looking ahead and hinting at further developments, the report could have acted as a spur to further reforms. By nature it was a milestone, marking the conclusion of one set of reforms and initiating a period of consolidation. Unfortunately, because it was allowed to stand unchallenged, it became, through its many re-printings and several editions, a millstone, a restraining influence on the development of geometry.

Clearly, Godfrey was eager to proceed with what he saw as the next step in geometry teaching, but this was not to be, for early in 1924 he was to die of bronchial pneumonia. A. W. Siddons wrote in his obituary [24] that "it is only those who knew him intimately who can appreciate how much of the reform of mathematical teaching in the last twenty years is really due to his influence . . . he seemed likely to be a dominant influence on the mathematical teaching of the country for many years to come".

We can only speculate on the directions geometry teaching in England would have taken had Godfrey lived.

Notes and References

1. For example: "Nearly twenty years ago, as a young teacher, – came to the conclusion that methods of teaching mathematics were out of date. Little had been changed for hundreds of years." Pioneers of new maths – 1, *Mod maths and science* (British Thornton newsletter) No. 1 (1970).
2. C. Godfrey, Methods of intuition and experiment in secondary schools, *Proc. 5th I.C.M. (1912)*, Vol. II, 641 (Cambridge: Cambridge University Press, 1913).
3. Previously schools had used 7-figure tables, and a questionnaire circulated in 1911 to 551 secondary schools showed that 7-figure tables were still used in the upper classes of about 15% of the schools. Interesting comments on the change-over from 7- to 4-figure tables are contained in A. W. Siddons, Progress, *Math. Gaz.* **20** (1936).

4. D. E. Smith, Intuition and experiment in mathematical teaching in the secondary schools, *Proc. 5th I.C.M. (1912)*, Vol. II, 622 (Cambridge: Cambridge University Press, 1913).

5. The moves which led to this decision are amusingly described by Siddons in the article referred to in [3] above.

6. C. Godfrey, The teaching of mathematics in English public schools for boys, *Proc. 4th I.C.M. (1908)*, Vol. III. Lincei (Rome, 1909).

7. Professor Perry, an engineer and a former master at Clifton College, inaugurated a discussion at the 1901 meeting of the British Association in Glasgow by reading a paper attacking the mathematical teaching of the period. He maintained that the study of mathematics began because it was useful, continues because it is useful and was valuable because of its usefulness. For these reasons it should be taught so as to appear useful and pupils should learn something of its applications. A summary of his main points is to be found in *Teaching mathematics in secondary schools*, HMSO (1958), and some indication of his effect overseas in *A history of mathematical education in the United States and Canada*, 32nd yearbook of the NCTM (1970).

   Oliver Heaviside dismissed as "fudge and fiddlesticks" "the prevalent idea (that) you must understand the reason why first, before you proceed to practice". Mathematical understanding, he claimed, came with practice – and could not precede it (cf. H. Whitney, Are we off the track in teaching mathematical concepts? in *Developments in Mathematical Education* (Cambridge, 1973).

8. *Nature, Lond.* **16** (January 1902). This letter led to the establishment in 1902 of the Teaching Committee of the Mathematical Association.

9. *Math. Gaz.* **2** (1903), reprinted **55**, 392, (March 1971) 239–240.

10. C. Godfrey, Geometry teaching: The next step, *Mathl Gaz.* **X**, (1920) 20–24.

11. G. St. L. Carson, The educational value of geometry, in *Teaching of Mathematics in the United Kingdom*, Vol. I. (HMSO, 1912).

12. Compare the emphasis placed on Haeckel's Law of Recapitulation by René Thom in his two papers 'Modern' mathematics: an educational and philosophical error?, *Am. Scient.* **59**, 6 (1971) 695–699 and Modern mathematics: Does it exist?, in *Developments in Mathematical Education* (Cambridge, 1973).

13. C. Godfrey, The Board of Education circular on the teaching of geometry, *Math. Gaz.* **5**, (1910) 195–200.

14. C. J. L. Wagstaff, Should we continue to teach geometry?, *Math. Gaz.* **9** (1917) 38–40.

15. W. J. Dobbs, Mathematics in secondary schools, *Math. Gaz.* **9** (1917) 40–42.

16. Compare the findings of the Jeffery Committee almost thirty years later (see, for example, *Teaching Mathematics in Secondary Schools* (HMSO, 1958).

17. W. J. Dobbs, The teaching of geometry and trigonometry, *Math. Gaz.* **5** (1913), 139–146, 167–170.

18. W. J. Dobbs, *A School Course in Geometry* (Longmans, 1913).

19. H. Freudenthal, What groups mean in mathematics and what they should mean in mathematical education, in *Developments in Mathematical Education* (Cambridge, 1973).

20. *Math. Gaz.* **12** (1924). Baker's criticism is reprinted in *Math. Gaz.* **55**, 392 (March 1971), 146–152.

21. *The Teaching of Geometry in Schools*, 4th Edn., (Bell, 1944). 1.

22. *Math. Gaz.* **10**, 24 (1920). Account of discussion following the presentation of Godfrey's paper Geometry teaching: The next step (see [10]).

23. D. A. Quadling, The mathematics of SMP, in *SMP: The First Ten Years* (Cambridge: Cambridge University Press, 1972).

24. *Math. Gaz.* **12** (1924) 137–138. An account of Godfrey's work in other branches of mathematics is given in A. G. Howson, Charles Godfrey and the reform of mathematical education, *Educ. Stud. Math.* (to appear).

6.7

The Place of Geometry in a Mathematical Education

J. V. ARMITAGE

It is a commonplace that pictures leave a deeper impress on the human mind than abstractions and, in spite of its abstract nature, that is also true of many parts of mathematics. A proof of a theorem or the construction of a counter-example is often suggested by pictorial considerations, and that is true of algebra and analysis as well as of obviously geometrical subjects like topology. (For entertaining examples see Professor J. E. Littlewood's book [1].) There is obviously a connection between a developed geometrical intuition of that kind and the kind of spatial experience which ought to be furnished by a school mathematics course. No one would deny its importance and the writers of textbooks on 'modern' mathematics are well aware of the need to stimulate geometrical intuition. What has been called in question is the relation between that kind of geometrical intuition and the formal treatment of the Euclidean plane based, albeit insecurely, on Euclid's *Elements*. Indeed, the traditional treatment has been abandoned with indecent haste and without serious question. In Part I of this paper we make a plea for a re-consideration of the case for some kind of formal geometry as an indispensable part of the education of the more mathematically able children.

Geometry has fared badly in university syllabus changes during the past twenty years. Although algebraic geometry, differential geometry and topology are the subject of intensive and exciting research and are increasingly an essential part of the equipment of workers in other fields, most undergraduates learn scarcely any of the first two and not much of the last, whilst projective geometry, once an essential constituent of any advanced course, is relegated to the status of a peripheral option, if it is taught at all. Euclidean geometry is seen as a special part of some more general theory, for example the theory of sesquilinear forms, and that is that. But does that represent an adequate account of geometry? And what is the best preparation for the topology and geometry which is taught? We deal briefly with these questions in Part II.

First published in *Mathematical Gazette* **57** (December 1973), pp. 267–278.

The present paper is not intended to be an exhaustive survey of the problem, but rather a stimulus to further discussion. One hopes that school syllabuses are moving towards a synthesis of the best features of 'traditional' and 'modern' and that the claims of formal geometry will not be ignored. Part I is based on lectures given to several branches of the Association, to which the author is indebted for their invitations and encouragement. Part II is based very loosely on a discussion on "the place of geometry in university teaching" held at the 2nd International Congress of Mathematical Education, Exeter 1972. That discussion took place under the chairmanship of Professor D. Rees. Readers who actually recall the discussion will realise that the author has not scrupled to follow the Snark's example:

> "But the Judge said he never had summed up before;
>   So the Snark undertook it instead,
> And summed it so well that it came to far more
>   Than the Witnesses ever had said."
>   *(Lewis Carroll, The Hunting of the Snark, Fit the Sixth.)*

1. The Place of Geometry in Schools

1.1. The Decline of Geometry

It is over 100 years since the Association for the Improvement of Geometrical Teaching was founded and it is sad that, in spite of its subsequent efforts [2], we are now faced with the subject's extinction in schools, or at any rate with an evanescent substitute which is almost worse than extinction. The reformers rightly attacked the caricature of geometry which was all too often presented in the classroom, "but", as Sir Philip Sidney wrote in defence of poetry, "shall the abuse of a thing make the right use odious?" There were good teachers and good textbooks (a comparatively recent one being [3]). My theme, then, is the defence of geometry; not merely the development of spatial intuition, but formal geometry presented as a deductive science of physical space.

At the outset, I must make it clear that I have in mind the top 30% of what used to be the grammar school population. If my defence is valid for a bigger proportion of the school population, then I shall be delighted. That is not to say that I am interested only in the future specialist; indeed, in one sense it is more important that academically gifted pupils whose main interests lie elsewhere should encounter real mathematics at O level. Nor must it be assumed that I wish to ignore the needs of the majority of pupils. For them a perceptual treatment of geometry, as developed for example in [4], is more appropriate. Something of my attitude may be inferred from the advice to the reader in the Introduction to [1]: "did you get anything out of geometry (at school)?" What else can one teach at school which conveys so authentically the flavour and excitement of real mathematics?

1.2. The Case Against Geometry

(a) One suspects that the real reason for the abolition of geometry is that it is difficult. Certainly many pupils found it incomprehensible, because they could not enter into the spirit of the formal game, and unattractive, because any pleasure to be derived from problem solving was destroyed by the demands imposed by the need to write out the solution in a particular way. There is no royal road in geometry, as Menaechamus said to Alexander (or Euclid to Ptolemy), but now we look for the broad paths in education (and where they lead is well known).

(b) Then geometry is held to be useless. To the question "What shall I get by learning these things?", Euclid replied "give him three pence, since he must make gain out of what he learns". In an age where everything has its price, we are satisfied neither with the three pence nor the answer.

(c) It is well known that Euclid's treatment is not rigorous (he offers no adequate account of the topology of the plane, for example) and the complete description, as magisterially worked out by Hilbert [5], is clearly too hard for schools. Because it cannot be done properly, so the argument goes, it had better not be done at all.

(d) Geometry used to be defended on the ground that it offered a training in logic. That never was the best defence and, like its use in the defence of Latin, it has proved to be an ineffective one. On a related theme, it is argued that proof can be learnt in algebra.

(e) School geometry (in the sense of Euclid) is a part of linear algebra and therefore is best approached that way. Accordingly one should abandon all those riders and get on with groups and linear transformations. This particular approach to geometry is inferred from contemporary university courses and one of the tenets of the curriculum reform movement has (rightly) been the belief that school mathematics should reflect the mathematics of mathematicians.

(f) It will be seen that most of those objections are connected with the phenomenon of 'modern' mathematics in schools. Curriculum reform has led to the introduction of many undoubtedly attractive topics (networks, matrices, probability) as well as some doubtful ones (statistics) and some of unbelievable banality (some of the work on sets). Something had to go and, for the reasons outlined above, that had to be formal geometry.

1.3. Responsio

(a) Mathematics is an attractive subject because it is difficult. Obstinacy and the ability to concentrate are a necessary part of the equipment of a mathematician and, in other contexts, they are desirable qualities for life in general. Geometry is a good discipline in which to encounter difficulties for the first time, because its problems lend themselves to pictorial exploration and investigation and that combination of intuition and logic which is of the essence of mathematics. Hilbert wrote that a good problem should possess these characteristics. It should be clear and easy to comprehend; so that it attracts. It should be difficult in order to entice, yet not inaccessible, lest it mock our efforts. It should be significant and a guide-post in the tortuous paths to hidden truths.

Here is an example. Let ABC be a triangle and let D, E, F be points in BC, AC, AB respectively (and between A and B, etc. – we don't need a betweenness axiom to know what that means). Then it is easy to see that the circles AFE, BDF, CED are concurrent. (In the classroom one should begin with the special case in which D, E, F are the feet of the altitudes, then weaken the hypotheses. That in itself is good mathematics.)

We can vary the problem by considering a line DEF meeting the sides of a general triangle. Then we have four lines and can omit one line at a time to obtain four triangles. Each triangle has a circumcircle and it is not too difficult for bright pupils (I tried it recently with a fourth form) to show that those four circles are concurrent. One can now generalise. Five lines determine, in general, five four-line configurations, each of which gives rise to a point of concurrency as before. These five points are concyclic. Then every six-line configuration determines, in general, six five-line configurations, each of which determines a circle. These six circles are concurrent, and so on.

Obviously the proofs quickly get much harder (cf. Coxeter [6], pp. 258–262, for a proof of Clifford's Theorem, based on a chain of theorems due to Cox), but pupils can readily see the beauty of the whole chain of theorems and so experience that sense of wonder which ought to be a chief end of education. Nor is the result a quaint geometrical curiosity unfit for serious study. It ties up with linear algebra and inversive geometry and has only recently been the subject of a paper by a theoretical physicist [7].

Difficulty is acceptable when it offers such rich rewards.

(b) The real answer to this is Euclid's own; those who expect education to produce a return in kind for financial investment will always seek to weigh and measure its fruits and ultimately impoverish the human spirit. But if one believes that the intellect and the ability to reason should be developed for their own sakes, then geometry, like classics, has a value which cannot be measured.

But even in practical terms geometry is useful in the sense that it has applications in surveying, navigation, engineering, photography and physics and the question of trustworthiness is of immediate importance and leads to proof. In any case the question of what will prove to be useful in 'real life' is not easy to answer in school terms and some of the skills and disciplines developed through geometry may turn out to be useful, especially when one remembers the variety of employments which our rootless society may require people to take up.

There is a further point in favour of the game, or useless, aspect of geometry. As part of education it is necessary to put the learner in active as well as passive situations. Indeed that is one of the main reasons for the change in emphasis to child-centred activity. Where immediate usefulness is paramount, rapid and accurate memorising is best. In order to learn by discovery, it is necessary to take part in game-like activity (not play!) and Euclidean (formal) geometry is best, because it makes constant reference to intuitive ideas about space, it is the least arbitrary and the richest in mathematical meaning (cf. [8]).

(c) This point is well dealt with in [8]. Questions such as the completeness or independence of one's system of axioms need not enter into a treatment of geometry conceived as a deductive science of space. Arguments which rely on the special features of a picture are, of course, suspect, but the picture is really used to suggest an argument and to support a platonic view of mathematics in which triangles, circles and so on are regarded as having an independent existence.

(d) The axiomatic view of mathematics, in which one lays down certain axioms, without further specification of the objects to which they are to apply, and develops their consequences by applying the laws of logic, has been widely adopted and has made tremendous progress in the solution of concrete problems. Yet the laws of logic provide only a mechanism for mathematics and one feels that propositions are true in a stronger sense than that of mere deduction.

So whilst one does learn logic from geometry (and from a study of logic, come to that) one learns a great deal more. It is not primarily a question of logic, but rather of inspiration and seeing that things must be so. Writing out a logical proof is important, both to convince oneself and others, but to reduce mathematics to logic is like reducing music to little black dots on a piece of appropriately ruled paper.

If geometry is presented as a deductive science of physical space, then it is also closely related to mechanics and physics: Euclid is the progenitor of Newton and Einstein as well as of Hilbert. Indeed, Euclid was the model *par excellence* for other fields of knowledge, for example Spinoza's *Ethics proved in geometrical order* (as will be familiar to all those who move readily through the World of Jeeves).

One recalls Einstein's own indebtedness to geometry when he wrote, concerning the theorem on the concurrency of the altitudes of a triangle, ". . . it is *marvellous* that man is capable at all to reach such a degree of certainty and purity in thinking as the Greeks showed us for the first time to be possible in geometry . . .".

But must this idea of a deductive science come from geometry; why not treat geometry informally and learn about proof in algebra? Certainly one can learn about proof in combinatorics (traversable networks, electric circuits) or arithmetic (tests for divisibility) and so one should. But the proposed substitutes are probably as difficult as geometry if done properly, and one quickly reaches a stage where they are decidedly more difficult. As far as pure algebra is concerned, it is a mistake to spend a lot of time verifying that group axioms are satisfied in trivial examples, whilst algebra proper is really too hard for a pre-sixth-form course.

As far as informal, non-deductive geometry is concerned, here is a quotation from the distinguished analyst Nevanlinna [9]. After discussing the relation between perceptual (informal) and conceptual (formal) geometry, he writes as follows: "One should make the transition to deductive procedure early on at school. For as a collection of empirical laws geometry is an impoverished theory. Its deeper significance is achieved only by the fact that its intrinsic logical structure can be explained completely."

(e) Certainly Euclidean geometry is a special case of more general theories and its most significant theorems are nothing other than the expression of relations in those theories. Thus, for example, the theory of invariants and covariants leads to an interpretation of the altitude theorem in terms of the linear dependence of covariants associated with the equations of the altitudes [10]. Again, the basic metrical theorems of geometry are a consequence of the definition of the Euclidean plane in terms of a certain bilinear form on a real vector space. But to say that a theorem is nothing other than a special case of something else is not an argument for removing the more special theory from the student's development. The significance of the choice of definitions still lies in the geometrical interpretation and the most unrelenting algebraist still appears to derive most pleasure from the interpretation of the most familiar geometrical theorems. The value of imagery lies in its relation to previous experience and that experience seems to be enfeebled when it is not accompanied by some kind of formal treatment.

It is by no means obvious that the best preparation for university studies is to encounter a watered-down version at school. Perhaps the best preparation for reading Bourbaki is to learn how to do hard exercises (an integral part of Bourbaki), rather than to learn about structures prematurely. Expressed in different terms, my point is that it is better to have such a good understanding of Pythagoras' Theorem that one can extend it by analogy to infinite dimensional spaces than to be taught about Hilbert spaces without acquiring understanding. To see the advantage of the former, see the article by N. Youd, written whilst he was a schoolboy, in [11]. (Also quoted in [12], where there are many other illustrations of the power of informed geometrical intuition.)

Those who think that Euclidean geometry in its own right is completely worked out might study [13] with pleasure and profit.

(f) The foregoing defence of geometry claims that geometry should be taught because an informed geometrical intuition is essential throughout mathematics and because a formal treatment of the subject introduces the pupil to the mathematics of mathematicians in a way in which no other elementary topic can. So one can reject formal geometry only if one does not aspire to that particular ambition. The writer believes that some children should be introduced to proof and the idea of a deductive system and therefore a place must be found for it, at least in the syllabus appropriate to a minority of the school population.

The positive part of the geometrical content is outlined in the next section. It will be seen to be more ambitious than anything undertaken hitherto, though many of the modern topics should make the treatment more streamlined. However, some topics will have to be omitted and from a typical modern O level syllabus I should omit so-called statistics, most of 'sets' (apart from the language and its use in introductory probability), ideas of structure (though one hopes that teachers would use those ideas to inform the whole of their teaching) and much of the work on scales of notation and so-called topology (retaining networks but omitting equivalence, which is difficult to do meaningfully unless one has a notion of map or continuous function).

1.4. An Outline Syllabus in Geometry

The ultimate aim of the school course in geometry should be:

(a) familiarity with different kinds of space;
(b) the notion of a group of transformations;
(c) (in particular) the structure of affine and Euclidean space in two and three dimensions and an appreciation of the *pleasure* to be derived from re-discovering geometrical propositions and proving them;
(d) some understanding of the relation between geometry and the physical world—for example in mechanics, special relativity, or in a geometrical treatment of the variational principles appropriate to a treatment of Snell's Law (cf. [14, 15, 9, 16]).

Geometry should be presented as a description of the real world (whatever that is) from clearly stated assumptions about the nature of space, in much the same way as physics should be presented. (In my experience, the more able pupils enjoyed such philosophical speculations.)

I assume that up to the age of 12 or 13 pupils will have made models, learnt how to draw, practised surveying, done navigation and encountered geographical maps and that they will have learnt about symmetry, rotations, etc. from handling models and working with plane figures. Starting from there, I would present the familiar old-fashioned material in the context of the following themes of ideas. Needless to say, it should not be presented as a grind, but as an exciting adventure.

1. Notions of congruence and distance (including congruence on a sphere), explicit use of real numbers and a clear statement of the relation between congruence of figures and

reflection and rotation. The triangle inequality and distance properties associated with reflection and isosceles triangles.

2. Notion of parallelism, parallelograms, vectors and translations. Comparison of direction at different points, angle sum of triangle, analogous questions on spheres.

3. Group structure, in conjunction with work on matrices. (Rotations, reflections and translations and their representation.)

4. Circles, the classical theorems on angles and tangents. (The angle properties by the traditional proof, the tangent properties by reflection, once symmetry is established. Though ideally one would like to present a wide range of alternative proofs at all stages.)

5. Similar triangles. To do this properly (as in Euclid V and VI; see [17]) is fascinating; I tried it once with a bright class and at least one girl understood it. However, perhaps one should be content with the case of a 'rational ratio' and assume the general real case. Of course, this work ties in with the work on linear transformations and is important for an understanding of the real number system. It is instructive to consider the analogous results for triangles on spheres.

6. Some solid geometry should be done and I should like to see perspective drawing discussed in some detail, partly as an introduction to projective geometry (cf. the remarks in [18]).

7. Introduction to nomograms as a useful tool and as an application of similarity and group structure.

The foregoing concludes O level (probably including some of the Additional Mathematics work). Properly integrated with algebra, it should not be too burdensome.

I envisage an orderly treatment with strong assumptions, based on physics and clearly stated, but thereafter proper proofs (yet without undue rigour). In examinations, questions would be set in which either the method to be used would be suggested or in which a variety of approaches would be allowed, provided that the candidate stated his assumptions clearly and correctly. (That last suggests a Tower of Babel, but there are essentially only two approaches and the sequences are fairly standard.)

At A level, I should like to see an approach to two- and three-dimensional affine and Euclidean spaces, based on linear algebra and including the conics and simple properties of spheres, cones and cylinders. Projective geometry, taken up from perspective drawing, could be developed synthetically and algebraically. The work should be linked with work in mechanics and statistics (and those subjects would have to be modified to fit in with the pattern), cf. [12].

If all students had that background, one could begin the university course with a survey of geometry in the style of Coxeter's book [6] and quickly move on to a course of linear algebra and geometry (as in Kaplansky's book [19]). The topology courses could start off with more examples drawn from the students' experience (notably the projective plane) and there would also be obvious advantages in basis free calculus.

The sixth-form course in geometry outlined above is similar to that offered in the MEI Pure Mathematics together with the Special Paper options on linear algebra and projective geometry. It is also similar to work in the SMP Further Mathematics syllabus. Whether work at that level will be possible in future depends on school reorganisation proposals now being canvassed. But one hopes that study in depth will still be possible in our schools.

2. The Place of Geometry in University Teaching

Up to about 1950, projective geometry occupied a central position in the structure of most university mathematics syllabuses. During the next decade it was almost completely ousted (there were a few honourable pockets of resistance) by demands made on time by new material in pure mathematics and pressure from applied mathematicians to include other, more 'useful' topics. In particular linear algebra took pride of place, since it fulfilled both conditions. At the present time there is evidence of a wish to re-establish geometry in some form, partly for aesthetic reasons and partly because a geometrical language and, inseparable from it, a geometrical way of thinking pervade much of modern mathematics. As examples drawn from postgraduate as well as undergraduate courses, we mention the language of modern algebraic geometry as it attempts to realise Kronecker's (other) dream of a unified treatment of geometry and arithmetic, the theory of differentiable manifolds and virtually the whole of topology, the connection between non-Euclidean geometry and modular forms, the language of functional analysis and of probability spaces, the foundations of quantum mechanics and relativity, the classical groups and crystallography, the application of finite geometries to problems in experimental design. All those demand an ability to think geometrically and in particular, to be able to use the geometrical language, one must have a feeling for the underlying geometrical analogies, even when intuitive ideas are at best suggestive and at worst misleading. One ought also to remember that classical algebraic geometry is still a source of inspiration in abstract algebraic geometry and not only as a source of translation exercises into the language of schemes. (See [20] for the passages for translation and [21] for the dictionary.)

At the undergraduate level, linear algebra ousted geometry and yet is impoverished without it. The case is succinctly made in the Preface to [19]:

"Linear algebra, like motherhood, has become a sacred cow. It is taught everywhere; it is reaching down into the high schools and even the elementary schools; it is jostling calculus for the right to be taught first.

Yet all is not well. The courses and books all too often stop short just as the going is beginning to get interesting. And classical geometry, linear algebra's twin sister, is a bridesmaid whose chance of getting near the altar becomes ever more remote. Generations of mathematicians are growing up who are on the whole splendidly trained, but suddenly find that, after all, they do need to know what a projective plane is."

On a related theme, we quote from the Introduction to [22]:

"Students have virtually no geometric experience of much significance between their fresh-man study of conics and quadrics and some graduate course devoted, more likely than not, to proving the Riemann-Roch Theorem in six easy lessons using sheaves and this sometimes without having ever beheld a curve possessed of a singular point."

Moreover, it is misleading to dismiss algebraic geometry as a branch of commutative algebra or elementary differential geometry as a part of calculus (without pictures). Perhaps the desire to draw pictures is a condescension to a psychological disability, though even so it may represent an essential stage in the development of mathematical understanding – a point of view cogently argued by Thom in [8]. Again, one needs geometrical intuition to pick out the significant theorems in algebra or calculus; mere manipulation of symbols is as unproductive in higher mathematics as it is for school pupils in the study of analytical geometry. It is salutary to recall another preface (they are usually easier than the main text), namely that of Cartan's book [23]: "The eminent services that have already been and will in the future be brought about by the absolute differential calculus of Ricci and Levi-Civita should not prevent us from avoiding the more exclusively formal calculus where the wealth of indices hide a geometric truth which is often very simple." (And of course, if one reads a bit further one sees that Cartan relied a great deal on Euclidean geometry.) To clinch the argument, although it is frequently asserted and undeniably true that Descartes improved geometry by the introduction of algebra, it was no less his aim to illuminate algebra by means of geometry. In the same vein, at the Princeton University Bicentennial Conference on the Problems of Mathematics, Lefschetz said that "to me algebraic geometry is algebra with a kick".

In Exeter, the effect of over-abstraction and the neglect of geometry on future teachers was discussed at length. Whilst there is no doubt that, for a mathematician, the best way to define a Euclidean space is from the point of view of real inner-product spaces, it is quite useless to do so if the student has no appreciation of the physical significance of the ideas involved. Nor is there much to be gained from a course on measures in abstract spaces if the student cannot evaluate a double integral or find the volume of a cone. So the student should be helped to see how his more advanced studies build on geometrical foundations acquired at school and the intending teacher should be encouraged to reflect on what he is about to teach in the light of what he has recently learnt. (Of course this argument begs the whole question of the relation between elementary courses and more advanced ones. It also presupposes that most university courses ought not to be designed primarily in terms of the expertise in research of the staff.)

It is clear that one can no longer indulge in the luxury of a geometry course that is construed independently of the rest of the mathematical curriculum. It should be related both to subjects being studied concurrently and to what has gone before and what comes after. With these criteria in mind, the following outline syllabus is suggested for consideration, as offering a satisfying geometrical education in itself and, at the same time, setting geometry within the context of a wider mathematical education.

Preliminary course (an expansion of ideas in Coxeter's book [6]). Review of Euclid's *Elements* and Hilbert's *Foundations*. Distance geometry, leading to a formal study of transformation geometry in the plane and space, groups of isometries, crystallographic groups, subgroups of SO(3), the Platonic solids. The Euclidean group (including similarities), Euclidean invariants, stereographic projection etc., Möbius (linear fractional) transformations and inversion. Real affine geometry, the affine group, invariance questions and classification of conics and quadrics. Klein's definition of a geometry. Real projective geometry, models of the projective plane, standard theorems on conics, relation with Euclidean and non-Euclidean geometries, topology of the projective plane. Basic theorems of elementary differential geometry of curves and surfaces up to (say) Bonnet's existence theorem for surfaces.

Such a course would be complementary to a standard first course on linear algebra and affords a supply of examples for more advanced work in algebra and topology, as well as covering an introduction to the background of differential geometry essential to physics or engineering (or biology, *d'après* Thom).

In addition to standard treatments of topology and differential geometry, the student might then be offered an option on algebraic curves on the lines, say, of Walker's book [24] or more ambitious treatments where appropriate. This could be taken as far as Bezout's Theorem, with Pascal's Theorem on the inscribed hexagon appearing as a corollary, and a preliminary study of elliptic curves, with reference to topology and arithmetic. Such a course fulfils the criteria hinted at above and affords a starting point for more advanced courses in algebraic geometry, for example on diophantine geometry, which must surely be one of the most aesthetically satisfying subjects in the whole of mathematics.

3. Conclusion

The author is well aware that much of the foregoing is debatable if not inflammatory, but he hopes that it will serve to restore formal geometry to some pupils in schools and to all mathematics students in universities.

For Plato, geometry was probably synonymous with mathematics, yet one can still echo his incomparable defence of our subject: "Geometry leads the soul towards truth."

References

1. J. E. Littlewood, *A Mathematician's Miscellany* (Methuen, 1953).
2. Mathematical Association reports on the teaching of geometry in schools (Bell, 1923, 1939, 1953).
3. L. Roth, *Modern Elementary Geometry* (Nelson, 1948).
4. School Mathematics Project, *Books A-II* (Cambridge University Press, 1968–1972).
5. D. Hilbert, *Grundlagen der Geometrie*, 7th Edn. (Leipzig, Berlin: Teubner 1930); published in English translation (Chicago: Open Court, 1930).
6. H. S. M. Coxeter, *Introduction to Geometry* (Wiley, 1961).
7. M. S. Longuet-Higgins, Clifford's chain and its analogues in relation to the higher polytopes, *Proc. R. Soc.* A **330** (1972), 443–466.
8. R. Thom, Les mathematiques 'modernes': une erreur pedagogique et philosophique?, *L'âge de la science* **3**, 3 (Paris: Dunod Ed.,) shorter version in English in *Am. Scient.* **59** (1971), 695–699.
9. R. Nevanlinna, *Space, Time and Relativity* (Addison-Wesley, 1968).
10. N. Bourbaki, *Eléments d'Histoire des Mathématiques* (Paris: Hermann, 1969).
11. N. Youd, An original solution of a problem in calculus, *Math. Spectrum* **3** (1970–71), 17–21.
12. T. J. Fletcher, *Linear Algebra Through its Applications* (Van Nostrand, 1972).
13. M. T. Powell & I. A. Tyrrell, A theorem in circle geometry, *Bull. Lond. Math. Soc.* (1971).
14. H. S. M. Coxeter, *Twelve Geometric Essays* (Southern Illinois University Press, 1968).
15. E. A. Milne, *Kinematic Relativity* (Oxford University Press).
16. R. P. Feynmann, *Lectures on Physics* (Addison-Wesley, 1970).
17. T. L. Heath, *Euclid's Elements* (Cambridge University Press, 1908); reprinted (New York: Dover, 1956).
18. H. F. Baker, The teaching of geometry in schools, *Math. Gaz.* **12**, 170 (May 1924); reprinted in the Centenary issue **LV**, 392 (March 1971), 146–152.
19. I. Kaplansky, *Linear Algebra and Geometry* (Boston: Allyn and Bacon 1969).
20. J. G. Semple & L. Roth, *Algebraic Geometry* (Oxford University Press, 1949).
21. D. Mumford, *Introduction to Algebraic Geometry* (in preparation), first three chapters. Harvard Lecture Notes.
22. W. E. Jenner, *Rudiments of Algebraic Geometry* (Oxford University Press, 1963).
23. E. Cartan, *Leçons sur la Géométrie des Espaces de Riemann* (Paris: Gauthier-Villars, 1928).
24. R. J. Walker, *Algebraic Curves* (Princeton University Press, 1930).

Appendices

Appendix I

Report of the M.A. Committee on Geometry

The following report, preliminary and subject to revision, has been drawn up by a Committee of the Mathematical Association, consisting of representatives from a large number of public schools, especially of those near London. It is the outcome of many meetings, and of prolonged deliberation on the more drastic of the changes proposed. It affords striking evidence, if any were needed, of the fact that mathematical teachers are neither unconscious of, or indifferent to, the condition of things so forcibly depicted at the last meeting of the British Association. It is gratifying to note, *en passant*, that the counsels of the Committee were on the whole pervaded by singular unanimity.

Similar reports will shortly be issued on the teaching of Arithmetic and Algebra. The reports taken as a whole will represent a body of opinion which cannot be ignored, and should have a wholesome effect upon the future of mathematical teaching in this country.

It is hoped that the readers of the *Gazette* will make a serious study of the reports. All criticisms and suggestions will receive careful consideration at the hands of the Committee. They should, in the first instance, be sent to the Secretary, Mr. A. W. Siddons Harrow School Middlesex. It is very desirable that mathematical masters and others should fully avail themselves of this opportunity of placing on record their views as to the proposed changes; and, it is hardly necessary to add, that the hands of the Committee will be greatly strengthened by finding that the mass of their colleagues in this profession endorse the proposals.

It has long been felt that changes were inevitable, and we hope that the suggestions made by the Committee will be approved by those who have given time and thought to the question.

Res ipsa loquitur, judices, quae semper valet plurimum. [The facts speak for themselves, gentlemen, which always carries the most weight. (Ed.)]

REPORT OF THE M.A. COMMITTEE ON GEOMETRY

The Committee make the following suggestions under the head of Geometry.

First published in *Mathematical Gazette* 2 (May 1902), pp. 167–172.

Introductory and Experimental Course

It is desirable.

1. That a first introduction to Geometry should not be formal but experimental, with use of instruments and numerical measurements and calculations.

2. That Public Schools in their entrance examinations should set a fair proportion of questions requiring the use of instruments, and the obtaining of numerical results from numerical data by measurements from accurately drawn figures; and that in their entrance scholarship examinations the same principle should be recognised.

3. That elementary geometry papers, in examinations such as University Local Examinations, the Examinations of the College of Preceptors, Oxford Responsions and the Cambridge Previous Examination, should contain some questions requiring the practical use of instruments.

Division of the formal course into two parallel courses of (i) Theorems, (ii) Constructions

4. Since pupils will have been already familiarised with the principal constructions of Euclid before they begin their study of formal geometry, it is desirable that the course of constructions should be regarded as quite distinct from the course of theorems. The two courses will probably be studied side by side, but great freedom should be allowed to the teacher as to the order in which he takes the different constructions.

With regard to Constructions

5. The course of constructions should be regarded as a *practical* course, the constructions being accurately made with instruments, and no construction, or proof of a construction, should be deemed invalid by reason of its being different from that given in Euclid, or by reason of its being based on theorems which Euclid placed after it.

With regard to Theorems

6. The Committee propose, with a view to making the course of theorems independent of methods of construction, that no proof of a theorem should be considered invalid by reason of an assumption that a line or angle may be divided into any number of equal parts or that a line may be drawn from any point in any assigned direction and of any assigned length, or that any figure may be duplicated or placed in any position.

7. It is not proposed to interfere with the logical order of Euclid's series of theorems; – in other words, it is not proposed to introduce any order of theorems that would render invalid Euclid's proof of any proposition.

8. As far as possible, proofs of theorems should be based on first principles, and long chains of dependent propositions should be avoided.

9. *Proof* of *congruence* by superposition, and, in particular, proof of symmetry about a line by folding should be considered fundamental methods of proof.

10. Connected theorems should be associated together in the pupil's mind. e.g. (I. 13, 14, 15), (I. 4, 8, 26), (VI. 4, 5, 6, 7), and, in particular, a theorem and its converse, when true, should always be so associated.

The importance of Riders

11. In pass examinations it is desirable that the system should be gradually introduced of requiring that a candidate, in order to secure a pass should evince some power besides that of being able to write out bookwork.

12. It is desirable that, when possible, in an examination in geometry there should be a paper of exercises, including the practical use of instruments, and the solution of riders.

ORDER OF TEACHING THE EARLIER BOOKS

13. The Committee recommend the following general order in teaching the *theorems* of the first three books, and think that examiners should be requested to recognise this order:-

Book I.
Book III. to 32 inclusive.
Book II.
Book III. 35 to the end.

14. The detailed order must depend largely on some of the definitions, and on the methods of proof admitted in certain propositions. The Committee consider it desirable in the teaching of the subject that Books I. and III. taken more or less concurrently.

DETAILED SUGGESTIONS

15. That definitions should not be taught *en bloc* at the beginning of each book, but that each definition should be introduced when required.

BOOK I

(i) Theorems

16. That 13, 14, 15 be taken first.

17. That a very short proof of 13 should be considered satisfactory.

18. That 7 be omitted.

19. That, when two triangles are congruent, the equal parts should be written side by side in two columns:

$$\text{e.g in } \triangle \text{s } ABC, DEF,$$
$$AB = DE,$$
$$AC = DF,$$
$$\text{and} \quad \angle BAC = \angle EDF;$$
$$\therefore \text{ the } \triangle \text{s are congruent.}$$

20. That 8 be proved by placing the triangles in opposition.

21. That proofs of 24 by 19, which are incomplete, should be amended; but that proof by 20 should be preferred.

22. That 26 be proved by superposition.

23. That, in connection with I. 4, 8, 26, the following proposition be introduced:-
   Two right-angled triangles which have their hypotenuses equal, and one side of one equal to one side of the other, are congruent. (This can be proved by placing the triangles in opposition with their equal sides coincident, and applying I. 5 and 26.)

24. That the following propositions be introduced:

(1) *The locus of points equidistant from two given points is the perpendicular bisector of the line joining the given points.*

(2) *The locus of points equidistant from two given intersecting straight lines is the pair of bisectors of the angles contained by the given lines.*

25. That Playfair's axiom is preferable to Euclid's 12th axiom.

26. That, in dealing with angles connected with parallel lines, triangles, and polygons, illustration by rotation is desirable.

27. That it should be proved (for commensurables) that the area of a parallelogram is measured by the product of the measures of its base, and height, and the area of a triangle by half this product. (Cf. § 47.)

(ii) Constructions

28. That 1 be replaced by 22.

29. That 2, 3 be omitted.

30. That, in 45, the figure should first be reduced to a triangle by an application of I. 38.

BOOK II

31. It is proposed that this book should be taken after III. 32; suggestions with regard to this book will be found below.

BOOK III

32. That there should be a preliminary discussion of such fundamental properties of the circle as the following:

Every diameter is a line of symmetry: i.e. if the figure be folded about a diameter, the two portions of the circle coincide: with the corollaries:

(1) the two ends of every chord drawn perpendicular to the diameter are equidistant from it and from any point in it. Such a pair of points may be called symmetrically opposite points with regard to the diameter, or one may be called the image of the other in the diameter.

(2) The line of centres (i.e. the line drawn through the centres) of two circles is a line of symmetry. Hence, if the circles cut in a point off the line of centres, they also cut in the symmetrically opposite point; and, if the circles touch, either internally or externally, the point of contact must be on the line of centres; also, if two circles have the same centre, they either do not meet at all, or coincide entirely.

33. That 2, 4, 5, 6, 10, 11, 12, 13 be omitted.

34. That the last parts of 7 and 8 be omitted.

35. That the 'limit' definition of a tangent be allowed.

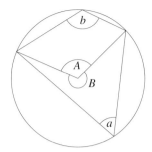

36. That 16, 18, 19 be replaced by the proposition, *The tangent at any point of a circle and the radius to the point of contact are at right angles to one another*: with the corollary, *One and only one tangent can be drawn at any point of a circle.*

37. That in 20, 21, 22 the use of angles greater than two right angles be allowed (ambiguity is rendered impossible if the angles are lettered instead of the points).

38. That 23, 24 be omitted.

39. That 26, 27 be stated as one proposition, and be proved by superposition, and that the equality of the sectors be proved as a corollary.

BOOK II

40. That the following definitions of a *rectangle* and a *square* be accepted:
A *rectangle* is a parallelogram which has one of its angles a right angle.
A *square* is a rectangle which has two adjacent sides equal.

41. That those proofs are preferable which do not make use of the diagonal.

42. That illustration from Algebra ought to be given where such is possible.

43. That 8, 9, 10 be omitted.

44. That the use of the signs + and − be allowed.

BOOK IV

45. That all propositions be omitted, as formal propositions, except 2, 3, 4, 5, 10, and that these be taken with earlier books; the rest of the book being treated as exercises in geometrical drawing.

BOOK VI

46. During the preliminary course of geometry it should be pointed out that drawings of different sizes can present exactly the same appearance, and that, in such drawings, the corresponding lines are proportional and the corresponding angles equal. Practical problems in heights and distances can in this way be solved graphically by quite young pupils, and are found most interesting exercises.

47. The committee suggest that the study of Book VI. should be preceded by

(1) A theory of measurement of lengths of lines and areas of rectangles for cases in which the lines and the sides of the rectangles are commensurable. (Cf. § 27.)

(2) All algebraical treatment of ratio and proportion for commensurables.

48. That an ordinary school course should not be required to include incommensurables; – in other words that in such a course all magnitudes of the same kind should be treated as commensurable.

49. This limitation would necessarily involve a change of proof in VI. 1 and 33; for VI. 1, either of the following methods of proof might be adopted:

(1) Let *ABC*, *DEF* be two △s whose bases *BC*, *EF* are commensurable and whose heights are equal.

Place the △s so that the bases *BC*, *EF* are in the same straight line and so that the △s are on the same side of the line.

BC, *EF* being commensurable, have a common measure. Let *YZ* be a common measure, and let *BC* and *EF* contain *YZ* *p* times and *q* times respectively.

Place *YZ* on *BCEF*; take any point *X* in *AD*, and join *XY*, *XZ*.

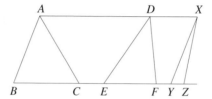

Since the △s *ABC*, *DEF* have equal heights, *AD* is parallel to *BF* (a proof of this should be given if there is no proposition or corollary to refer to).

Hence △*ABC* is *p* times the △*XYZ*, and △*DEF* is *q* times the △*XYZ*;

$$\therefore \triangle ABC : \triangle DEF = p : q$$
$$= BC : EF$$

i.e etc., etc.   (Q.E.D.)

(2) Let *ABC*, *DEF* be two △s of the same height.

Draw *AG*, *DH* perpendicular to *BC*, *EF* respectively, then *AG* = *DH*. Now the area of a triangle is measured by half the product of the measures of its base and height (see § 27);

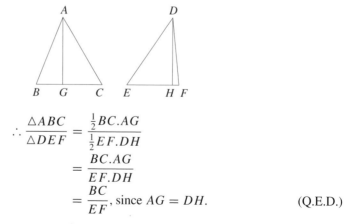

$$\therefore \frac{\triangle ABC}{\triangle DEF} = \frac{\frac{1}{2}BC.AG}{\frac{1}{2}EF.DH}$$
$$= \frac{BC.AG}{EF.DH}$$
$$= \frac{BC}{EF}, \text{ since } AG = DH.$$ (Q.E.D.)

50. The Committee desire to call attention to the method of proof of an extension of VI. 2 in the A.I.G.T. Geometry.

51. In stating the conditions for the similarity of two triangles *ABC*, *DEF*,

$$\frac{BC}{EF} = \frac{CA}{FD} = \frac{AB}{DE}$$

(or *BC* : *EF* = *CA* : *FD* = *AB* : *DE*) is preferable to

$$\frac{AB}{BC} = \frac{DE}{EF}, \frac{BC}{CA} = \frac{EF}{FD}, \text{ and } \frac{CA}{AB} = \frac{FD}{DE}.$$

52. The expression *corresponding sides* is preferable to the expression *homologous sides*.

53. In connection with the formal course, as soon as the proposition that equiangular triangles are similar has been proved, the sine, cosine, and tangent can be defined (if this has not been done earlier in the experimental course). In order to make the meanings and importance of these functions sink deeply into the pupil's mind, numerical examples should be given on right-angled triangles (heights and distances); these should be worked with the help of four-figure tables.

54. The expression for the area of a triangle, and of a parallelogram, in terms of two sides and the included angle, may be introduced simultaneously with the propositions concerning areas in Book VI.

55. In accordance with the spirit of the above proposals, the Committee suggest that the following proposition be adopted:

If two triangles (or parallelograms) have one angle of the one equal to one angle of the other, their areas are proportional to the areas of the rectangles contained by the sides about the equal angles.

If the $\triangle$s *ABC*, *DEF* have $\angle B = \angle E$, their areas are in the ratio $\frac{AB.BC}{DE.EF}$.
Draw *AG*, *DH* perpendicular to *BC*, *EF* respectively.

$$\text{Then } \frac{\triangle ABC}{\triangle DEF} = \frac{\frac{1}{2} BC \cdot AG}{\frac{1}{2} EF \cdot DH} = \frac{BC \cdot AG}{EF \cdot DH}.$$

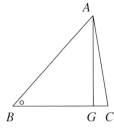 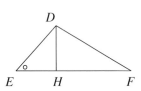

Also, since $\angle B = \angle E$ and $\angle G = \angle H$, the $\triangle$s *ABG*, *DEH* are similar;

$$\therefore \frac{AG}{DH} = \frac{AB}{DE};$$

$$\therefore \frac{\triangle ABC}{\triangle DEF} = \frac{BC \cdot AB}{EF \cdot DE}.$$

(Q.E.D.)

56. 19 follows immediately in the form, *The areas of similar triangles are proportional to the squares on corresponding sides.*

57. 20 can be deduced as in Euclid.

58. 22 follows at once.

59. 14, 15, 16, 17, 21, 23, 24, 26, 27, 28, 29, 32 should be omitted.

60. All statements of ratio may be made in fractional form and the sign = used instead of : : (as has been done above).

61. In the ordinary school course, reciprocal proportion should be dropped and compounding replaced by multiplying.

Appendix II

Euclidean Propositions

In the articles by Rouse Ball, Bertrand Russell and Thomas Heath and elsewhere in the book, Euclidean propositions are referred to simply by their book and proposition numbers. Their statements are given below in Heath's translation of the *Elements*.

Book I

Prop. 1: On a given straight line to construct an equilateral triangle.

Prop. 4: If two triangles have the two sides equal to two sides respectively, and have the angles contained by the equal straight lines equal, they will also have the base equal to the base, the triangle will be equal to the triangle, and the remaining angles will be equal to the remaining angles respectively, namely those which the equal sides subtend.

Prop. 5: In isosceles triangles the angles at the base are equal to one another, and, if the equal straight lines be produced further, the angles under the base will be equal to one another.

Prop. 6: If in a triangle two angles be equal to one another, the sides which subtend the equal angles will also be equal to one another.

Prop. 7: Given two straight lines constructed on a straight line (from its extremities) and meeting in a point, there cannot be constructed on the same straight line (from its extremities), and on the same side of it, two other straight lines meeting in another point and equal to the former two respectively, namely each to that which has the same extremity with it.

Prop. 8: If two triangles have the two sides equal to two sides respectively, and have also the base equal to the base, they will also have the angles equal which are contained by the equal straight lines.

Prop. 12: To a given infinite straight line, from a given point which is not on it, to draw a perpendicular straight line.

Prop. 13: If a straight line set upon a straight line make angles, it will make either two right angles or angles equal to two right angles.

Prop. 14: If with any straight line, and at a point on it, two straight lines not lying on the same side make the adjacent angles equal to two right angles, the two straight lines will be in a straight line with one another.

Prop. 15: If two straight lines cut one another, they make the vertical angles equal to one another.

Prop. 16: In any triangle, if one of the sides be produced, the exterior angle is greater than either of the interior and opposite angles.

Prop. 19: In any triangle the greater angle is subtended by the greater side.

Prop. 20: In any triangle two sides taken together in any manner are greater than the remaining one.

Prop. 22: Out of three straight lines, which are equal to three given straight lines, to construct a triangle.

Prop. 24: If two triangles have the two sides equal to two sides respectively, but have the one of the angles contained by the equal straight lines greater than the other, they will also have the base greater than the base.

Prop. 26: If two triangles have the two angles equal to two angles respectively, and one side equal to one side, namely, either the side adjoining the equal angles, or that subtending one of the equal angles, they will also have the remaining sides equal to the remaining sides and the remaining angle to the remaining angle.

Prop. 29: A straight line falling on parallel straight lines makes the alternate angles equal to one another, the exterior angle equal to the interior and opposite angle, and the exterior angles on the same side equal to two right angles.

Prop. 32: In any triangle, if one of the sides be produced, the exterior angle is equal to the two interior and opposite angles, and the three interior angles of the triangle are equal to two right angles.

Prop. 38: Triangles which are on equal bases and in the same parallels are equal to one another.

Prop. 44: To a given straight line to apply, in a given rectilineal angle, a parallelogram equal to a given triangle.

Prop. 45: To construct, in a given rectilineal angle, a parallelogram equal to a given rectilineal figure.

Prop. 47: In right-angled triangles the square on the side subtending the right angle is equal to the squares on the sides containing the right angle.

Prop. 48: If in a triangle the square on one of the sides be equal to the squares on the remaining two sides of the triangle, the angle contained by the remaining two sides of the triangle is right.

Book II

Prop. 2: If a straight line be cut at random, the rectangle contained by the whole and both of the segments is equal to the square on the whole.

Prop. 14: To construct a square equal to a given rectilineal figure.

Book III

Prop. 20: In a circle the angle at the centre is double of the angle at the circumference, when the angles have the same circumference as base.

Prop. 21: In a circle the angles in the same segment are equal to one another.

Prop. 22: The opposite angles of quadrilaterals in circles are equal to two right angles.

Prop. 26: In equal circles equal angles stand on equal circumferences, whether they stand at the centres or at the circumferences.

Prop. 27: In equal circles angles standing on equal circumferences are equal to one another, whether they stand at the centres or at the circumferences.

Prop. 31: In a circle the angle in the semicircle is right, that in a greater segment less than a right angle, and that in a less segment greater than a right angle; and further the angle of the greater segment is greater than a right angle, and the angle of the less segment less than a right angle.

Prop. 32: If a straight line touch a circle, and from the point of contact there be drawn across, in a circle, a straight line cutting the circle, the angles which it makes with the tangent will be equal to the angles in the alternate segments of the circle.

Book IV

Prop. 2: In a given circle to inscribe a triangle equiangular with a given triangle.

Prop. 3: About a given circle to circumscribe a triangle equiangular with a given triangle.

Prop. 4: In a given triangle to inscribe a circle.

Prop. 5: About a given triangle to circumscribe a circle.

Prop. 10: To construct an isosceles triangle having each of the angles at the base double of the remaining one.

Book V

Def. 4: Magnitudes are said to have a ratio to one another which are capable, when multiplied, of exceeding one another.

Def. 5: Magnitudes are said to be in the same ratio, the first to the second and the third to the fourth, when, if any equimultiples whatever be taken of the first and third, and any equimultiples whatever of the second and fourth, the former equimultiples alike exceed, and alike are equal to, or alike fall short of, the latter equimultiples respectively taken in corresponding order.

Book VI

Prop 1: Triangles and parallelograms which are under the same height are to one another as their bases.

Prop 2: If a straight line be drawn parallel to one of the sides of a triangle, it will cut the sides of the triangle proportionally; and, if the sides of the triangle be cut proportionally, the line joining the points of section will be parallel to the remaining side of the triangle.

Prop. 4: In equiangular triangles the sides about the equal angles are proportional, and those are corresponding sides which subtend the equal angles.

Prop. 5: If two triangles have their sides proportional, the triangles will be equiangular and will have those angles equal which the corresponding sides subtend.

Prop. 6: If two triangles have one angle equal to one angle and the sides about the equal angles proportional, the triangles will be equiangular and will have those angles equal which the corresponding sides subtend.

Prop. 7: If two triangles have one angle equal to one angle, the sides about other angles proportional, and the remaining angles either both less or both not less than a right angle, the triangles will be equiangular and will have those angles equal, the sides about which are proportional.

Prop. 17: If three straight lines be proportional, the rectangle contained by the extremes is equal to the square on the mean; and, if the rectangle contained by the extremes be equal to the square on the mean, the three straight lines will be proportional.

Prop. 25: To construct one and the same figure similar to a given rectilineal figure and equal to another given rectilineal figure.

Prop. 27: Of all the parallelograms applied to the same straight line and deficient by parallelogrammic figures similar and similarly situated to that described on the half of the straight line, that parallelogram is greatest which is applied to the half of the straight line and is similar to the defect.

Prop. 28: To a given straight line to apply a parallelogram equal to a given rectilineal figure and deficient by a parallelogrammic figure similar to a given one: thus the given rectilineal figure must not be greater than the parallelogram described on the half of the straight line and similar to the defect.

Prop. 29: To a given straight line to apply a parallelogram equal to a given rectilineal figure and exceeding by a parallelogrammic figure similar to a given one.

Prop. 33: In equal circles angles have the same ratio as the circumferences on which they stand, whether they stand at the centres or at the circumferences.

Book X

Prop. 1: Two unequal magnitudes being set out, if from the greater there be subtracted a magnitude greater than its half, and from that which is left a magnitude greater than its half, and if this process be repeated continually, there will be left some magnitude which will be less than the lesser magnitude set out.

Prop. 9: The squares on straight lines commensurable in length have to one another the ratio which a square number has to a square number; and squares which have to one another the ratio which a square number has to a square number will also have their sides commensurable in length. But the squares on straight lines incommensurable in length have not to one another the ratio which a square number has to a square number; and squares which have not to one another the ratio which a square number has to a square number will not have their sides commensurable in length either.

Book XII

Prop. 1: Similar polygons inscribed in circles are to one another as the squares on the diameters.

Prop. 2: Circles are to one another as the squares on the diameters.

Prop. 3: A pyramid which has a triangular base is divided into two pyramids equal and similar to one another, similar to the whole and having triangular bases, and into two equal prisms; and the two prisms are greater than the half of the whole pyramid.

Prop. 4: If there be two pyramids of the same height which have triangular bases, and each of them be divided into two pyramids equal to one another and similar to the whole, and into two equal prisms, then, as the base of the one pyramid is to the base of the other pyramid, so will all the prisms in the one pyramid be to all the prisms, being equal in multitude, in the other pyramid.

Prop. 5: Pyramids which are of the same height and have triangular bases are to one another as the bases.

Prop. 6: Pyramids which are of the same height and have polygonal bases are to one another as the bases.

Prop. 7: Any prism which has a triangular base is divided into three pyramids equal to one another which have triangular bases.

Prop. 10: Any cone is a third part of the cylinder which has the same base with it and equal height.

Prop. 16: Given two circles about the same centre, to inscribe in the greater circle an equilateral polygon with an even number of sides which does not touch the lesser circle.

Prop. 17: Given two spheres about the same centre, to inscribe in the greater sphere a polyhedral solid which does not touch the lesser sphere at its surface.

Prop. 18: Spheres are to one another in the triplicate ratio of their respective diameters.

Book XIII

Prop. 10: If an equilateral pentagon be inscribed in a circle, the square on the side of the pentagon is equal to the squares on the side of the hexagon and on that of the decagon inscribed in the same circle.

Princeton U-Store, NJ
Mon 20 Oct 2003 ~ $32.00
$40 list   + 1.92 tax
              ‾‾‾‾‾‾‾‾‾
                33.92